T0327660

FUNDAMENTALS OF OPTICAL FIBER SENSORS

WILEY SERIES IN MICROWAVE AND OPTICAL ENGINEERING

KAI CHANG, Editor
Texas A&M University

A complete list of the titles in this series appears at the end of this volume.

FUNDAMENTALS OF OPTICAL FIBER SENSORS

ZUJIE FANG
KEN K. CHIN
RONGHUI QU
HAIWEN CAI

A JOHN WILEY & SONS, INC., PUBLICATION

For general information on our other products and services or for technical support, please contact our Customer Care Department within the United States at (800) 762-2974, outside the United States at (317) 572-3993 or fax (317) 572-4002.

Wiley also publishes its books in a variety of electronic formats. Some content that appears in print may not be available in electronic formats. For more information about Wiley products, visit our web site at www.wiley.com.

Library of Congress Cataloging-in-Publication Data

Fundamentals of optical fiber sensors / Zujie Fang ... [et al.].
  p. cm.
 ISBN 978-0-470-57540-6 (hardback)
 1. Optical fiber detectors.   2. Fiber optics.   I. Fang, Zujie, 1942–
TA1815.F86 2012
681'.25–dc23

                                            2012005812

ISBN: 9780470575406

CONTENTS

PREFACE

Since the inventions of the laser and optical fiber in the 1960s, optical fiber communications and related technologies have been a great success story. The means and the way of human communication have been changed dramatically all over the world, and the standard of people's enjoyment of information is raised substantially. Relying on optical communication and computer science and technology, the function of information in various aspects of human life has reached inestimable importance. Stimulated and advanced by optical communication, research and development of fiber and related devices have made tremendous progress as well. A huge industry has emerged and boomed, including related materials, various types of optical fibers, devices and components, apparatus, machines, systems, networks, and their applications.

Optical fiber sensors, another important application of the optical fiber, have also experienced fast development, and attracted wide attention in fundamental scientific research as well as in practical applications. Sensing in the information system is often likened to human sense organs. Optical fiber can not only transport information acquired by sensors at a high speed and in large volume, but it can also play the role of a sensing element itself. In addition, compared with electric and other types of sensors, fiber sensor technology has unique merits, such as immunity from electromagnetic interference, being waterproof, and resistance to chemical corrosion. It has advantages over conventional bulky optical sensors, such as the combination of sensing and signal transmission, smaller size, and the possibility of building distributed systems. Fiber sensor technology has been used in various areas of industry, transportation, communication, security, and defense, as well as in people's daily life. Its importance has been growing with the advancement of the technology and the expansion of the scope of its application.

A large number of research papers and technical materials on optical fiber sensors have been published in professional journals, stimulated by progress in the sciences and demand for applications. A few monographs and textbooks have also been published. They have played their respective roles in academic and technical development of the field, each with its individual features and special applicable scopes. In view of a great deal of new progress—including that made by the authors of this volume—in the area of optical fiber sensors, which have emerged in recent years, there appears to be a lack of a comprehensive and updated textbook for senior undergraduate and graduate students, as well as a convenient reference resource for scientists and engineers working in the field. This book aims at serving this need—to students: explaining with clarity and exploring in depth the physical principles of optical fibers sensors; to workers in the field: making practical applications of the devices readily and conveniently accessible. The optical fiber sensors involve quite a large number of fields of science and technology, including optics, materials, electronics, and computing. This book puts the emphasis on their structures and optical characteristics and explains their physical mechanisms by using clear figures and basic formulas. The book cannot cover all aspects of the technology; detailed references are listed for interested readers.

This book consists of seven chapters. After the introduction in Chapter 1, the fundamental principles of optical fibers are reviewed in Chapter 2, including the electromagnetic theory and ray optics of optical fibers. Chapter 3 is engaged in fiber sensitivity and fiber devices. Chapter 4 describes fiber gratings of various structures and their application in sensor technology. Chapter 5 reviews the distributed fiber sensors, based on elastic and inelastic optical scattering in fibers. Chapter 6 introduces fiber sensors of special interest, including fiber gyroscopes, fiber hydrophones, Faraday effect sensors, and sensors based on surface plasmons. Chapter 7 is devoted to the extrinsic fiber Fabry–Perot interferometer sensors. The appendices give mathematical formulas used in the text, fundamentals of elasticity and polarization optics, and some data sheets of fibers and fiber devices.

This book will be published simultaneously in Chinese, by Science Press, Beijing, China, and in English in the Wiley Series in Microwave and Optical Engineering.

The authors of this book—Zujie Fang, Ronghui Qu, and Haiwen Cai—are professors of Shanghai Institute of Optics and Fine Mechanics, Chinese Academy of Sciences (SIOM/CAS); Ken K. Chin is Professor of Physics of New Jersey Institute of Technology (NJIT), USA.

The authors acknowledge the guidance and editorial support of Kai Chang, as well as the generous support of Dr. George Georgiou of NJIT, and helpful work of Dr. Qing Ye and Dr. Zhengqing Pan of SIOM, without which the publication of this book would not have been possible.

ZUJIE FANG
KEN K. CHIN
RONGHUI QU
HAIWEN CAI

CHAPTER 1

INTRODUCTION

Optical fiber is a thready material that makes use of optical total internal reflection (TIR) to guide light waves. The fiber loss was predicted low enough to transmit optical signals for long distances in the 1960s, and the low loss silica fiber was fabricated in the 1970s. Since then, optical fibers have been used in telecommunication systems in tremendous amounts and with great success. Their applications in sensor and other science and technology fields are also developed quickly, playing increasingly important roles in various fields.

1.1 HISTORICAL REVIEW AND PERSPECTIVE

The optical fiber was proposed and fabricated earlier in the 1920s [1,2], demonstrating light propagation in a glass waveguide based on the principle of TIR. The invention of optical fiber broke the limitation of the straight propagation of light. Fibers with cladding were invented later to reduce propagation loss, caused by the outer medium of air, for the earlier fibers without cladding. This improvement resulted in practical applications using optical fibers, such as image transmissions in bundles [3]. It was predicted theoretically in 1966 by K.C. Kao in his initiative

Fundamentals of Optical Fiber Sensors, First Edition.
Zujie Fang, Ken K. Chin, Ronghui Qu, and Haiwen Cai.
© 2012 John Wiley & Sons, Inc. Published 2012 by John Wiley & Sons, Inc.

paper that optical fiber with extremely low loss could be realized, and its application to telecommunications was proposed [4]. Soon after, a fiber with loss down to a few tens dB/km was fabricated [5]. Combined with lasers, especially the semiconductor laser, which came into being in the same decade [6–8], the fiber technology gave an impetus to the emergence of optical communication technology, which is now recognized as one of the bases of the modern information society.

A great number of improvements to fiber performance were achieved. It was found that silica fiber is the most suitable material for low-cost and high-quality fabrication technologies. Matched with the development of semiconductor lasers, three low loss bands at 850, 1,300, and 1,550 nm were fully exploited. Fiber dispersion properties were also investigated in detail. The single mode fiber showed much lower dispersion, superior to multimode fibers. In the 1980s, optical fiber communication systems were built for practical telecommunication applications [9]. An important milestone in optical communications was the success of wavelength division multiplexing (WDM) technology, especially dense wavelength division multiplexing (DWDM) in the 1990s, and the issuance of a series of international telecommunication union standards. Optical fiber communication technology, combined with the Internet, has tremendously changed the state of telecommunication all over the world.

Stimulated by the development of optical fiber technologies, a variety of optical devices and components have been developed. Apart from the semiconductor lasers and the photodetectors in different wavelength bands, optical modulators, switches, and WDM filters, and so on, are widely used for high-speed data transportation and networking. It is found that optical fiber is not only a long-distance transportation medium but also a good material of optical devices with special functions, especially for fiber amplifiers, which play a key role for DWDM technology.

The optical fiber is very useful—with unique features in sensor technologies—not only for signal transportation but also as a sensing element itself. Optical fiber sensors (OFS) have obvious merits over electrical sensors and bulky optical sensors. Some fiber sensors show unsubstitutable features, such as fiber optic gyros, nonlinear optical scattering sensors, and fiber gratings. Optical fiber interferometers can be used in many areas, while keeping the merits of high precision and high sensitivity.

Optical fiber sensor technology has grown into a large-scale industry, with its research and development becoming a trending field.

International conferences on OFS have been held since 1983 every year or so, with several hundred attendants. The OFS sessions are also listed in related international conferences and topic meetings, such as the Optical Fiber Communication Conference, Photonics East and West, International Symposium on Test and Measurement, and Society of Photo-Optical Instrumentation Engineers (SPIE) conferences. Many topical books have been published [10–14]; a great number of research papers can be read in journals, among them some review papers [15–19], which are also referred to in this book.

OFS have found varied applications in human social activities and daily living, from industrial production to cultural activities, from civil engineering to transportation, from medicine and health care to scientific research, and from residence security to national defence. OFS are used widely in manufacturing automation, production quality control; in oil well, tank, and pipeline monitoring, power system monitoring, and communication network monitoring; in building status monitoring and seismological observation; in navigation and vehicle status monitoring; in metrology and scientific instruments; in antiterrorist activities and intrusion alarming; and in many military applications.

The R&D and industry of optical fiber sensors are becoming major stimulants to the economy. According to estimates and forecasts of the Optoelectronics Industry Development Association, the average annual growth rate of fiber optic sensor revenue is about 63% during the period 2005–2010 [20].

1.2 CLASSIFICATIONS OF OPTICAL FIBER SENSORS

A sensor is considered an indispensable part of an information system. In automatically controlled equipment, sensors provide feedback signals for controlling operations; in industrial and civil engineering, sensors indicate basic conditions, such as stress and strain, vibrations, and temperature changes; in applications of security, military, and antiterrorism, they sound alarms; in health care, they are used to detect and transmit biochemical information.

Many kinds of sensors have been invented and developed. Most of them can be categorized into two types—electric and optical sensors—which have their respective merits and demerits. Soon after the invention of fiber, it was found that fiber itself possesses functions to sense external physical changes; its sensitivities were exploited to develop a variety of devices and sensors. Fiber sensors carry out a twofold function: acquisition of information and transport of the signals. The OFS shows unique merits, including the following:

- Small size and weight
- Environmental robustness, water- and moist-proof
- Immunity to electromagnetic interference and radio frequency interference
- Capability of remote sensing and distributed sensing
- Safe and convenient; integration with signal transportation
- Capability of multiplexing and multiparameter sensing
- Large bandwidth and higher sensitivity
- Lower cost and economic effectiveness

To construct a fiber sensor application system, various optical devices are needed, just like in optical fiber communication systems. As fiber technologies have grown into a vast industry, and a universal means in research and development as well, almost all optical components and devices find their homologs in optical fiber technology. Devices and components used in sensor technologies can be classified with some overlaps as follows.

According to functions, fiber devices are conventionally divided into active and passive devices, although their division is not exact. Generally, the former can generate or alter optical signals by some electrical methods, such as lasers, amplifiers, modulators, switches, and so on. The passive devices have no electrical means to alter optical signals; their main function is to define paths of optical signals and to configure various optical fiber systems. Couplers, connectors, collimators, attenuators, isolators, circulators, polarization controllers, and wavelength division multiplexers are the most important passive devices in fiber communications and fiber sensors.

According to materials and structures, fiber devices and sensors are roughly categorized into intrinsic and extrinsic devices. If the material that plays the main role is the fiber itself, it is called an intrinsic fiber device; whereas an extrinsic fiber device uses other optical materials incorporated with the fiber. For example, a fiber Mach–Zehnder interferometer is a typical intrinsic fiber device composed of fiber couplers and two fiber sections as the interference beams. Its extrinsic counterpart is lithium niobate electro-optic waveguide modulator connected with input and output fiber pigtails. Many bulky optical materials are used to fabricate components and devices in fiber technology, including silica, glass with different compositions, various crystals, semiconductors, polymer, and so on.

The main parameters of a light wave include amplitude, frequency, phase, polarization state, and intensity, which is the square of its amplitude. All of them can carry information, and thus can be used as sensor parameters. According to the parameters, OFS are divided into two categories: intensity-modulated sensors and phase-modulated sensors. For example, a sensor by detecting fiber bending loss is an intensity-modulated sensor, whereas a sensor based on birefringence caused by fiber bending is a phase-modulated sensor. Generally speaking, intensity-modulated sensors cost less, whereas phase-modulated sensors provide higher sensitivity and higher precision.

According to sensing elements, similar to the ordinary optical devices, fiber sensors can be categorized into intrinsic and extrinsic sensors. The former is a sensor that makes use of fiber's sensitivity to environmental conditions, whereas the latter is based on the sensitivity of materials other than fiber. For example, an electric current sensor can be made of fiber by its Faraday effect, so can some crystals with higher Faraday effect coefficients, connected with fibers as input and output leads. In most extrinsic fiber sensors the fiber plays a role of signal transportation, which is an important function in practical applications, and also a reflection of the fiber sensor's merits. Some extrinsic fiber sensors make use of the combined effects of fiber and other optical components, such as the fiber Fabry–Perot interferometer sensors. This book focuses mostly on intrinsic fiber sensors, with some discussion on extrinsic sensors.

According to the properties of sensed parameters, OFS are categorized into many types. The measurands include the following:

1. *Geometrical*: position, displacement, distance, thickness, move/stop signaling, liquid level, and so on.
2. *Mechanical*: strain, stress, pressure, and so on.
3. *Dynamical*: velocity, acceleration, angular velocity, fluid velocity, flow rate, vibration frequency and amplitude, and so on.
4. *Physical*: temperature, electric current, voltage, magnetic field, sound, ultrasonic and acoustic parameters, and so on.
5. *Chemical/biochemical*: flammable gases, toxic gases, specimen analysis, chemical etching detection, and so on.
6. *Miscellaneous*: break detection, fiber losses, intrusion detection, fire alarming, and so on.

The classification of OFS is based on different mechanisms, including the following:

1. *Basic effects of materials*: photoelastic effect and thermal photo effect (strain-induced and thermally induced refractive index change), thermal expansion of materials in optical path.
2. *Fiber interferometers*: Mach–Zehnder interferometers, Michelson interferometers, Fabry–Perot interferometers, Sagnac interferometers, Fizeau interferometer, and so on.
3. *Polarization dependences*: polarization maintaining fiber interferometers, strain-induced birefringence of the fiber, Faraday effect, and so on.
4. *Gratings and filters*: fiber gratings, spectral dependence of fiber couplers, wavelength converters, Doppler effect, and so on.
5. *Nonlinear optical effect and scatterings*: Rayleigh scattering, Raman scattering, Brillouin scattering, Kerr effect, self-phase modulation and cross-phase modulation, and so on.
6. *Mode coupling*: mode coupling by evanescent field, axial mode coupling, and so on.
7. *Loss-related mechanism*: fiber attenuation, end coupling, fiber bending loss, and so on.
8. *Aided with transducers*: various mechanical structures to convert the measurands to parameters of sensor elements.
9. *Aided with external materials*: reactants and fluorescence.

This list cannot cover all kinds of OFS, neither can this book. Among them the most important fiber sensors are introduced and analyzed in the following chapters.

1.3 OVERVIEW OF THE CHAPTERS

This book is intended to introduce basic OFS with emphases on their principles and physics, to provide helpful fundamentals to students and graduate students of the specialty, and to readers working in research and development. Its contents include seven chapters, introduced as follows.

Chapter 2 gives the fundamentals of optical fibers. After a brief introduction in Section 2.1, the electromagnetic theory of conventional step-index optical fibers is provided in Section 2.2. The gradient index fiber is analyzed in Section 2.3 based on ray optics, and discussed briefly by wave optics. Section 2.4 introduces some special fibers, including the rare-earth-doped active fiber and the double-cladding fiber, the polarization maintaining fiber, and the photonic crystal fiber.

Chapter 3 is devoted to fiber sensitivities and related devices. Fiber sensitivities to strains and temperature changes induced by different conditions are discussed in Section 3.1, which is the basis of the sensing unit of fiber itself, and is used to develop related fiber devices. Section 3.2 introduces fiber couplers, mainly the 2×2 and 3×3 directional couplers. Axial mode coupling is also discussed briefly. Fiber loops based on the couplers are analyzed in Section 3.3, such as the fiber Sagnac loop, fiber ring resonator, fiber Mach–Zenhder interferometer, and fiber Michelson interferometer. These devices have been used widely in fiber sensor systems. Section 3.4 discusses the polarization characteristics of fiber, especially the polarization state evolution in propagation under different conditions. Section 3.5 introduces several polarization devices used in the communication and sensor systems.

Chapter 4 focuses on fiber gratings, which play very active roles in fiber sensor technology. Their basic structures and fabrication processes, as well as photosensitivity are introduced in Section 4.1. Section 4.2 is devoted to the theory of fiber grating and the design methods of various gratings. Several special fiber gratings are analyzed in Section 4.3, such as multisection fiber gratings, chirped fiber Bragg gratings, tilt fiber Bragg gratings, and polarization maintaining fiber gratings. Section 4.4 introduces sensitivities of fiber gratings, their applications in sensor technologies, and the related technical issues.

Chapter 5 introduces distributed OFS, which utilize the effects of light scattering in fibers. Section 5.1 gives a brief introduction of elastic scattering (Rayleigh scattering) and inelastic scatterings (Raman scattering and Brillouin scattering). The distributed fiber sensors based on Rayleigh scattering is discussed in Section 5.2, including optical time domain reflectometer (OTDR), polarization OTDR, phase-sensitive OTDR, and optical frequency domain reflectometer. Section 5.3 introduces sensors based on Raman scattering and its application to temperature sensing, that is, the distributed anti-Stokes Raman thermometry. Section 5.4 discusses sensors based on Brillouin scattering, which are sensitive to both strain and temperature. Two sensors are introduced: Brillouin OTDR and Brillouin optical time domain analyzer. They make use of spontaneous and stimulated Brillouin scatterings, respectively. Several other sensors are briefly introduced in Section 5.5, including those composed of fiber loops, based on low coherence technology, and those based on speckle effect in fibers.

Chapter 6 reviews several fiber optic sensors with special applications in four sections: gyroscopes, hydrophones, fiber Faraday sensor, and sensors based on surface plasmon waves respectively. These sensors have unique features and special applications. Since there have been professional monographs published, this book does not go into detail, but gives basic concepts and principles, and discusses the basic technical issues for practical applications.

Chapter 7 focuses on the extrinsic fiber Fabry–Perot (F-P) interferometer (EFFPI) sensors, which have a heart element of F-P cavity composed of a fiber facet and a mirror such as the diaphragm fiber optic sensor (DFOS). The structures and principles of EFFPI sensors are introduced in Section 7.1. The theoretical analysis based on Gaussian beam is presented in Section 7.2. Section 7.3 describes the basic characteristics and performance of EFFPI sensors. The last section introduces more applications of the EFFPI sensor and discusses some technical issues.

Appendix 1 provides mathematics formulas useful in fiber sensor analyses. Appendices 2 and 3 give the fundamentals of elasticity and polarization optics, respectively. Appendix 4 lists the specifications of related materials and device products, which are used frequently.

REFERENCES

1. Baird JL. British Patent 285,738, 1928. Quoted from Agrawal GP. *Nonlinear Fiber Optics*. San Diego, CA: Elsevier Science, 2004.

2. Hancell CW. US Patent 1751584, 1930. Quoted from Agrawal GP: *Nonlinear Fiber Optics*. San Diego, CA: Elsevier Science, 2004.

3. Van Heel ACS. A new method of transporting optical image without aberration. *Nature* 1954; 173: 39–39.

4. Kao KC, Hockham GA. Dielectric-fiber surface waveguides for optical frequencies. *IEE Proceedings 1966*. Reprinted in IEE Proceedings 1986; 113(Pt. J): 191–198.

5. Kapron FP, Keck DB, Maurer RD. Radiation losses in glass optical waveguides. *Applied Physics Letters* 1970; 17: 423–425.

6. Schalow AL, Townes CH. Infrared and optical masers. *Physics Review* 1958; 112: 1940–1949.

7. Maiman TH. Stimulated optical radiation in ruby. *Nature* 1960; 187: 493–494.

8. Hall RN, Fenner GE, Kingsley JD, Soltys TJ, Carlson RO. Coherent light emission from GaAs junctions. *Physics Review Letters* 1962; 9: 366–368.

9. Kaminow IP, Koch TL. *Optical fiber Telecommunications IIIA*. San Diego, CA: Academic Press, 1997.

10. Culshaw B, Dakin J (Ed.). *Optical Fiber Sensors*. London: Artech House, 1988.

11. Krohn DA. *Fiber Optic Sensors-Fundamentals and Applications*. Second Edition. Research Triangle Park, NC: Instrument Society of America, 1992.

12. Grattan LS, Meggitt BT (Ed.). *Optical Fiber Sensor Technology*. Berlin: Springer, 2000.

13. Goure JP, Verrier I. *Optical Fiber Devices*. Bristol: Institute of Physics Publishing, 2002.

14. Yin S, Ruffin PB, Yu FTS. *Fiber Optic Sensors*. Second Edition. Boca Raton, London, New York: CRC Press; Taylor & Francis Group, 2008.

15. Giallorenzi TG, Bucaro JA, Dandridge A, Sigel GH, Jr., Cole JH, Rashleigh SC, Priest RG. Optical fiber sensor technology. *IEEE Journal of Quantum Electronics* 1982; 18: 626–665.

16. Pitt GD, Extance P, Neat RC, Batchelder DN, Jones RE, Barnett JA, Pratt RH. Optical-fiber sensors. *IEE Proceedings* 1985; 132(Pt. J): 214–248.

17. Rogers A. Distributed optical-fiber sensing. *Measurement Science and Technology* 1999; 10: R75–R99.

18. Lee B. Review of the present status of optical fiber sensors. *Optical Fiber Technology* 2003; 9: 57–79.

19. Culshaw B. Optical fiber sensor technologies: opportunities and—perhaps—pitfalls. *Journal of Lightwave Technology* 2004; 22: 39–50.

20. Krohn D. Market opportunities and standards activities for optical fiber sensors. *Optical Fiber Sensors Conference 2006*, Hamden, CT, paper FB1.

CHAPTER 2

FUNDAMENTALS OF OPTICAL FIBERS

Chapter 2 focuses on the fundamentals of optical fibers, which is the basis of fiber sensors and fiber devices involved in this book. Section 2.1 gives a brief introduction to the fiber, including its structure and fabrication, basic characteristics, and classifications of the main fiber productions. Section 2.2 expounds the electromagnetic theory of step-index fiber, which is the theoretical basis of the fiber waveguide. Gradient-index fibers are analyzed in Section 2.3 by ray optics, which gives a clearer and more explicit view on the light propagation in fiber. Section 2.4 introduces briefly special optical fibers, including rare-earth-doped fibers and double-cladding fibers (DCFs), polarization-maintaining fibers (PMFs), and photonic crystal fibers (PCFs).

2.1 INTRODUCTION TO OPTICAL FIBERS

2.1.1 Basic Structure and Fabrication of Optical Fiber

Optical fiber is a thready material with a circular cross section that makes use of optical total internal reflection (TIR) to guide light waves. Figure 2.1(a) is a typical structure of a step-index fiber, which consists of a core with refractive index n_1 and a cladding layer with index n_2

Fundamentals of Optical Fiber Sensors, First Edition.
Zujie Fang, Ken K. Chin, Ronghui Qu, and Haiwen Cai.
© 2012 John Wiley & Sons, Inc. Published 2012 by John Wiley & Sons, Inc.

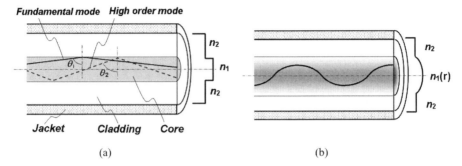

Fundamental mode   High order mode

Jacket   Cladding   Core

(a)   (b)

Figure 2.1 Schematic illustration of optical fibers: (a) step-index fiber and (b) gradient index fiber.

slightly lower than n_1. The light propagating inside the core will be totally reflected at the interface between the core and cladding layer when the incident angle is larger than the critical angle θ_c, and will be well confined in the core. According to Snell's law, the critical angle is determined by the refractive indexes of the core and cladding:

$$\theta_c = \arcsin(n_2/n_1). \tag{2.1}$$

The optical ray with the maximum incident angle θ_1 near $\pi/2$ is called the fundamental mode, whereas rays with smaller angle θ_2, but still larger than θ_c, can also propagate in the fiber and are called high-order modes, as depicted in Figure 2.1(a). The characteristics of optical fibers and their mode properties have been analyzed in detail in books [1–14], and will be discussed in Section 2.2 in this chapter.

Figure 2.1(b) is another kind of fiber, called the gradient index (GRIN) fiber, in which the core index descends with the radial distance r, expressed as

$$n(r) = \begin{cases} n_1[1 - \Delta(r/a)^p] & \text{(for } r \leq a), \\ n_1[1 - \Delta] & \text{(for } r > a), \end{cases} \tag{2.2}$$

where p is a positive real number, a is the core radius, and $\Delta = (n_1 - n_2)/n_1$. In the GRIN fiber, the optical ray turns toward the axis, where the peak index n_1 is located, and takes a waved path.

The fabrication of silica fiber is based mainly on modified chemical vapor deposition (MCVD) technology [15–17]. In the process, SiO_2 of extremely high purity is produced from pure $SiCl_4$, which is the basic material of microelectronic industry, and sintered to a fused silica preform at a temperature of about 1,600°C. In the processing that follows, the preform is drawn into a fiber with diameter ~0.1 mm at high

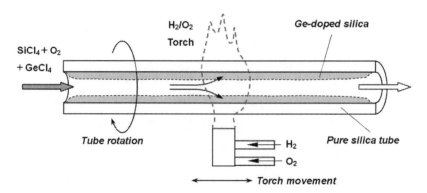

Figure 2.2   Schematic diagram of MCVD process for silica fiber preform.

temperature. To obtain the index difference between core and cladding, the index is adjusted by doping some special impurities into the pure silica, such as GeO_2 and P_2O_5. Figure 2.2 shows a schematic diagram of MCVD processing for Ge-doped silica fiber performs. The doping level is designed roughly by a linear interpolation of the component indexes, such as $n_{Si/Ge} = (1 - f)n_{SiO_2} + f n_{GeO_2}$ for Ge-doped silica. The atomic fraction f of germanium should usually be controlled to be around 3% to get $\Delta n = n_1 - n_2$ at an amount of ~ 0.003 for conventional single-mode fibers (SMF). Another useful method of optical fiber fabrication is called vapor-phase axial-deposition (VAD) method [18], which has the advantages of lower cost and higher productivity. The optical fiber can also be composed of a pure silica core and a silica cladding doped with some lower index elements, such as B_2O_3. Pure silica fiber is considered to have higher long-term reliability.

The bare drawn fiber is coated by a plastic jacket and packaged into a cable to enhance its strength. Many kinds of optical cables have been developed, such as the single-fiber cable, multifiber cable, armored cable, submarine cable, and so on. Most of them are for telecommunication, but are also necessary and useful for fiber sensor technology. Figure 2.3 is a schematic cross-section of a typical fiber cable. Notice that the fibers are usually cased loosely in respective plastic tubes to avoid stress on the fibers during packaging and paving.

2.1.2   Basic Characteristics

2.1.2.1   *Transmission Loss*   Transmission loss is one of the basic characteristics of an optical fiber. Even in fully pure silica, Rayleigh scattering loss still exists due to thermal movement of molecules, which

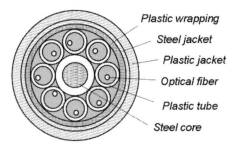

Figure 2.3 Schematic cross-section of a typical fiber cable.

is inversely proportional to λ^4 [10]. Infrared loss will be dominant in the longer wavelength band, thus a low-loss window is formed in the wave band of 1–2 μm, and the lowest loss appears at 1,550 nm for silica fiber, as shown in Figure 2.4 [19]. It is customary to express the fiber loss in units of dB/km by using the relation

$$\alpha_{dB} = \frac{10}{L} \lg \frac{P_0}{P_L},\tag{2.3}$$

where P_0 is the power launched into a fiber, and P_L is the power transmitted through the fiber with length L. It is shown that some high loss

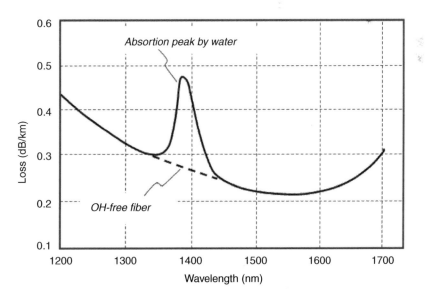

Figure 2.4 Typical loss spectrum of silica fibers. (Reprinted with permission from reference [20].)

peaks exist in the loss spectrum because the silica contains some un-wanted impurities, such as water and OH radicals, which results in loss peaks of close to 1.39 μm. With improvements in processing technology, water molecules have been removed as much possible and the corresponding loss peak almost disappears, as shown in Figure 2.4 by a dashed line [20].

2.1.2.2 Modes It is important to understand that the condition of the incident angle being larger than the critical angle for guiding light in fiber is just a necessary condition, not a sufficient condition. The propagating light must satisfy phase conditions at the boundary between core and cladding, that is, the phase shift of the light wave between successive reflections keeps an integer multiple of 2π. The requirement results in one of the basic characteristics of a guided wave: only with discrete angles can the light beams propagate in the fiber. Among them, the light with the smallest angle to the axis is termed the fundamental mode, and others are high-order modes. When the core radius is small enough and/or the index step is low enough, only the fundamental mode can propagate inside. Such a fiber is called SMF; the other type of fiber being a multimode fiber (MMF). The mode characteristics depend also on wavelength. A fiber can be an SMF for longer wavelengths, but becomes an MMF for shorter wavelengths. Thus a special term of cutoff wavelength is used to define a single-mode wave band of a fiber. The term 'cladding mode' is used when the incident angle θ is smaller than the critical angle, the light is refracted to the cladding layer, but is still reflected at the outer boundary to the air. Radiation mode happens when the incident angle is smaller than the critical angle at the outer boundary and light radiates into the air.

In the electromagnetic theory, the light propagating in the fiber is described as

$$E(t, z) = E_0 \exp[-\alpha z/2 + j(\beta z - \omega t)], \tag{2.4}$$

where $\alpha = (\alpha_{dB}/10)\ln 10$ is the attenuation coefficient in e-based logarithmic scale, $\beta = n_{eff}k_0$ is the propagation constant with vacuum wave vector $k_0 = 2\pi/\lambda$ and effective index n_{eff}. The effective index is a function of fiber structure and working wavelength, and will be deduced in Section 2.2; geometrically it can be regarded as $n_{eff} \sim n_1 \sin\theta$. Different modes correspond to individual propagation constants, and to different field distributions in the transverse cross section of the fiber, which comes from the solution of Helmholtz equations. This will be analyzed in Section 2.2.

2.1.2.3 Dispersion   Another important characteristic of the fiber is its chromatic dispersion [4], namely, the dependence of refractive index on the optical frequency ω, which causes optical signal pulse broadening in fiber communication, and also brings about sensed signal impairment in fiber sensors. The dispersion is mainly attributed to two factors; one of them is material dispersion [21]. The other is the effect of the waveguide on the propagation constant, including the dispersion between modes and the intramodal waveguide dispersion. For MMF, different modes possess different discrete effective indexes, corresponding to propagating rays with different angles and different group velocity v_g in z-direction, called the modal dispersion. For a single mode, the effective index is also a function of wavelength, determined by a Helmholtz equation. Generally, the effect of modal dispersion is much more serious than the intramodal dispersion. This is one of the reasons why SMFs are mostly used in communications.

The dispersion is described by the Taylor expansion of β over the optical frequency:

$$\beta(\omega) = \beta_0 + \beta_1(\omega - \omega_0) + \frac{1}{2}\beta_2(\omega - \omega_0)^2 + \cdots, \qquad (2.5)$$

where ω_0 is the central frequency, and $\beta_m = (\partial^m \beta / \partial \omega^m)|_{\omega=\omega_0}$. The first two coefficients are as follows:

$$\beta_1 = \frac{n_{\text{eff}}}{c} + \frac{\omega}{c}\frac{\partial n_{\text{eff}}}{\partial \omega} = \frac{n_{\text{eff}}}{c} - \frac{\lambda}{c}\frac{\partial n_{\text{eff}}}{\partial \lambda} = \frac{n_g}{c} = \frac{1}{v_g}, \qquad (2.6a)$$

$$\beta_2 = \frac{\partial \beta_1}{\partial \omega} = \frac{\partial}{\partial \omega}\frac{n_g}{c} = \frac{-\lambda^2}{2\pi c^2}\frac{\partial n_g}{\partial \lambda}. \qquad (2.6b)$$

It is shown that group velocity v_g is a function of wavelength λ, which is called group velocity dispersion (GVD). The GVD causes pulse delay and broadening in propagation, which are the problems most people pay close attention to; a dispersion parameter is thus commonly used for group delay per unit length and per unit line-width in ps/(km nm):

$$D = \frac{\mathrm{d}}{\mathrm{d}\lambda}\frac{1}{v_g} = -\frac{\omega}{\lambda}\beta_2 = \frac{\lambda}{c}\frac{\mathrm{d}^2 n_{\text{eff}}}{\mathrm{d}\lambda^2}. \qquad (2.7)$$

The variations of refractive index n and group refractive index n_g of fused silica with wavelength λ are shown in Figure 2.5 [4]. Spectra of the dispersion parameter D of typical single-mode silica fibers are depicted in Figure 2.6, including contributions from waveguide dispersion

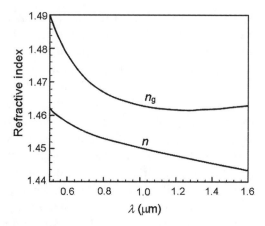

Figure 2.5 Variation of refractive index and group index of silica with wavelength. (Reprinted with permission from reference [4].)

(curve A), from material dispersion (curve B), and from total dispersion (curve C) [5,22].

It is shown in the figures that for conventional single-mode silica fibers, the minimum dispersion is obtained at 1,300 nm range, whereas the lowest loss is located at 1550 nm, where the dispersion parameter is about −17 ps/(km nm). The International Telecommunication Union (ITU) has standardized this kind of fiber, defined as ITU-T G.652 fiber [23].

As an optical component, a fiber's numerical aperture (NA) is an important parameter, which is defined as $NA = \sin\theta_h$, where θ_h is the half-angle of the emitting beam from the fiber's facet. For MMFs, the

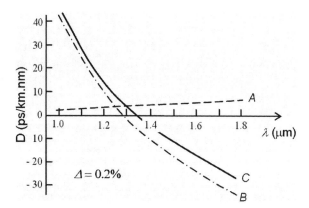

Figure 2.6 Dispersion spectrum of silica fiber: waveguide dispersion (A); material dispersion (B); and total dispersion (C). (Reprinted with permission from reference [5].)

maximum emitting angle is determined by the indexes of core and cladding; the numerical aperture is deduced by Snell's law to be

$$NA = \left(n_1^2 - n_2^2\right)^{1/2} \approx n_1 \sqrt{2\Delta}, \tag{2.8}$$

where $\Delta = (n_1 - n_2)/n_1$, as defined in Equation (2.2). For MMFs, NA is typically 0.175 ± 0.015 [24]. For SMFs the core diameter is so small that the emitting beam is not only determined by the indexes of core and cladding, but also depends on mode field distribution by a Fourier transform relation between the far-field pattern and the near-field pattern, which will be discussed further in Section 2.2. For conventional SMFs, NA is about 0.11–0.12.

2.1.3 Classifications of Optical Fibers

Developed for over 50 years, fiber technology has grown into a large-scale industry. Various fibers are commercially available; and new kinds are being investigated in labs for future practical applications. All kind of fibers may be classified as follows with some overlaps.

2.1.3.1 According to Materials Silica fibers are used most widely, as described above. Others include

- compound glass fiber, with several compositions, such as silicate, phosphate, and fluoride;
- plastic optical fiber (POF), which has attractive features, such as a large aperture, working in visible band, and costs less;
- infrared fiber, which works in mid-infrared band;
- crystal fiber; and
- others.

Generally, different materials are suitable for different working wavelength bands, and provide also various functions with their different features.

2.1.3.2 According to Structures From index distribution, step-index fibers and GRIN fibers as described in Figure 2.1 are the main structures. Double-cladding fibers (DCF), photonic crystal fibers (PCF) and other microstructure fibers are the most attractive types, developing fast in recent years. Polarization maintaining fibers (PMF) are useful in many applications. Large-aperture fibers and fiber bundles are widely used for lightening, endoscopy, and energy transportation.

2.1.3.3 *According to Function and Performance*   Fibers for optical signal transportation are manufactured in large quantities and used widely over the world; these include SMFs and MMFs. To satisfy requirements of optical signal transportation, fibers with modified dispersion properties are developed. Among them, the dispersion-shifted fiber (DSF, $D = 0$ at 1,550 nm) and nonzero dispersion-shifted fiber (NZDSF, $D \sim -4$ ps/km.nm at 1,550 nm) are standardized by ITU as ITU-T G.653 and ITU-T G.655 fibers respectively [25,26]. A small amount of dispersion is left in NZDSF to mitigate the nonlinear four-wave mixing effect and to avoid cross-talk between communication channels in dense wavelength division multiplexing (DWDM) systems. To meet demand in access network applications, such as Fiber-To-The-Home (FTTH), a bending loss insensitive fiber has been developed, standardized as ITU-T G. 657 fiber [27]. Generally, SMFs are preferred for communications and sensors; however, MMFs are still widely used in many applications. The fibers form the basis of optical fiber sensor technologies, not only for sensed signal transportation but also for sensing by the fiber itself. Basic specifications of typical fibers are listed in Appendix 4.

Apart from light transmission, fibers posses other important functions. Fibers with some rare-earth ions doped in their core material can amplify the propagating light when pumped optically. Among them, the erbium-doped silica fiber (EDF) is the most important one and is used to build fiber amplifiers (EDFA) and fiber lasers.

For some applications, such as in distributed fiber sensors, higher optical nonlinear effect plays an important role; consequently fibers with high nonlinear coefficient are developed.

In addition, there are special functional fibers, such as the dispersion compensation fiber, liquid core fiber, image transmission fiber, and so on. Some of these special fibers will be introduced in Section 2.4.

2.2 ELECTROMAGNETIC THEORY OF STEP-INDEX OPTICAL FIBERS

Light propagation in fiber is briefly explained in Section 2.1 by geometric optics, which addresses its basic principles and characteristics. Because fiber size is related to the order of optical wavelength, it is necessary to treat the fiber as a dielectric waveguide, and to get insights into wave evolution in optical fibers by using electromagnetic theory. The theoretical analyses in this section concentrate on conventional step-index fibers, which are the most often used fibers; the GRIN fiber will be discussed in the next section. The electromagnetic theories of optical fibers are given in quite a number of publications, with more detailed analyses and with different emphases [1–8]; references [28,29] give comprehensive theoretical analyses on the electromagnetic theory of optical fibers. As such, this section would focus only on giving a brief introduction.

2.2.1 Maxwell Equations in Cylindrical Coordinates

To obtain comprehensive understanding of light propagation in the fiber, we start with the fundamental theory of electromagnetic fields, that is, Maxwell equations, which were derived from the basic experimental facts including Coulomb's law, Ampere's law, and Faraday's law, written in MKS units as

$$\nabla \cdot \boldsymbol{D} = \rho_f, \quad \nabla \times \boldsymbol{H} = \frac{\partial \boldsymbol{D}}{\partial t} + \boldsymbol{J}, \quad \nabla \cdot \boldsymbol{B} = 0, \quad \nabla \times \boldsymbol{E} = -\frac{\partial \boldsymbol{B}}{\partial t},$$

$$(2.9a)$$

and in a medium with dielectric constant ε and permeability μ:

$$\boldsymbol{D} = \varepsilon \boldsymbol{E}, \qquad \boldsymbol{B} = \mu \boldsymbol{H}. \qquad (2.9b)$$

In a medium with no free charge ρ_f and no current $\boldsymbol{J}$, an electromagnetic wave equation is derived as

$$\nabla^2 \boldsymbol{E} - \frac{n^2}{c^2} \frac{\partial^2 \boldsymbol{E}}{\partial t^2} = 0, \qquad (2.10)$$

where $c = (\mu_0 \varepsilon_0)^{-1/2}$ is the velocity of light in vacuum and $n = (\mu \varepsilon / \mu_0 \varepsilon_0)^{1/2} \sim (\varepsilon / \varepsilon_0)^{1/2}$ is the refractive index of the medium with $\mu \sim \mu_0$ in light wave band. The refractive index is real for media with neither loss nor gain. For a single-frequency light wave with angular frequency of ω, Equation (2.10) is written as the Helmholtz equation:

$$\nabla^2 \boldsymbol{E} + n^2 k_0^2 \boldsymbol{E} = 0, \qquad (2.11)$$

where $k_0 = \omega / c$ is the vacuum wave vector. The electromagnetic wave is derived as

$$\boldsymbol{E}(r, t) = \mathrm{Re}\{\boldsymbol{E}(r) \exp[j(\boldsymbol{k} \cdot \boldsymbol{r} - \omega t)]\}, \qquad (2.12a)$$

$$\boldsymbol{H}(r, t) = \mathrm{Re}\{\boldsymbol{H}(r) \exp[j(\boldsymbol{k} \cdot \boldsymbol{r} - \omega t)]\}. \qquad (2.12b)$$

The phase factor in the above exponential term is taken such that it goes with the wave spatially, not temporally as $[j(\omega t - \boldsymbol{k} \cdot \boldsymbol{r})]$; this is a convenient notation in optics, though there is no difference physically. This notation will continue to be used unless otherwise stated. For a light wave propagating in fiber with its axis set in z-direction, the electric field and magnetic field are written as

$$\boldsymbol{E}(x, y, z, t) = \mathrm{Re}\{\boldsymbol{E}(x, y)\exp[j(\beta z - \omega t)]\}, \qquad (2.12c)$$

$$\boldsymbol{H}(x, y, z, t) = \mathrm{Re}\{\boldsymbol{H}(x, y)\exp[j(\beta z - \omega t)]\}, \qquad (2.12d)$$

where β is the wave vector component in z-direction, termed the propagation constant. The electric field and magnetic field are combined as an electric-magnetic wave with a transverse field distribution of $\boldsymbol{E}(x, y)$ and $\boldsymbol{H}(x, y)$, which satisfy Maxwell equations with following forms for their three components in Cartesian coordinates:

$$\frac{\partial^2 E_i}{\partial x^2} + \frac{\partial^2 E_i}{\partial y^2} + \beta_t^2 E_i = 0 \qquad (i = x, y, z), \qquad (2.13a)$$

$$\frac{\partial^2 H_i}{\partial x^2} + \frac{\partial^2 H_i}{\partial y^2} + \beta_t^2 H_i = 0 \qquad (i = x, y, z), \qquad (2.13b)$$

where $\beta_t^2 = n^2 k_0^2 - \beta^2$ is the transverse wave number. As is well known, the electromagnetic wave in free space is a transverse wave without components in propagation direction. In contrast, the z-components not only exist in dielectric waveguides, but also play an important role because the z-components stand for the guided wave propagating in the transverse direction, implying the effect of waveguide, whereas the transverse components of the field can be expressed as functions of z-components. Therefore, z-components of the electric field and magnetic field are solved first.

A regular step-index fiber is axially symmetric with the refractive index distribution shown in Figure 2.7(a). The wave equations of E_z and H_z are written in cylindrical coordinates as follows:

$$\frac{\partial^2 E_z}{\partial r^2} + \frac{1}{r}\frac{\partial E_z}{\partial r} + \frac{1}{r^2}\frac{\partial^2 E_z}{\partial \varphi^2} + \beta_t^2 E_z = 0,$$

$$\frac{\partial^2 H_z}{\partial r^2} + \frac{1}{r}\frac{\partial H_z}{\partial r} + \frac{1}{r^2}\frac{\partial^2 H_z}{\partial \varphi^2} + \beta_t^2 H_z = 0. \qquad (2.14)$$

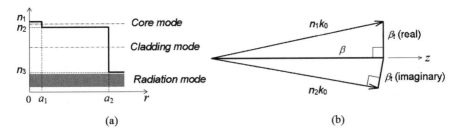

Figure 2.7 (a) Index distribution in a step-index fiber; (b) relations between β and β_t.

The transverse components of the fields are obtained by the following relations deduced from Maxwell equation (2.9):

$$E_r = \frac{j}{\beta_t^2}\left(\beta\frac{\partial E_z}{\partial r} + \frac{\omega\mu}{r}\frac{\partial H_z}{\partial\varphi}\right), \qquad (2.15a)$$

$$E_\varphi = \frac{j}{\beta_t^2}\left(\frac{\beta}{r}\frac{\partial E_z}{\partial\varphi} - \omega\mu\frac{\partial H_z}{\partial r}\right), \qquad (2.15b)$$

$$H_r = \frac{j}{\beta_t^2}\left(\beta\frac{\partial H_z}{\partial r} - \frac{\omega\varepsilon}{r}\frac{\partial E_z}{\partial\varphi}\right), \qquad (2.15c)$$

$$H_\varphi = \frac{j}{\beta_t^2}\left(\frac{\beta}{r}\frac{\partial H_z}{\partial\varphi} + \omega\varepsilon\frac{\partial E_z}{\partial r}\right). \qquad (2.15d)$$

Equations (2.14) for a fiber with circularly uniform refractive index are solved by means of variable separation:

$$E_z(r, \varphi)[\text{and } H_z(r, \varphi)] = R(r)\Theta(\varphi). \qquad (2.16)$$

The equation is decomposed into

$$\frac{d^2\Theta}{d\varphi^2} + \nu^2\Theta = 0, \qquad (2.17)$$

$$\frac{d^2R}{dr^2} + \frac{1}{r}\frac{dR}{dr} + \left(\beta_t^2 - \frac{\nu^2}{r^2}\right)R = 0, \qquad (2.18)$$

where v is an integer, including zero. The refractive index in step-index fibers is uniform in three regions, as depicted in Figure 2.7(a), resulting in constant β_t for each region.

The transverse wave number $\beta_t = \sqrt{n_i^2 k_0^2 - \beta^2} = \sqrt{n_i^2 - n_{\text{eff}}^2} k_0$ ($i = 1, 2, 3$) can be either a real number or an imaginary number, as shown in Figure 2.7(b). Three mode types—core mode, cladding mode, and radiation mode—are classified according to the location of the effective index as shown in Figure 2.7(a).

The solution of Equation (2.17) is

$$\Theta(\varphi) = \begin{cases} \cos v\varphi, \\ \sin v\varphi, \end{cases} \tag{2.19}$$

which is identical for the three regions. Equation (2.18) is a Bessel equation with standard solutions. Because singularities at $r = 0$, and at $r = \infty$ have to be avoided, the general solutions are thus written as follows.

For core modes ($n_1 k_0 > \beta > n_2 k_0 > n_3 k_0$),

$$
\begin{aligned}
R(r) &= A J_v(\beta_{t1} r) & (a_1 \geq r \geq 0), \\
R(r) &= B_1 K_v(\beta_{t2} r) + B_2 I_v(\beta_{t2} r) & (a_2 \geq r > a_1), \\
R(r) &= C K_v(\beta_{t3} r) & (r > a_2).
\end{aligned}
\tag{2.20a}
$$

For cladding modes ($n_1 k_0 > n_2 k_0 > \beta > n_3 k_0$),

$$
\begin{aligned}
R(r) &= A J_v(\beta_{t1} r) & (a_1 \geq r \geq 0), \\
R(r) &= B_1 J_v(\beta_{t2} r) + B_2 N_v(\beta_{t2} r) & (a_2 \geq r > a_1), \\
R(r) &= C K_v(\beta_{t3} r) & (r > a_2).
\end{aligned}
\tag{2.20b}
$$

where $\beta_{t1} = \sqrt{n_1^2 k_0^2 - \beta^2}$, $\beta_{t2} = \sqrt{|n_2^2 k_0^2 - \beta^2|}$, and $\beta_{t3} = \sqrt{|n_3^2 k_0^2 - \beta^2|}$. By usual notations, J_v and N_v are the vth-order Bessel functions of the first and second kinds; I_v and K_v are the modified Bessel functions of the first and second kinds. The third region is usually the air with $n_3 = 1$. When the incident angle at the boundary between the cladding and the surrounding medium is less than the critical angle, the optical beam is radiation mode with $\beta < n_3 k_0$ and $R(r) = C H_v(\beta_{t3} r)$, where H_v is the Hankel function.

2.2.2 Boundary Conditions and Eigenvalue Equations

The coefficients in expression (2.20), A, B, and C, are to be determined by boundary conditions. The boundary conditions of the electromagnetic field for physical states of no surface charge and no surface current are generally expressed as $E_{1t} = E_{2t}$, $\varepsilon_1 E_{1n} = \varepsilon_2 E_{2n}$, $H_{1t} = H_{2t}$, and $H_{1n} = H_{2n}$, where subscript t stands for the tangential components, and n stands for the normal components. The boundary conditions at $r = a_1$ and $r = a_2$ are written as

$$E_z(a_1)|_{\text{core}} = E_z(a_1)|_{\text{cladding}}, \qquad E_z(a_2)|_{\text{cladding}} = E_z(a_2)|_{\text{air}}, \quad (2.21\text{a})$$

$$\varepsilon_1 E_r(a_1)|_{\text{core}} = \varepsilon_2 E_r(a_1)|_{\text{cladding}}, \quad \varepsilon_2 E_r(a_2)|_{\text{cladding}} = \varepsilon_3 E_r(a_2)|_{\text{air}},$$
$$(2.21\text{b})$$

$$E_\varphi(a_1)|_{\text{core}} = E_\varphi(a_1)|_{\text{cladding}}, \qquad E_\varphi(a_2)|_{\text{cladding}} = E_\varphi(a_2)|_{\text{air}}, \quad (2.21\text{c})$$

$$H_z(a_1)|_{\text{core}} = H_z(a_1)|_{\text{cladding}}, \qquad H_z(a_2)|_{\text{cladding}} = H_z(a_2)|_{\text{air}}, \quad (2.21\text{d})$$

$$H_r(a_1)|_{\text{core}} = H_r(a_1)|_{\text{cladding}}, \qquad H_r(a_2)|_{\text{cladding}} = H_r(a_2)|_{\text{air}}, \quad (2.21\text{e})$$

$$H_\varphi(a_1)|_{\text{core}} = H_\varphi(a_1)|_{\text{cladding}}, \qquad H_\varphi(a_2)|_{\text{cladding}} = H_\varphi(a_2)|_{\text{air}}, \quad (2.21\text{f})$$

It is proved that four of the six conditions at each boundary are independent, whereas the other two are compatible.

For a light beam confined in the fiber core, the second boundary is too far away to bring about effects; therefore, a two-region model can replace the three-region model, and $R(r)$ is rewritten as

$$R(r) = \begin{cases} A J_\nu(\beta_{t1} r) & (r \le a = a_1), \\ B K_\nu(\beta_{t2} r) & (r \ge a, \quad n_1 k > \beta > n_2 k), \end{cases} \quad (2.22\text{a})$$

$$R(r) = \begin{cases} A J_\nu(\beta_{t1} r) & (r \le a = a_1), \\ B H_\nu(\beta_{t2} r) & (r \ge a, \quad \beta < n_2 k). \end{cases} \quad (2.22\text{b})$$

Since solutions for z-components of electric and magnetic fields are derived, propagation constant β and coefficients A and B can then be solved by substituting them into the boundary conditions. By denoting

$u = \beta_{t1}a = \sqrt{n_1^2 k_0^2 - \beta^2}\, a$, $w = \beta_{t2}a = \sqrt{|n_2^2 k_0^2 - \beta^2|}\, a$, $x_1 = \beta_{t1}r$, and $x_2 = \beta_{t2}r$, two degenerate modes are obtained, expressed as

$$
E_z^{(x)} = \begin{cases} A J_v(x_1) \cos v\varphi & (r \leq a), \\ B K_v(x_2) \cos v\varphi & (r \geq a), \end{cases}
$$

$$
H_z^{(x)} = \begin{cases} C J_v(x_1) \sin v\varphi & (r \leq a), \\ D K_v(x_2) \sin v\varphi & (r \geq a), \end{cases} \tag{223}
$$

$$
E_z^{(y)} = \begin{cases} A J_v(x_1) \sin v\varphi & (r \leq a), \\ B K_v(x_2) \sin v\varphi & (r \geq a), \end{cases}
$$

$$
H_z^{(y)} = \begin{cases} C J_v(x_1) \cos v\varphi & (r \leq a), \\ D K_v(x_2) \cos v\varphi & (r \geq a). \end{cases} \tag{224}
$$

It is noticed that a $\pi/2$ phase shift exists between Θ_H and Θ_E to obey Maxwell equation (2.9). The two sets of modes are for x-polarization and y-polarization, respectively. A basic relation between u and w exists:

$$
u^2 + w^2 = (n_1^2 - n_2^2)k_0^2 a^2 = V^2, \tag{2.25}
$$

where $V = \sqrt{n_1^2 - n_2^2}\, k_0 a = n\sqrt{2\Delta}k_0 a$ is an important parameter termed the *normalized frequency*.

The boundary conditions of z-direction components of the electric field and magnetic field require $E_z(a)|_{\text{core}} = E_z(a)|_{\text{cladding}}$ and $H_z(a)|_{\text{core}} = H_z(a)|_{\text{cladding}}$, leading to a relation of $B/A = D/C = J_v(u)/K_v(w)$. By denoting the ratio by q, the electric and magnetic field components are deduced from (2.15) as follows:

$$
E_r^{(x)} = j \begin{cases} \dfrac{1}{\beta_{t1}} \left[A\beta J_v'(x_1) + C \dfrac{v\omega\mu_0}{x_1} J_v(x_1) \right] \cos v\varphi, \\[3mm] \dfrac{-q}{\beta_{t2}} \left[A\beta K_v'(x_2) + C \dfrac{v\omega\mu_0}{x_2} K_v(x_2) \right] \cos v\varphi, \end{cases} \tag{2.26a}
$$

$$
E_\varphi^{(x)} = j \begin{cases} \dfrac{-1}{\beta_{t1}} \left[A\dfrac{v\beta}{x_1} J_v(x_1) + C\omega\mu_0 J_v'(x_1) \right] \sin v\varphi, \\[3mm] \dfrac{q}{\beta_{t2}} \left[A\dfrac{v\beta}{x_2} K_v(x_2) + C\omega\mu_0 K_v'(x_2) \right] \sin v\varphi, \end{cases} \tag{2.26b}
$$

$$
H_r^{(x)} = j \begin{cases} \dfrac{1}{\beta_{t1}} \left[A \dfrac{\nu\omega\varepsilon_1}{x_1} J_\nu(x_1) + C\beta J_\nu'(x_1) \right] \sin\nu\varphi, \\[3mm] \dfrac{-q}{\beta_{t2}} \left[A \dfrac{\nu\omega\varepsilon_2}{x_2} K_\nu(x_2) + C\beta K_\nu'(x_2) \right] \sin\nu\varphi, \end{cases} \tag{2.26c}
$$

$$
H_\varphi^{(x)} = j \begin{cases} \dfrac{1}{\beta_{t1}} \left[A\omega\varepsilon_1 J_\nu'(x_1) + C \dfrac{\nu\beta}{x_1} J_\nu(x_1) \right] \cos\nu\varphi, \\[3mm] \dfrac{-q}{\beta_{t2}} \left[A\omega\varepsilon_2 K_\nu'(x_2) + C \dfrac{\nu\beta}{x_2} K_\nu(x_2) \right] \cos\nu\varphi \end{cases} \tag{2.26d}
$$

and

$$
E_r^{(y)} = j \begin{cases} \dfrac{1}{\beta_{t1}} \left[A\beta J_\nu'(x_1) - C \dfrac{\nu\omega\mu_0}{x_1} J_\nu(x_1) \right] \sin\nu\varphi, \\[3mm] \dfrac{-q}{\beta_{t2}} \left[A\beta K_\nu'(x_2) - C \dfrac{\nu\omega\mu_0}{x_2} K_\nu(x_2) \right] \sin\nu\varphi, \end{cases} \tag{2.27a}
$$

$$
E_\varphi^{(y)} = j \begin{cases} \dfrac{1}{\beta_{t1}} \left[A \dfrac{\nu\beta}{x_1} J_\nu(x_1) - C\omega\mu_0 J_\nu'(x_1) \right] \cos\nu\varphi, \\[3mm] \dfrac{-q}{\beta_{t2}} \left[A \dfrac{\nu\beta}{x_2} K_\nu(x_2) - C\omega\mu_0 K_\nu'(x_2) \right] \cos\nu\varphi, \end{cases} \tag{2.27b}
$$

$$
H_r^{(y)} = j \begin{cases} \dfrac{-1}{\beta_{t1}} \left[A \dfrac{\nu\omega\varepsilon}{x_1} J_\nu(x_1) - C\beta J_\nu'(x_1) \right] \cos\nu\varphi, \\[3mm] \dfrac{q}{\beta_{t2}} \left[A \dfrac{\nu\omega\varepsilon}{x_2} K_\nu(x_2) - C\beta K_\nu'(x_2) \right] \cos\nu\varphi, \end{cases} \tag{2.27c}
$$

$$
H_\varphi^{(y)} = j \begin{cases} \dfrac{1}{\beta_{t1}} \left[A\omega\varepsilon J_\nu'(x_1) - C \dfrac{\nu\beta}{x_1} J_\nu(x_1) \right] \sin\nu\varphi, \\[3mm] \dfrac{-q}{\beta_{t2}} \left[A\omega\varepsilon K_\nu'(x_2) - C \dfrac{\nu\beta}{x_2} K_\nu(x_2) \right] \sin\nu\varphi, \end{cases} \tag{2.27d}
$$

where the prime of Bessel functions stands for derivatives to the argument $x_{1,2}$, $\mu_1 = \mu_2 = \mu_0$ for nonmagnetic medium in optical frequency band, and the upper and lower lines are for region ($r \leq a$) and ($r > a$), respectively. By using conditions (2.21b) and (2.21c) we obtain

$$
A\beta \left[\frac{\varepsilon_1 J_\nu'(u)}{u J_\nu(u)} + \frac{\varepsilon_2 K_\nu'(w)}{w K_\nu(w)} \right] \pm C\nu\omega\mu_0 \left[\frac{\varepsilon_1}{u^2} + \frac{\varepsilon_2}{w^2} \right] = 0 \tag{2.28}
$$

and

$$Av\beta \left[\frac{1}{u^2} + \frac{1}{w^2} \right] \pm C\omega\mu_0 \left[\frac{J'_\nu(u)}{uJ_\nu(u)} + \frac{K'_\nu(w)}{wK_\nu(w)} \right] = 0, \qquad (2.29)$$

where sign "+" in front of coefficient C corresponds to the x-polarization mode, whereas sign "–" corresponds to the y-polarization mode. Then, we arrive at an equation [6]:

$$\left[\frac{J'_\nu(u)}{uJ_\nu(u)} + \frac{K'_\nu(w)}{wK_\nu(w)} \right] \left[\frac{\varepsilon_1 J'_\nu(u)}{uJ_\nu(u)} + \frac{\varepsilon_2 K'_\nu(w)}{wK_\nu(w)} \right] = \nu^2 \left(\frac{1}{u^2} + \frac{1}{w^2} \right) \left(\frac{\varepsilon_1}{u^2} + \frac{\varepsilon_2}{w^2} \right).$$

$$(2.30a)$$

An equivalent equation is derived from conditions (2.21c) and (2.21e) [3,7]:

$$\left[\frac{J'_\nu(u)}{uJ_\nu(u)} + \frac{K'_\nu(w)}{wK_\nu(w)} \right] \left[\frac{\varepsilon_1 J'_\nu(u)}{uJ_\nu(u)} + \frac{\varepsilon_2 K'_\nu(w)}{wK_\nu(w)} \right] = \frac{\nu^2 \beta^2}{\omega^2 \mu_0} \left(\frac{1}{u^2} + \frac{1}{w^2} \right)^2.$$

$$(2.30b)$$

Referring to the definition of u and w, Equations (2.30a) and (2.30b) are the eigenvalue equations to determine the propagation constant as $\beta = \beta(V)$.

Because Bessel functions $J_\nu(u)$ are in forms of decayed periodic oscillation, the eigen equation generally has multiple roots for a certain integer ν, which are labeled by integer $l = 1, 2, \ldots$. It means that only discrete values of β hold for the guided modes; and a pair of integers, ν and l, are used to specify a mode. This is the basic property of guided modes, reflecting the phase matching requirement at the boundary of a waveguide. β can take continuously changing values only for radiation modes.

2.2.3 Weakly Guiding Approximation, Hybrid Modes, and Linear Polarized Modes

For conventional fibers used in fiber communication and fiber sensor technology, the index step between core and cladding is small, typically around 0.003. It does not cause obvious errors in solving the eigen equation to take an approximation of $\varepsilon_1 \approx \varepsilon_2$, called *weakly guiding approximation* [29]. Equation (2.30a) is then simplified as

$$\left[\frac{J'_\nu(u)}{uJ_\nu(u)} + \frac{K'_\nu(w)}{wK_\nu(w)} \right] = \pm\nu \left(\frac{1}{u^2} + \frac{1}{w^2} \right). \qquad (2.31)$$

Its solutions are discussed for three cases: (1) $\nu = 0$; (2) $\nu \neq 0$ with sign "$-$"; (3) $\nu \neq 0$ with sign "$+$." For $\nu = 0$, solution of Equation (2.17) turns into $\Theta(\varphi) = const.$; and coefficients in expressions (2.23) and (2.24) fall into one of two choices: $A = B = 0$ or $C = D = 0$. The former corresponds to transverse electric (TE) mode, and the later to transverse magnetic (TM) mode. They have the same eigen equation of

$$\frac{J_0'(u)}{u J_0(u)} + \frac{K_0'(w)}{w K_0(w)} = 0. \tag{2.32a}$$

By using following identities of Bessel functions (See Appendix 1):

$$J_\nu'(x) = \frac{\nu}{x} J_\nu(x) - J_{\nu+1}(x) = -\frac{\nu}{x} J_\nu(x) + J_{\nu-1}(x),$$

$$K_\nu'(x) = \frac{\nu}{x} K_\nu(x) - K_{\nu+1}(x) = -\frac{\nu}{x} K_\nu(x) - K_{\nu-1}(x),$$

Equation (2.32a) is rewritten as

$$\frac{J_1(u)}{u J_0(u)} + \frac{K_1(w)}{w K_0(w)} = 0. \tag{2.32b}$$

For $\nu \neq 0$, the eigen equation is rewritten as

$$\frac{J_{\nu-1}(u)}{u J_\nu(u)} - \frac{K_{\nu-1}(w)}{w K_\nu(w)} = 0, \tag{2.32c}$$

$$\frac{J_{\nu+1}(u)}{u J_\nu(u)} + \frac{K_{\nu+1}(w)}{w K_\nu(w)} = 0, \tag{2.32d}$$

for sign "$-$" and for sign "$+$" in (2.31), respectively.

It is seen from Equation (2.28) and (2.29) that in case of $\nu \neq 0$ neither A nor C is equal to zero. That is, the guided light wave is neither a TE mode, nor a TM mode, which is essentially different from the transverse electromagnetic wave in free space. The modes are termed the *hybrid mode* and denoted by *HE* modes and *EH* modes for cases (2.32c) and (2.32d), respectively.

Formulas (2.28) and (2.29) give the proportionality between the electric and magnetic field amplitudes: $C = (\beta/\omega\mu_0)A$ for *HE* modes of x-polarization and *EH* modes of y-polarization; and $C = -(\beta/\omega\mu_0)A$ for the other two mode types, respectively.

By redefining the subscript as [6]: $m = 1$ for *TM* and *TE* modes with $\nu = 0$, $m = \nu - 1$ for *HE* modes with $\nu > 0$, and $m = \nu + 1$ for

EH modes with $\nu > 0$, the three eigen equations—(2.32b), (2.32c), and (2.32d)—can be rewritten in a unified form:

$$\frac{u J_{m-1}(u)}{J_m(u)} + \frac{w K_{m-1}(w)}{K_m(w)} = 0. \tag{2.33}$$

The identical eigen equation indicates that *HE*, *EH*, *TE*, and *TM* modes have an identical propagation constant β for the same mode orders of *m* and *l*. In other words, they are degenerate under weakly guiding approximation. The degenerate modes can constitute the *linear polarized mode*, denoted by LP_{ml}, whose polarization properties are discussed hereafter.

The eigen equations are to be solved numerically for different fiber structure parameters and wavelength. To show the propagation constants as functions of the normalized frequency V, a parameter, called the normalized effective index, is defined as

$$b = \frac{n_{\text{eff}}^2 - n_2^2}{n_1^2 - n_2^2} = \frac{\beta^2 - n_2^2 k_0^2}{(n_1^2 - n_2^2)k_0^2} = \frac{w^2}{V^2}. \tag{2.34}$$

The effective index is then expressed as $n_{\text{eff}}^2 = b(n_1^2 - n_2^2) + n_2^2$. The curves of $b \sim V$ are depicted in Figure 2.8 for several lowest order modes.

It is noticed that the propagation constant β of a certain mode decreases with the normalized frequency V decreasing, until β reaches $n_2 k_0$, where $b = 0$, $w = 0$, and $u = V$. It means that the mode does not exist below the corresponding value of V_c; in other words, the mode is

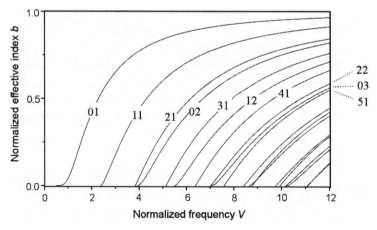

Figure 2.8 Normalized effective index varied with normalized frequency.

cut off at that point. For a certain fiber with fixed structure parameters, the normalized frequency V decreases with wavelength increasing. As wavelength increases, the existing mode number decreases, until only the fundamental mode exists. In the range that is far from cutoff, the propagation constant and the effective index can be approximated as

$$\beta \approx n_1 k_0 - u^2/(2n_1 k_0 a^2), \tag{2.35a}$$

$$n_{\text{eff}} \approx n_1 - u^2/(2n_1 k_0^2 a^2). \tag{2.35b}$$

This expression is useful in discussing the fiber characteristics [33].

The fundamental mode exists without wavelength limitation, as its cutoff frequency is $V_c = 0$, corresponding to an infinitive λ. The fundamental mode is one with the lowest integers of m and l, that is, LP_{01} or HE_{11} mode, corresponding to the first root of the eigen equation:

$$\frac{J_0(u)}{u J_1(u)} - \frac{K_0(w)}{w K_1(w)} = 0. \tag{2.36}$$

The conditions of mode cutoff for higher order modes are deduced from the eigen equation under condition of $w = 0$. The limiting forms of $K_m(w)$ for $w \to 0$ are expressed as $K_m(w) = 2^{m-1}(m - 1)!/w^m$ for $m \geq 1$; and $K_0(w) \approx \ln(2/\gamma w)$ with Euler constant $\gamma \cong 1.781$. The cut-off condition is written as

$$\frac{V J_{m-1}(V)}{J_m(V)} = \frac{-w K_{m-1}(w)}{K_m(w)} \approx -\frac{w^2}{2(m-1)} \to 0,$$

that is, $$J_{m-1}(V) = 0. \tag{2.37}$$

The cutoff condition of mode LP_{11} is the first zero of $J_0(V)$, that is, $V_c = 2.4048$. This is the condition of SMF: below this point only the LP_{01} mode can propagate in the fiber. The equations to determine the cutoff frequency for the several lowest order modes under weakly guiding approximation are listed in Table 2.1.

NB: LP_{2l} mode and $LP_{0(l+1)}$ mode have different eigen equations, but the same cutoff frequency, because $J_{-1}(x) = -J_1(x)$.

2.2.4 Field Distribution and Polarization Characteristics

On the basis of the eigen equation and its solution $\beta = \beta_{vl}$, the field distributions of individual modes are obtained, expressed as follows:

Table 2.1   Cut-Off Frequencies of Low Order Modes

LP Mode	Hybrid mode	Equation	V_c	Modes cut-off
LP_{01}	HE_{11}	$J_0(V) = 0$ $(l = 1)$	2.4048	LP_{11} and higher modes
LP_{11}	$TM_{01}/TE_{01}/HE_{21}$	$J_1(V) = 0$ $(l = 1)$	3.8317	LP_{21}, LP_{02}, and higher modes
LP_{21}, LP_{02}	$HE_{12}/HE_{31}/EH_{11}$	$J_2(V) = 0$ $(l = 1)$	5.1356	LP_{31} and higher modes
LP_{31}	HE_{41}/EH_{21}	$J_0(V) = 0$ $(l = 2)$	5.5201	LP_{12} and higher modes
LP_{12}	$TM_{02}/TE_{02}/HE_{22}$	$J_3(V) = 0$ $(l = 1)$	6.3802	LP_{41} and higher modes
LP_{41}	HE_{51}/EH_{31}	$J_1(V) = 0$ $(l = 2)$	7.0156	LP_{22}, LP_{03}, and higher modes

$HE_{\nu l}$ modes ($\nu \neq 0$):

$$E_z^{(x)} = jA\frac{\beta_{t1}}{\beta}\begin{cases} J_\nu(\beta_{t1}r)\cos\nu\varphi, \\ qK_\nu(\beta_{t2}r)\cos\nu\varphi, \end{cases} \qquad E_z^{(y)} = jA\frac{\beta_{t1}}{\beta}\begin{cases} J_\nu(\beta_{t1}r)\sin\nu\varphi, \\ qK_\nu(\beta_{t2}r)\sin\nu\varphi, \end{cases}$$

$$\tag{2.38a}$$

$$H_z^{(x)} = -jA\frac{\beta_{t1}}{\omega\mu_0}\begin{cases} J_\nu(\beta_{t1}r)\sin\nu\varphi, \\ qK_\nu(\beta_{t2}r)\sin\nu\varphi, \end{cases} \qquad H_z^{(y)} = -jA\frac{\beta_{t1}}{\omega\mu_0}\begin{cases} J_\nu(\beta_{t1}r)\cos\nu\varphi, \\ qK_\nu(\beta_{t2}r)\cos\nu\varphi, \end{cases}$$

$$\tag{2.38b}$$

$$E_r^{(x)} = A\begin{cases} (1/u)J_{\nu-1}(\beta_{t1}r)\cos\nu\varphi, \\ (q/w)K_{\nu-1}(\beta_{t2}r)\cos\nu\varphi, \end{cases} \qquad E_r^{(y)} = A\begin{cases} (1/u)J_{\nu-1}(\beta_{t1}r)\sin\nu\varphi, \\ (q/w)K_{\nu-1}(\beta_{t2}r)\sin\nu\varphi, \end{cases}$$

$$\tag{2.38c}$$

$$E_\varphi^{(x)} = A\begin{cases} (1/u)J_{\nu-1}(\beta_{t1}r)\sin\nu\varphi, \\ (q/w)K_{\nu-1}(\beta_{t2}r)\sin\nu\varphi, \end{cases} \qquad E_\varphi^{(y)} = A\begin{cases} (1/u)J_{\nu-1}(\beta_{t1}r)\cos\nu\varphi, \\ (q/w)K_{\nu-1}(\beta_{t2}r)\cos\nu\varphi. \end{cases}$$

$$\tag{2.38d}$$

$EH_{\nu l}$ modes ($\nu \neq 0$):

$$E_z^{(x)} = jA\frac{\beta_{t1}}{\beta}\begin{cases} J_\nu(\beta_{t1}r)\cos\nu\varphi, \\ qK_\nu(\beta_{t2}r)\cos\nu\varphi, \end{cases} \qquad E_z^{(y)} = jA\frac{\beta_{t1}}{\beta}\begin{cases} J_\nu(\beta_{t1}r)\sin\nu\varphi, \\ qK_\nu(\beta_{t2}r)\sin\nu\varphi, \end{cases}$$

$$\tag{2.39a}$$

$$H_z^{(x)} = -j\frac{A\beta_{t1}}{\omega\mu_0}\begin{cases} J_\nu(\beta_{t1}r)\sin\nu\varphi, \\ qK_\nu(\beta_{t2}r)\sin\nu\varphi, \end{cases} \qquad H_z^{(y)} = j\frac{A\beta_{t1}}{\omega\mu_0}\begin{cases} J_\nu(\beta_{t1}r)\cos\nu\varphi, \\ qK_\nu(\beta_{t2}r)\cos\nu\varphi, \end{cases}$$

$$\tag{2.39b}$$

$$E_r^{(x)} = A\begin{cases} (1/u)J_{\nu+1}(\beta_{t1}r)\cos\nu\varphi, \\ (q/w)K_{\nu+1}(\beta_{t2}r)\cos\nu\varphi, \end{cases} \qquad E_r^{(y)} = A\begin{cases} (1/u)J_{\nu+1}(\beta_{t1}r)\sin\nu\varphi, \\ (q/w)K_{\nu+1}(\beta_{t2}r)\sin\nu\varphi, \end{cases}$$

$$\tag{2.39c}$$

$$E_\varphi^{(x)} = A \begin{cases} (1/u)J_{v+1}(\beta_{t1}r)\sin v\varphi, \\ (q/w)K_{v+1}(\beta_{t2}r)\sin v\varphi, \end{cases} \qquad E_\varphi^{(y)} = A \begin{cases} (1/u)J_{v+1}(\beta_{t1}r)\cos v\varphi, \\ (q/w)K_{v+1}(\beta_{t2}r)\cos v\varphi, \end{cases}$$

(2.39d)

TM modes:

$$E_z^{(\text{TM})} = jA\frac{\beta_{t1}}{\beta} \begin{cases} J_0(\beta_{t1}r), \\ qK_0(\beta_{t2}r), \end{cases} \qquad H_z^{(\text{TM})} = 0, \qquad (2.40a)$$

$$E_r^{(\text{TM})} = A \begin{cases} (1/u)J_1(\beta_{t1}r), \\ -(q/w)K_1(\beta_{t2}r), \end{cases} \qquad H_r^{(\text{TM})} = 0, \qquad (2.40b)$$

$$E_\varphi^{(\text{TM})} = 0, \qquad H_\varphi^{(\text{TM})} = A\frac{\omega\varepsilon_0}{\beta} \begin{cases} (n_1^2/u)J_1(\beta_{t1}r), \\ (n_2^2q/w)K_1(\beta_{t2}r). \end{cases} \qquad (2.40c)$$

TE modes:

$$E_z^{(\text{TE})} = 0, \qquad H_z^{(\text{TE})} = -jA\frac{\beta_{t1}}{\omega\mu_0} \begin{cases} J_0(\beta_{t1}r), \\ qK_0(\beta_{t2}r), \end{cases} \qquad (2.41a)$$

$$E_r^{(\text{TE})} = 0, \qquad H_r^{(\text{TE})} = -A\frac{\beta}{\omega\mu_0} \begin{cases} (1/u)J_1(\beta_{t1}r), \\ (q/w)K_1(\beta_{t2}r), \end{cases} \qquad (2.41b)$$

$$E_\varphi^{(\text{TE})} = A \begin{cases} (1/u)J_1(\beta_{t1}r), \\ (q/w)K_1(\beta_{t2}r), \end{cases} \qquad H_\varphi^{(\text{TE})} = 0. \qquad (2.41c)$$

NB: The variables $q, u,$ and w are functions of mode order (m, l), determined by the eigen equation. Constant A is to be normalized by optical power, leaving a ratio factor $(j\beta_{t1}/\beta)$ in expressions of E_z and H_z. In each of the expressions, the upper line is for core region and the lower line is for cladding. H_r and H_φ are omitted in (2.38) and (2.39) for simplicity. Figure 2.9 shows the electric and magnetic lines for several lowest order modes.

It is seen that *TM* modes are radially polarized waves, and *TE* modes are azimuthally polarized waves. The direction of field vectors are functions of transverse positions (r, φ) for most of the modes. It is necessary to investigate their polarization state in Cartesian coordinates. On

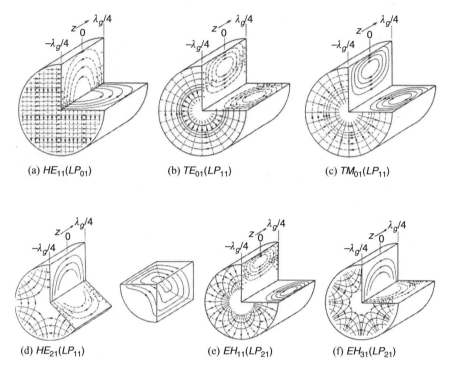

(a) $HE_{11}(LP_{01})$ (b) $TE_{01}(LP_{11})$ (c) $TM_{01}(LP_{11})$

(d) $HE_{21}(LP_{11})$ (e) $EH_{11}(LP_{21})$ (f) $EH_{31}(LP_{21})$

Figure 2.9 Electric (solid lines) and magnetic (dashed lines) fields of modes. (Reprinted with permission from reference [6].)

the basis of the transform relations between cylindrical and Cartesian coordinates:

$$E_x = E_r \cos\varphi - E_\varphi \sin\varphi,$$
$$E_y = E_r \sin\varphi + E_\varphi \cos\varphi, \tag{2.42}$$

the electric fields of x-polarization mode in Cartesian coordinates are deduced as

$$E_x^{HE} = A_1 \begin{cases} (1/u)J_{\nu-1}\cos(\nu-1)\varphi, \\ (q/w)K_{\nu-1}\cos(\nu-1)\varphi, \end{cases} \quad E_y^{HE} = A_1 \begin{cases} (1/u)J_{\nu-1}\sin(\nu-1)\varphi, \\ (q/w)K_{\nu-1}\sin(\nu-1)\varphi, \end{cases}$$
$$\tag{2.43a}$$

$$E_x^{EH} = A_2 \begin{cases} (1/u)J_{\nu+1}\cos(\nu+1)\varphi, \\ (q/w)K_{\nu+1}\cos(\nu+1)\varphi, \end{cases} \quad E_y^{EH} = -A_2 \begin{cases} (1/u)J_{\nu+1}\sin(\nu+1)\varphi, \\ (q/w)K_{\nu+1}\sin(\nu+1)\varphi, \end{cases}$$
$$\tag{2.43b}$$

$$E_x^{TM} = A_3 \begin{cases} (1/u)J_1(x_1)\cos\varphi, \\ -(q/w)K_1(x_2)\cos\varphi, \end{cases} \quad E_y^{TM} = A_3 \begin{cases} (1/u)J_1(x_1)\sin\varphi, \\ -(q/w)K_1(x_2)\sin\varphi, \end{cases}$$
$$\tag{2.43c}$$

$$E_x^{\text{TE}} = A_4 \begin{cases} (1/u)J_1(x_1)\sin\varphi, \\ -(q/w)K_1(x_2)\sin\varphi, \end{cases} \qquad E_y^{\text{TE}} = -A_4 \begin{cases} (1/u)J_1(x_1)\cos\varphi, \\ -(q/w)K_1(x_2)\cos\varphi. \end{cases}$$

$$(2.43\text{d})$$

The fields of the y-polarization mode are obtained by exchanging x and y. Obviously, they can be merged in the same expressions as LP modes by mode index m with superscripts (x) and (y) for two degenerate polarization modes:

$$E_x^{(x)} \propto E_y^{(y)} = A \begin{cases} (1/u)J_m(\beta_{t1}r)\cos m\varphi, \\ (q/w)K_m(\beta_{t2}r)\cos m\varphi, \end{cases} \qquad (2.44\text{a})$$

$$E_y^{(x)} \propto E_x^{(y)} = A \begin{cases} (1/u)J_m(\beta_{t1}r)\sin m\varphi, \\ (q/w)K_m(\beta_{t2}r)\sin m\varphi. \end{cases} \qquad (2.44\text{b})$$

It is seen that the mode HE_{1l} (LP_{0l}) itself is a linear polarization mode; whereas the other modes are not polarized by their selves. However, a linear polarization mode can be composed of hybrid modes. For example, by taking $A_2 = A_1$, the sum of HE_{ml} and EH_{ml} modes gives an x-polarized LP mode. A proper combination of the TM, TE, and HE_{2l} modes gives the LP_{1l} mode. The most important and useful mode is the fundamental mode with $m = 0$ and $l = 1$, that is, EH_{11} (LP_{01}) mode, whose fields are depicted in Figure 2.9(a).

The *linear polarized modes* (LP_{ml}) are just composed of the degenerate hybrid modes with the same m, l. The LP modes can be derived from Helmholtz's equation in Cartesian coordinates [6–8]. To meet the cylindrical boundary conditions with Cartesian coordinates, some approximations are taken in deriving LP modes: the first is that the x- (or y-) component of the magnetic field is neglected in deriving the x- (or y-)direction electric field; the second is that approximation of $\beta \approx n_1 k_0 \approx n_2 k_0$ is assumed to simplify expressions of the LP modes. The usual deduction of LP modes is not repeated here; instead, the LP modes are regarded as the combination of the hybrid modes, so long as weak guiding approximation holds; and the same propagation constant of the degenerate modes ensures an identical phase factor of $(\beta z - \omega t)$ in propagation in the fiber.

The intensity distributions of LP_{11x} and LP_{11y} are shown schematically in Figure 2.10 with arrows indicating polarization directions. Similarly, Figure 2.11 shows that of LP_{21} modes. It is shown from the mode distributions that the two indexes, ν and l, are related to the node numbers of the electric and magnetic fields in fiber circumference and

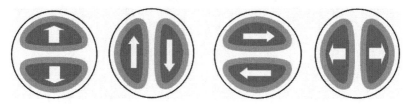

Figure 2.10 Optical intensity distributions of LP_{11} modes with electric field directions.

in its radius, respectively, indicating the standing waves in the two directions, respectively.

It is noticed that although all eigenmodes and their combinations are not exact transverse waves, except the TE modes with $E_z = 0$, the z-component of the electric field is quite small compared with the transverse components. The ratio of their amplitudes is estimated to be [8]

$$\frac{\beta_{t1}^2}{\beta^2} \approx \frac{2(n_1 - n_{\text{eff}})}{n_{\text{eff}}} \approx 2\Delta, \qquad (2.45)$$

which is much smaller than 1 in a weakly guiding fiber. For conventional SMFs with $(n_1 - n_2) \sim 0.003$, the ratio is around 0.004. Moreover, referring to the Poynting vector, E_z gives no contribution to z-direction energy flow; it represents just the standing wave transversely.

Transverse distributions of LP mode intensity depend on the normalized frequency V. Figure 2.12 shows radial distributions of the normalized power density of LP_{01} and LP_{11} modes, with the normalized effective index $b = 0.9$ for far away from cutoff, and $b = 0.1$ for near cutoff [6].

The fraction of mode power traveling within core is a useful parameter, expressed as [7]:

$$\Gamma = \frac{P_{\text{core}}}{P_{\text{total}}} = \frac{w^2}{V^2}\left[1 - \frac{J_m^2(u_{ml})}{J_{m+1}(u_{ml})J_{m-1}(u_{ml})}\right] \qquad (2.46a)$$

Figure 2.11 Optical intensity distributions of LP_{21} modes with electric field directions.

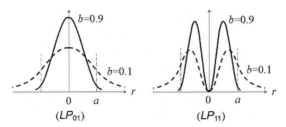

Figure 2.12 Normalized power density distributions of *LP* modes.

and complementarily as [3]

$$\Gamma_2 = \frac{P_{\text{clad}}}{P_{\text{total}}} = \frac{1}{V^2}\left[u^2 + \frac{w^2 J_m^2(u_{ml})}{J_{m+1}(u_{ml})J_{m-1}(u_{ml})}\right]. \qquad (2.46b)$$

2.2.5 Multimode Fiber and Cladding Modes

If the normalized frequency is larger than $V_c = 2.4048$ for a step-index fiber, it becomes an MMF. The core diameter of mass-produced MMFs is set to 62.5 μm, as standardized by ITU [24]; and the index step is also set larger than that of SMF, resulting in a much larger normalized frequency. It is obvious that much more guided modes can exist. The total mode number N is estimated based on the mode cutoff condition, $J_{m-1}(V) = 0$. The Bessel function for a large argument is expressed approximately as (A1.23)

$$J_{m-1}(x) \approx \sqrt{\frac{2}{\pi x}} \cos\left(x - \frac{m\pi}{2} + \frac{\pi}{4}\right). \qquad (2.47)$$

Therefore, the roots of the cutoff equation are obtained as $x_{ml} = (m + 2l)\pi/2 - 5\pi/4$, where the first root is numbered as $l = 1$; and the root spacing is approximated to be π. The number of mode index l for $u \leq V$ is then deduced to be $l_{\max} = V/\pi - m/2 + 5/4$; and the maximum of index m for $l = 1$ is $M \approx 2V/\pi$. The total number of modes is then calculated as [1]

$$N = 2 \times 2 \times \sum_{m=0}^{M}\left(\frac{V}{\pi} - \frac{m}{2}\right) = 4\left[\frac{MV}{\pi} - \frac{M(M+1)}{4}\right] \approx \frac{4V^2}{\pi^2}. \quad (2.48)$$

The two doublings are for two polarizations and for *HE* and *EH* modes. The number may be overestimated in region of low index l, but with a minor error for large V. It is often approximated further to

$N = V^2/2$, valid for $V > 20$ [13]. For conventional MMFs with a core diameter of 62.5 μm and $NA \sim 0.175$, the normalized frequency V reaches 23–26 in 1,300–1,500 nm band; and the total mode number reaches 800–1,100.

It is almost impossible to excite only one mode in a conventional MMF. The actual field pattern is a sum of all existing modes. For optical signal propagations the multiple modes cause serious pulse-width broadening due to the modal dispersion, as stated in Section 2.1, which will be discussed further in Section 2.3 in comparison with GRIN fibers. Generally the optical power tends to propagate inside the MMF's core and the percentage of power in the cladding is approximated to [13]

$$\Gamma_2 = P_{\text{clad}}/P_{\text{total}} = 2\sqrt{2}/3V. \tag{2.49}$$

Apart from the SMFs and conventional MMFs, fibers with a few modes are also developed, such as two-mode fibers with LP_{01} and LP_{11} modes, for certain applications [30].

Cladding Modes. In the above analyses, a mode with its effective index below n_2 would be regarded as radiation mode. Actually, the modes with effective index between n_2 and n_3 are cladding modes. In view of geometric optics, if the incident angle at the boundary between the cladding and the surrounding is larger than the critical angle $\theta_c = \sin^{-1}(n_3/n_2)$, the propagating light is confined in the cladding layer. Although cladding modes are usually with high loss, its effects play important roles in some cases and in some devices. To investigate cladding mode properties, a three-region model must be used instead of the two-region model. In addition, as the index of surrounding area is usually much smaller than the cladding, the weak guiding approximation will no longer hold; the electromagnetic theory for cladding modes should start from expressions (2.20) and boundary conditions (2.21). Therefore, the theoretical analysis of cladding modes is much more complicated. The detailed theoretical analyses for the cladding modes are given in reference [31].

Here, a simplified model is presented. Because the index step between core and cladding is usually about 0.003, much smaller than the step between cladding and surrounding medium, about 0.45 for the air, the cladding modes are regarded approximately as the modes existing in a multimode waveguide composed of the cladding and the small core with index n$_2$, surrounded by medium n_3. For the simplified two-region fiber, the above theoretical analysis and results are applicable, including the mode expressions (2.26) and (2.27) and the eigen equations

(2.30a) and (2.30b). By using relations of Bessel functions, (2.26) for $\nu > 0$ is deduced as [28]

$$E_r^{(1)} = \frac{jA\beta a}{2} \begin{cases} (1/u)[(1+p)J_{\nu-1} - (1-p)J_{\nu+1}]\cos\nu\varphi, \\ (q/w)[(1+p)K_{\nu-1} - (1-p)K_{\nu+1}]\cos\nu\varphi, \end{cases} \quad (2.50a)$$

$$E_\varphi^{(1)} = \frac{-jA\beta a}{2} \begin{cases} (1/u)[(1+p)J_{\nu-1} + (1-p)J_{\nu+1}]\sin\nu\varphi, \\ (q/w)[(1+p)K_{\nu-1} + (1-p)K_{\nu+1}]\sin\nu\varphi, \end{cases} \quad (2.50b)$$

where $p = (C/A)(\mu_0\omega/\beta)$ is denoted for the z-component ratio of the electric and magnetic fields. Under weakly guiding approximation, $p = 1$ corresponds for HE modes, and $p = -1$ for EH modes. Here p does not equal 1, nor -1. The eigen equation is rewritten as

$$\left(\frac{J_\nu'}{uJ_\nu} + \frac{K_\nu'}{wK_\nu}\right)\left(\frac{\varepsilon_2 J_\nu'}{uJ_\nu} + \frac{\varepsilon_3 K_\nu'}{wK_\nu}\right) = \nu^2\left(\frac{1}{u^2} + \frac{1}{w^2}\right)\left(\frac{\varepsilon_2}{u^2} + \frac{\varepsilon_3}{w^2}\right) \quad (2.51)$$

with $u = \sqrt{n_2^2 k_0^2 - \beta^2}a_2$, $w = \sqrt{\beta^2 - n_3^2 k_0^2}a_2$, and $V = \sqrt{n_2^2 - n_3^2}k_0 a_2$. For the conventional fiber, $a_2 = 62.5$ μm, it is calculated $V \approx 266$ at 1,550 nm. The mode characteristics of such a multimode waveguide show much difference in features from those under the weakly guiding approximation.

The TE and TM modes are obtained for $\nu = 0$, as can be seen from (2.23) and (2.24). However, their eigen equations are different:

$$\frac{J_\nu'}{uJ_\nu} + \frac{K_\nu'}{wK_\nu} = 0, \quad \text{for } TM \text{ modes}, \quad (2.52a)$$

$$\frac{\varepsilon_2 J_\nu'}{uJ_\nu} + \frac{\varepsilon_3 K_\nu'}{wK_\nu} = 0, \quad \text{for } TE \text{ modes}. \quad (2.52b)$$

It is indicated that the TM and TE modes are no longer degenerate, they have different effective indexes. The cases for hybrid modes are the same, that is, the $(\nu - 1)$th HE mode is no longer degenerate with the $(\nu + 1)$th EH mode. The electric fields for x-polarization mode in Cartesian coordinates are expressed as

$$E_x^{(x)} = \frac{jA\beta a}{2} \begin{cases} \dfrac{1}{u}[(1+p)J_{\nu-1}\cos(\nu-1)\varphi - (1-p)J_{\nu+1}\cos(\nu+1)\varphi], \\ \dfrac{q}{w}[(1+p)K_{\nu-1}\cos(\nu-1)\varphi - (1-p)K_{\nu+1}\cos(\nu+1)\varphi], \end{cases}$$

$$(2.53a)$$

$$E_y^{(x)} = \frac{jA\beta a}{2} \begin{cases} \frac{1}{u}[(1+p)J_{\nu-1}\sin(\nu-1)\varphi + (1-p)J_{\nu+1}\sin(\nu+1)\varphi], \\ \frac{q}{w}[(1+p)K_{\nu-1}\sin(\nu-1)\varphi + (1-p)K_{\nu+1}\sin(\nu+1)\varphi]. \end{cases}$$

$$(2.53b)$$

It is shown that the individual modes are not a linearly polarized mode, no matter $\nu = 1$, or $\nu \neq 1$. By denoting $h = 1 - \varepsilon_3/\varepsilon_2$, $X = J'_\nu/(uJ_\nu) + K'_\nu/(wK_\nu)$, and $\rho = K'_\nu/K_\nu$, eigen equation (2.51) is transformed to

$$X^2 - \frac{h\rho}{w}X - \left[\nu^2\left(\frac{1}{u^2} + \frac{1}{w^2}\right)^2 - \frac{h\nu^2}{w^2}\left(\frac{1}{u^2} + \frac{1}{w^2}\right)\right] = 0, \qquad (2.54)$$

and solved to be

$$X = \frac{h\rho}{2w} \pm \sqrt{\nu^2\left(\frac{1}{u^2} + \frac{1}{w^2}\right)^2 - \frac{h\nu^2}{w^2}\left(\frac{1}{u^2} + \frac{1}{w^2}\right) + \frac{h^2\rho^2}{4w^2}}. \qquad (2.55)$$

For low-order modes and far away from cutoff, $w \approx V \gg 1$, the approximation of $\rho \approx 1 + 1/2w \approx 1$ holds. For modes with $\nu = 1$, the eigen equation is reduced to

$$\frac{J'_1}{uJ_1} + \frac{K'_1}{wK_1} = -\left(\frac{1}{u^2} + \frac{\varepsilon_3/\varepsilon_2}{w^2}\right), \qquad (2.56a)$$

$$\frac{J'_1}{uJ_1} + \frac{K'_1}{wK_1} = \frac{1}{u^2} + \frac{1}{w^2}, \qquad (2.56b)$$

which correspond to HE and EH modes, respectively. It is seen that (2.56b) is with the same form as (2.32d), whereas (2.56a) can be simplified to

$$\frac{J_0}{uJ_1} - \frac{K_0}{wK_1} = \frac{h}{w^2}. \qquad (2.57)$$

Under the conditions, parameter p is approximated to $p = V^2/(V^2 - hu^2)$, and (2.53a) is written as

$$E_x^{(x)} \approx jA\beta a \begin{cases} \frac{1}{u}\left(J_0 - \frac{hu^2}{2V^2}J_2\cos 2\varphi\right), \\ \frac{q}{w}\left(K_0 - \frac{hu^2}{2V^2}K_2\cos 2\varphi\right). \end{cases} \qquad (2.58)$$

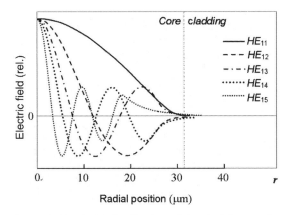

Figure 2.13   Field distribution of the cladding modes.

The second term is much smaller than the first in case $u \ll V$; basically, the low-order modes for $v = 1$ are linearly polarized. Figure 2.13 shows field distributions of several low-order cladding modes, simulated approximately by the simplified model with two regions.

The total number of the cladding modes is estimated to be about 36,000. Actually, a lot of cladding modes lose their energy to the radiation mode as propagation along the fiber. It is difficult to investigate the individual cladding mode separately; but they do determine characteristics of some fiber devices, such as long period fiber gratings. The above discussions are just for a qualitative understanding. References [8,31] give detailed theoretical deduction and analyses.

2.2.6 Propagation of Optical Pulses in Optical Fibers

It is shown in formulas (2.5) and (2.6) that the propagation velocity of optical waves in fiber is determined by the propagation constant β, which possesses the chromatic dispersion induced not only by the material dispersion, but also by the guiding effect, as shown in Figure 2.6. The dispersion causes different propagation velocities for different spectral compositions, resulting in changes of the pulse shape. To analyze the propagation of optical pulses, the electric field of the light wave is expressed as

$$\boldsymbol{E}(x, y, z, t) = \mathrm{Re}\{\boldsymbol{E}(x, y)A(z, t) \exp[j(\beta z - \omega t)]\}, \qquad (2.59)$$

where $A(z, t)$ describes the amplitude of the optical pulse. As a basic physical principle, the temporally varied optical wave corresponds necessarily to a finite spectral linewidth, governed by Fourier transform as

$$\tilde{A}(z, \omega) = \frac{1}{\sqrt{2\pi}} \int_{-\infty}^{\infty} A(z, t) \exp(j\omega t) dt. \tag{2.60}$$

To understand the evolution of optical pulses propagating in fibers, a pulse with Gaussian amplitude and chirped spectrum at the input is investigated as a typical example:

$$A(0, t) = A_0 \exp\left[-\frac{t^2}{2\tau^2}(1 + jC) - j\omega_0 t\right], \tag{2.61}$$

where ω_0 is the central frequency; C is the chirping parameter, which means the transient frequency in the pulse is time-dependent, expressed as $\omega_0 + (C/2\tau^2)t$; and $\tau/\sqrt{1 + C^2}$ is the $1/e$ width of pulse intensity $I(0, t) = |A(0, t)|^2$. Its spectrum is obtained as

$$\tilde{A}(0, \omega) = A_0\tau \exp\left[-\frac{(\omega - \omega_0)^2\tau^2}{2(1 + jC)}\right]. \tag{2.62}$$

The pulse amplitude evolves with z in the fiber with a dispersive propagation constant β, described by (2.5), which is expressed by an inverse Fourier transform as

$$A(z, t) = \frac{1}{\sqrt{2\pi}} \int_{-\infty}^{\infty} \tilde{A}(\omega) \exp[j(\beta z - \omega t)] d\omega. \tag{2.63}$$

By substituting (2.5) in the integral the amplitude waveform is deduced as

$$A(z, t) = A_0\tau\sqrt{\frac{1 + jC}{\tau^2 - j\beta_2 z(1 + jC)}} \exp\frac{-\hat{t}^2(1 + jC)}{2[\tau^2 - j\beta_2 z(1 + jC)]} e^{j(\beta_0 z - \omega_0 t)}, \tag{2.64}$$

where $\hat{t} = t - z/v_g$ is the time in a traveling coordinate. This means that the peak of pulse moves with the group velocity v_g, instead of the phase velocity $v_p = \omega_0/\beta_0$. The intensity waveform is obtained:

$$I(z, t) = I_0\tau^2\sqrt{\frac{1 + C^2}{(\tau^2 + \beta_2 zC)^2 + (\beta_2 z)^2}} \exp\left[\frac{-\hat{t}^2\tau^2}{(\tau^2 + \beta_2 zC)^2 + (\beta_2 z)^2}\right]. \tag{2.65}$$

It is shown that the pulse-width changes with the propagation distance z in case of $\beta_2 \neq 0$. It is deduced from (2.59) that the pulse-width may reach a minimum at

$$z_{\min} = \frac{-\tau^2 C}{(1 + C^2)\beta_2}, \tag{2.66}$$

which is a positive value if the chirping of input pulse meets condition of $C/\beta_2 < 0$. The pulse width reaches $\Delta t_{1/e} = \tau/\sqrt{1 + C^2}$ at $z_{\min}$, which is less than the input width, indicating that the pulse width is compressed in the range from 0 to $z_{\min}$. This is the basic mechanism of so-called prechirping technology of pulse-width narrowing.

After $z_{\min}$, or in case of $(C/\beta_2) > 0$, the pulse width will increase with the propagation distance. When the propagation distance z is large enough, the intensity is approximated as

$$I(z, t) \approx \frac{I_0 \tau^2}{\beta_2 z} \exp\left[\frac{-\hat{t}^2 \tau^2}{(1 + C^2)\beta_2^2 z^2}\right], \tag{2.67}$$

meaning that the pulse-width is proportional to z, written as $\Delta t_{1/e} = \sqrt{1 + C^2}|\beta_2|z/\tau$, no matter what the signs of dispersion and chirping are. It is noticed that the smaller the input pulse width, the larger the broadened width, which is reasonable as smaller input width means broader spectral width.

It is necessary to analyze also the spectral composition evolution in pulse duration. By extracting a phase factor Φ from the exponential term of (2.58), the transient frequency is derived as

$$\hat{\omega}(\hat{t}) = \frac{\partial \Phi(t)}{\partial t} = \omega_0 + \frac{\hat{t}[C\tau^2 + \beta_2 z(1 + C^2)]}{[(\tau^2 + \beta_2 z C)^2 + (\beta_2 z)^2]} \approx \omega_0 + \frac{\hat{t}}{\beta_2 z}, \tag{2.68}$$

where the approximation is for large enough z. It is shown that the transient frequency within the pulse duration changes in proportion to time $\hat{t}$, which is so-called redshift or blueshift, depending on the dispersion sign. Figure 2.14 gives a schematic illustration of pulse broadening with redshift. It is worth noticing that the total spectrum of the pulse does not change in the propagation though its composition distributes temporally along the pulse so long as no nonlinear optical effects exist.

For the input pulses with arbitrary waveforms, other than Gaussian type, the pulse amplitude may not be expressed by analytical functions. It has to be solved from time domain equations, taking into account the index dispersion, and even the nonlinear effect term and the loss term, written as [4]

$$\frac{\partial A}{\partial z} + \beta_1 \frac{\partial A}{\partial t} + j\frac{\beta_2}{2}\frac{\partial^2 A}{\partial t^2} + \frac{\alpha}{2}A = j\gamma |A|^2 A, \quad (2.69)$$

where $\beta_m = (\partial^m \beta/\partial\omega^m)_{\omega=\omega_0}$ are the dispersion factors, α stands for the loss described as the Beer–Lambert law, and $\gamma = n^{(2)}\beta_0/A_{\text{eff}}$ is a nonlinear parameter with nonlinear index coefficient $n^{(2)}$ and effective mode area A_{eff}. Equation (2.69) is usually called a nonlinear Schrodinger equation, and its revised version with more nonlinear effect terms forms the basis of the theory of ultrashort pulse laser optics [4].

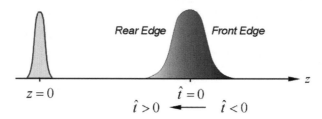

Figure 2.14 Schematic illustration of pulse broadening and spectral shift.

2.3 BASIC THEORY OF THE GRADIENT-INDEX OPTICAL FIBER

The basic concept of the optical waveguide comes from total internal reflection, which occurs at planar interfaces between two homogeneous media, and the light beam is idealized as an infinitively wide planar wave. Because optical waveguides, including fibers, may consist of small-size complicated structures, such as curved surfaces, and inhomogeneous media, it is necessary to describe the light as a transversely finite wave, termed an optical ray. It is the main task of this section to analyze the optical fiber by ray optics. The concept of ray optics is briefly introduced in Section 2.3.1; the basic properties of GRIN fiber are discussed in Section 2.3.2 by ray optics and in Section 2.3.3 by wave optics. The last subsection is devoted to briefly describing the characteristics of gradient index lens, in which the optical ray obeys the same equation as that in the GRIN fiber.

2.3.1 Ray Equation in Inhomogeneous Media

Ray optics may be regarded as advanced geometric optics. It is widely used to understand and explain phenomena related to light propagation. Its principle is deduced from the electromagnetic theory of light waves. Max Born gives a rigorous deduction in Reference [10].

In an inhomogeneous medium, the refractive index is a function of spatial position: $n = n(\mathbf{r})$, and the electromagnetic wave is no longer an exact plane wave. Expressions (2.12a) and (2.12b) are now used to describe a quasi-plane wave, but the wave vector $\mathbf{k}$ may not be a constant vector, that is, its unit vector is a function of spatial position: $\hat{k} = \mathbf{k}/|\mathbf{k}| = \mathbf{s}(\mathbf{r})$, and the spatial distributions of field amplitudes have to be taken into consideration.

Rewriting the phase factor as $\mathbf{k} \cdot \mathbf{r} = k_0 n(\mathbf{r})(\mathbf{s} \cdot \mathbf{r}) = k_0 S(\mathbf{r})$ with the optical path S, called the *eikonal*, optical fields are written as [10]

$$\mathbf{E}(\mathbf{r}) \exp[j(\mathbf{k} \cdot \mathbf{r} - \omega t)] = \mathbf{E}(\mathbf{r}) \exp j[k_0 S(\mathbf{r}) - \omega t], \quad (2.70a)$$

$$\mathbf{H}(\mathbf{r}) \exp[j(\mathbf{k} \cdot \mathbf{r} - \omega t)] = \mathbf{H}(\mathbf{r}) \exp j[k_0 S(\mathbf{r}) - \omega t]. \quad (2.70b)$$

Substituting them into Maxwell equations, we have

$$\nabla S \times \mathbf{H} + \varepsilon c \mathbf{E} = j\nabla \times \mathbf{H}/k_0, \quad \nabla S \times \mathbf{E} - \mu c \mathbf{H} = j\nabla \times \mathbf{E}/k_0,$$

$$\mathbf{E} \cdot \nabla S = i(\mathbf{E} \cdot \nabla \ln \varepsilon + \nabla \cdot \mathbf{E})/k_0, \quad \mathbf{H} \cdot \nabla S = i(\mathbf{H} \cdot \nabla \ln \mu + \nabla \cdot \mathbf{H})/k_0.$$

In the limit of geometric optics, the wavelength is much smaller than the transverse size of the light beam, that is, the wave vector is so large that the right-hand sides of the above four formulas can be neglected, resulting in

$$\nabla S \times \mathbf{H} + \varepsilon c \mathbf{E} = 0, \quad (2.71a)$$

$$\nabla S \times \mathbf{E} - \mu c \mathbf{H} = 0, \quad (2.71b)$$

$$\mathbf{E} \cdot \nabla S = 0, \quad (2.71c)$$

$$\mathbf{H} \cdot \nabla S = 0. \quad (2.71d)$$

By vector operation of the formulas we have

$$\nabla S \times (\nabla S \times \mathbf{E}) - \mu c \nabla S \times \mathbf{H} = 0$$

$$= [(\nabla S \cdot \mathbf{E})\nabla S - (\nabla S \cdot \nabla S)\mathbf{E}] + \varepsilon \mu c^2 \mathbf{E}$$

and, consequently, $\nabla S \cdot \nabla S - \varepsilon \mu c^2 - n^2(\mathbf{r})$, or in a simplified form

$$|\nabla S(\mathbf{r})| = n(\mathbf{r}), \quad (2.72)$$

which is called the *eikonal equation*. It is deduced also from formulas (2.71c) and (2.71d) that the gradient of the eikonal is perpendicular to both $\mathbf{E}$ and $\mathbf{H}$, that is, it coincides with the wave vector $\mathbf{k}$. Thus the wave front can be defined as $S(r) = \text{Const.}$, and $\nabla S = ns$ with the unit vector of $s = \nabla S/|\nabla S| = \nabla S/n$. The Poynting vector of the electromagnetic wave is obtained as

$$S = \langle E \times H \rangle = \frac{2c}{n^2} \langle w_e \rangle \nabla S, \tag{2.73}$$

where $\langle w_e \rangle = \langle E \cdot D \rangle / 2$ is the timed average of the electric energy density.

The optical ray is now defined as a spatial curve whose direction coincides with the Poynting vector at every point. By denoting the position vector of a point P in the ray as r, the equation of the ray is written as

$$n \frac{dr}{dl} = \nabla S, \tag{2.74}$$

where l is the arc length with $dl = \sqrt{dx^2 + dy^2 + dz^2}$, and the unit vector of the ray is $s = dr/dl$. The ray equation is further deduced as

$$\frac{d}{dl}\left(n\frac{dr}{dl}\right) = \frac{d}{dl}\nabla S = \frac{dr}{dl} \cdot \nabla(\nabla S)$$

$$= \frac{1}{n}\nabla S \cdot \nabla(\nabla S) = \frac{1}{2n}\nabla[(\nabla S)^2] = \frac{1}{2n}\nabla(n^2),$$

so that

$$\frac{d}{dl}\left(n\frac{dr}{dl}\right) = \nabla n. \tag{2.75}$$

This is the differential equation of determining the trace of ray in medium with index distribution of $n(r)$, called the *ray equation*.

In a homogeneous index medium the equation is simplified as $d^2r/dl^2 = 0$ and its solution is a straight line, as expected. Another example with practical interest is a medium with spherically symmetric index, which depends only on the distance r from a fixed center, $n = n(r)$, such as the atmospheric layer of the earth. It is deduced [9] that in such a medium we have

$$\frac{d}{dl}(r \times ns) = s \times ns + r \times \frac{d}{dl}(ns) = r \times \nabla n = r \times \frac{r}{r}\frac{dn}{dr} = 0$$

and thus

$$r \times (ns) = \text{Const.} \tag{2.76}$$

It is implied that the ray is in a plane, and satisfies the relation of $nr \sin \phi = nd = \text{Const.}$, where d is the perpendicular distance from the

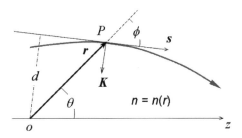

Figure 2.15 An optical ray in a medium with index $n = n(r)$.

origin to the tangent, as shown in Figure 2.15. Equation (2.76) is the trace analog of a moving particle under action of a central force obeying conservation of angular momentum.

In the polar coordinate of (r, θ), the position of P is a function of θ: $r = r(\theta)$, and the angle ϕ is expressed by the following relation:

$$\sin \phi = \frac{r(\theta)}{\sqrt{r^2(\theta) + (dr/d\theta)^2}} = \frac{C}{n(r)r(\theta)}, \qquad (2.77a)$$

where C is a constant; and a differential equation is given as

$$\frac{dr}{d\theta} = \frac{r}{C}\sqrt{n^2r^2 - C^2}. \qquad (2.77b)$$

In Figure 2.15 vector $K = ds/dl$ is called a curvature vector whose direction is in the principle normal of the ray, and whose magnitude is the reciprocal of curvature. From (2.75) we have $K = [\nabla n - s(dn/dl)]/n$, which implies that both the gradient of refractive index and the curvature vector are in the plane of the ray; and the ray bends toward the region of higher index.

By means of the optical ray, some basic theorems of geometrical optics are explained [10]:

1. Intensity law: the intensity of an optical beam is inversely proportional to its transverse area dA:

$$I_1 dA_1 = I_2 dA_2 = \text{Const.} \qquad (2.78)$$

2. Lagrange's integral invariant: due to $\nabla \times (ns) = \nabla \times (\nabla S) = 0$, we have

$$\oint ns \cdot dr = 0. \qquad (2.79)$$

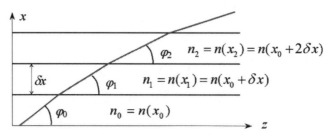

Figure 2.16   Modified Snell law.

3. The principle of Fermat: the optical length of a ray between two points P_1 and P_2, $\int_{P_1}^{P_2} n\,dl$, is shorter than any other curves joining the two points.

4. Snell's law holds at a curved surface, so long as the sizes of the incident beam and the boundary surface are larger enough than the wavelength.

5. Snell's law for a medium with spatially inhomogeneous index. Figure 2.16 shows a ray propagating in a medium with index $n = n(x)$, which is modeled as a multilayer medium with homogeneous index in each layer. The ray is refracted at each interface according to Snell's law: $n_i \cos \varphi_i = n_{i+1} \cos \varphi_{i+1}$. In the limitation of $\delta x \to 0$, the ray should obey an equation of

$$n(x) \cos \varphi(x) = n(0) \cos \varphi_0 = \text{Const.} \tag{2.80}$$

2.3.2 Ray Optics of GRIN Fiber

As described in Section 2.1, the GRIN fiber was invented at an early stage, and is still used widely. Its characteristics have been investigated in detail [3,9,32]. The refractive index has a cylindrical symmetric distribution in a GRIN fiber:

$$n^2(r) = \begin{cases} n_1^2[1 - 2\Delta(r/a)^p] & (r \leq a), \\ n_1^2[1 - 2\Delta] & (r > a), \end{cases} \tag{2.81a}$$

or approximately,

$$n(r) = \begin{cases} n_1[1 - \Delta(r/a)^p] & (r \leq a), \\ n_1[1 - \Delta] & (r > a), \end{cases} \tag{2.81b}$$

where $\Delta = [n^2(0) - n^2(a)]/2n^2(0) \approx [n(0) - n(a)]/n(0) = (n_1 - n_2)/n_1$. Note that r here is the radial position in the cylindrical coordinate, that is, $r = x\hat{i} + y\hat{j}$, not the distance to a fixed original as used above in Section 2.2.1. The power index p in (2.81) may take different values to express different index distributions, such as step-index with $p = \infty$, parabolic index with $p = 2$, and triangle index with $p = 1$.

For a conventional GRIN fiber, $\Delta n = n_1 - n_2 \ll n_1$, only paraxial rays need to be taken into consideration; then $dl \approx dz$, and ray equation (2.75) is simplified to be

$$\frac{d}{dz}\left(n\frac{dr}{dz}\right) = n\frac{d^2r}{dz^2} = \frac{dn}{dr}. \tag{2.82}$$

By substituting (2.81) the equation is written as

$$\frac{d^2r}{dz^2} = \frac{1}{n}\frac{dn}{dr} = -\Delta\frac{pr^{p-1}}{a^p} \tag{2.83}$$

and

$$\frac{d^2r}{dz^2} = -\alpha^2 r \tag{2.84}$$

for the parabolic index fiber with $p = 2$ and $\alpha = \sqrt{2\Delta}/a$. Because the fiber is cylindrically symmetric, it can be decomposed into two equations of $d^2x/dz^2 = -\alpha^2 x$ and $d^2y/dz^2 = -\alpha^2 y$ with solutions of

$$x(z) = x_0 \cos \alpha z, \tag{2.85a}$$

$$y(z) = y_0 \cos(\alpha z + \varphi). \tag{2.85b}$$

Its radial position r is

$$r^2(z) = x^2 + y^2 = r_0^2 + r_1^2 \cos(2\alpha z + \psi), \tag{2.86}$$

where $r_0^2 = \frac{1}{2}(x_0^2 + y_0^2)$, $r_1^2 = (r_0^4 - 4x_0^2 y_0^2 \sin^2 \varphi)^{1/2}$, and $\tan \psi = y_0^2 \sin 2\varphi/((x_0^2 + y_0^2 \cos^2 \varphi))$. It is shown that $r(z)$ varies between $r_{max} = \sqrt{r_0^2 + r_1^2}$ and $r_{min} = \sqrt{r_0^2 - r_1^2}$; and the ray goes in an elliptically helical line, as shown in Figure 2.17(a). The ray's angles to the fiber axis, $\theta_x = dx/dz$ and $\theta_y = dy/dz$, also change periodically on $2\pi/\alpha$, which is called the pitch length $l_p = 2\pi/\alpha = \pi a\sqrt{2/\Delta}$.

When no phase shift exists between x and y, that is, $\varphi = 0$, we have $r_1 = r_0$, $r_{min} = 0$, and

$$\frac{y(z)}{x(z)} = \frac{y_0}{x_0} = \text{Const.} \tag{2.87}$$

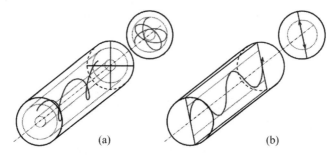

Figure 2.17   Rays in gradient index fiber: (a) helical ray and (b) meridional ray.

The ray is called the meridional ray passing through the axis, as shown in Figure 2.17(b). When the initial incident ray meets the condition of $\varphi = \pi/2$ and $x_0 = y_0$, resulting in $r_1 = 0$ and $r_{\min} = r_{\max}$, the ray goes in a circular helical line.

Formulas (2.85a), (2.85b) and (2.86) describe the path of optical rays. It is shown that the rays propagate in self-focused ways in parabolic index fibers. For fibers with refractive index distribution other than parabolic, the ray goes similarly in helical lines, although Equation (2.83) may not be solved analytically.

However, the helical lines have to meet some phase conditions, which leads to stationary electromagnetic field distributions in the transverse cross section of the fiber, and forms propagation modes in gradient-index fibers. To understand such propagation modes, the ray is described by a quasi-planar wave as

$$U(r) \propto \exp[-jk_0 S(r)] = \exp[-j(\kappa_r r + \nu\varphi + \beta z)]. \qquad (2.88)$$

The wave vector is composed of its three components: $\boldsymbol{k} = \kappa_r \hat{r} + \kappa_\varphi r \hat{\varphi} + \beta \hat{z}$. Among the components β is a constant with the same physical meaning as in wave optics; the azimuthal component is in the form of $\kappa_\varphi = \nu/r$ to ensure stationary circular phase conditions; and the radial one, κ_r, is determined by the relation

$$|\boldsymbol{k}|^2 = n^2(r)k_0^2 = \kappa_r^2 + \nu^2/r^2 + \beta^2. \qquad (2.89)$$

Different from the step-index fibers, κ_r is now a function of r; and has to meet the phase condition in radial direction [1]:

$$\int_{r_1}^{r_2} \kappa_r \mathrm{d}r = \int_{r_1}^{r_2} \sqrt{n^2(r)k_0^2 - \nu^2/r^2 - \beta_{\nu m}^2} \, \mathrm{d}r = \left(m + \frac{1}{2}\right)\pi, \qquad (2.90)$$

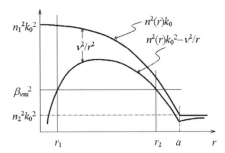

Figure 2.18 Schematic illustration of the wave vector in GRIN fibers.

where $m = 0, 1, \ldots M_v$; r_1 and r_2 are the two roots of equation

$$n^2(r)k_0^2 - v^2/r^2 - \beta_{vm}^2 = 0, \tag{2.91}$$

corresponding to the parameters r_{min} and r_{max} in the helical trajectory description. It is noticed that Equation (2.90) is rigorously deduced by WKB method [1] with factor $\pi/2$ in its right-hand side, which is often omitted since the mode number in usual MMFs is quite large. The equation determines the separate values of the propagation constant β with subscript index v and m.

Figure 2.18 gives a schematic picture of relations between the three components. The propagation constant β_{vm} of a GRIN fiber can then be solved from Equation (2.90), assisted by graphic method, shown in Figure 2.18.

The total mode number is then estimated as

$$\begin{aligned}
M &= 4 \sum_{v=0}^{v_{max}} \frac{1}{\pi} \int_0^{r_2(v)} \sqrt{k^2(r) - \beta_c^2 - v^2/r^2}\, dr \\
&\rightarrow \frac{4}{\pi} \int_0^{v_{max}} dv \int_0^{r_2(v)} \sqrt{k^2(r) - \beta_c^2 - v^2/r^2}\, dr,
\end{aligned} \tag{2.92}$$

where $\beta_c = n_{cl}k_0$ is the limited propagation constant corresponding to the cladding index n_{cl}, and the largest index v_{max} corresponds to $v_{max} = \sqrt{n^2(r)k_0^2 - \beta_c^2}r$. Factor 4 in (2.92) comes from two polarization modes, and clockwise and counterclockwise helical rays. Sum on v can be approximately replaced by integration, expressed in the second line of (2.92), to give an explicit expression:

$$M = \int_0^{\sqrt{n^2(r)k_0^2 - \beta_c^2}r} [n^2(r)k_0^2 - \beta_c^2]r\, dr. \tag{2.93}$$

Formula (2.93) can also be used to count the mode number M_β for propagation constants larger than a certain value β:

$$M_\beta = \int_0^{\sqrt{n^2(r)k_0^2 - \beta^2}r} [n^2(r)k_0^2 - \beta^2]r\,dr. \tag{2.94}$$

The integration can then be carried out for different refractive index distributions. For the power law index described by (2.81a) and (2.81b) we obtain

$$M = \frac{p}{p+2}\Delta n_1^2 k_0^2 a^2 = \frac{p}{p+2}\frac{V^2}{2}, \tag{2.95a}$$

$$M_\beta = \frac{p}{p+2}\Delta n_1^2 k_0^2 a^2 \left(\frac{n_1^2 k_0^2 - \beta^2}{2\Delta n_1^2 k_0^2}\right)^{\frac{2}{p}+1} = M\left(\frac{n_1^2 k_0^2 - \beta^2}{2\Delta n_1^2 k_0^2}\right)^{\frac{2}{p}+1}. \tag{2.95b}$$

The dependence of β on mode number can be given from the above formulas:

$$\beta = n_1 k_0 \left[1 - 2\Delta \left(\frac{M_\beta}{M}\right)^{\frac{p}{p+2}}\right]^{1/2}. \tag{2.96}$$

As discussed in Section 2.2.6, the propagation of optical pulsed signals is one of the most important characteristics, which is mainly determined by group velocity, defined as $v_g = (\partial\beta/\partial\omega)^{-1}$. The group velocity is deduced from (2.96) [9]:

$$v_g \approx \frac{c}{n_1}\left[1 - \frac{p-2}{p+2}\Delta\left(\frac{M_\beta}{M}\right)^{\frac{p}{p+2}}\right]. \tag{2.97}$$

It is shown that for parabolic index fiber, $p = 2$, the group velocity does not depend on mode numbers, meaning that the modal dispersion is minimized in the parabolic index fiber, which is its unique merit over other MMFs. Reference [1] gives analyses of GRIN fiber with other index distributions, such as cosecant-type functions, showing similar properties.

Basic propagation characteristics of multimode step-index fibers are obtained by using formulas (2.95)–(2.97) with $p = \infty$. The total mode number is obtained from (2.95a): $M = V^2/2$, which coincides approximately with (2.44). Formulas (2.96) and (2.97) are changed to

$$\beta = n_1 k_0 \left[1 - 2\Delta \left(\frac{M_\beta}{M}\right)\right]^{1/2} \tag{2.98}$$

and

$$v_g \approx \frac{c}{n_1} \left[1 - \Delta \left(\frac{M_\beta}{M} \right) \right] \xrightarrow{M_B = M} = \frac{c}{n_1} (1 - \Delta). \tag{2.99}$$

The last expression corresponds to an averaged group velocity when the optical energy is distributed uniformly over the multiple modes. It is an approximation just for qualitative understanding. The actual averaged group velocity depends on the energy distribution over the modes of input signal.

2.3.3 Wave Optics of GRIN Fiber

To understand optical propagation in GRIN fibers rigorously, it is necessary to have as a starting point the fundamental electromagnetic theory, that is, to solve Maxwell equations in a medium with spatially inhomogeneous refractive index distributions. Moreover, as introduced briefly in Section 2.1 and Section 2.2, the dispersion characteristics can be adjusted by using specially designed index distribution; for special fibers, such as DSF and dispersion compensation fibers, precise calculations of optical fields in such designed fibers are needed. A brief and primary description of wave optics for the GRIN fiber is introduced here in this section.

The Helmholtz equation for inhomogeneous medium is written as

$$\nabla^2 E + \frac{\omega^2 \mu \varepsilon}{c^2} E + \nabla(E \cdot \nabla \ln \varepsilon) = 0. \tag{2.100}$$

Because the difference between core and cladding is usually so small that the weakly guiding approximation is valid, it is reasonable that the gradient of dielectric constant in (2.100) can be neglected. Then we use variable separation, similar to (2.16), to solve (2.100) with the refractive index $n = n(r)$, expressed by (2.81), which is symmetric cylindrically, but not constant radially. Rigorously, a two-district model has to be used, meaning that the boundary conditions between the core and the cladding have to be considered. However an approximated model with index distribution extending to the infinitive is used to give simplified discussions and to investigate the paraxial mode field, as done in [33]. That is to say, if the mode field decays fast with r, and it is little enough to be neglected at the boundary between core and cladding, the equation is regarded as a problem without a boundary; and a transverse wave, similar to that in free space, is expected, which is different from the hybrid modes in step-index fibers. Therefore, a linear polarization mode can be applied directly. To be solved in cylindrical coordinates, the polarized field is written by means of variable separation as

$$E_x(r, \varphi)[\text{or } E_y(r, \varphi)] = R(r)\Theta(\varphi). \tag{2.101}$$

The azimuthal function obeys the same equation as (2.17) and has the same solution as (2.19). Then Helmholtz equation for radial function is written as

$$\frac{\partial^2 R}{\partial r^2} + \frac{1}{r}\frac{\partial R}{\partial r} + \left[n^2(r)k_0^2 - \beta^2 - \frac{\nu^2}{r^2}\right]R = 0. \tag{2.102}$$

For a parabolic index of (2.81) with $p = 2$, the equation has a special solution in the form of a Laguerre–Gaussian function:

$$R(r) = \left(\frac{r}{\sigma}\right)^\nu L(r)\exp\frac{-r^2}{2\sigma^2}, \tag{2.103}$$

where σ is a constant to be determined, and ν is the same as that in (2.17). Substituting it into Equation (2.102), and replacing the argument by $\xi = r^2/\sigma^2$, the equation is transformed to

$$\xi L''(\xi) + (1 + \nu - \xi)L'(\xi) + \mu L(\xi) = 0, \tag{2.104}$$

where $\mu = (\sigma^2/4)[n_1^2 k_0^2(1 - 2\Delta r^2/a^2) + r^2/\sigma^4 - \beta^2 - 2(\nu + 1)/\sigma^2]$. It is seen if $\sigma^2 = a/(n_1 k_0\sqrt{2\Delta})$ is taken, μ will not depend on r. Furthermore, if it is a nonnegative integer $\mu = l$, Equation (2.104) is a known Laguerre equation with a solution of Laguerre polynomials:

$$L_l^{(\nu)}(x) = \frac{e^x}{l!x^\nu}\frac{d^l}{dx^l}(e^{-x}x^{\nu+l}). \tag{2.105}$$

The lowest order Laguerre polynomials are listed as

$$L_0^{(\nu)}(x) = 1, \tag{2.106a}$$

$$L_1^{(\nu)}(x) = 1 + \nu - x, \tag{2.106b}$$

$$L_2^{(\nu)}(x) = (\nu + 1)(\nu + 2)/2 - (\nu + 2)x + x^2/2. \tag{2.106c}$$

The optical field of mode (ν, l) is now expressed as

$$E_{\nu l}(r, \varphi) = \frac{1}{\sigma\sqrt{\pi}}\left[\frac{l!}{(l+\nu)!}\right]^{1/2}\left(\frac{r}{\sigma}\right)^\nu L_l^{(\nu)}\left(\frac{r^2}{\sigma^2}\right)\exp\frac{-r^2}{2\sigma^2}\begin{cases}\cos\nu\varphi, \\ \sin\nu\varphi,\end{cases} \tag{2.107}$$

where term $[l!/(l+v)!]^{1/2}/(\sqrt{\pi}\sigma)$ is a normalized factor. The corresponding propagation constant is obtained as

$$\beta_{vl}^2 = n_1^2 k_0^2 - 2(2l + v + 1)\sqrt{2\Delta}n_1 k_0 a^{-1}. \qquad (2.108)$$

The modes with $v = 0$ are written as

$$E_{0l}(r, \varphi) = \frac{E_0}{\sigma\sqrt{\pi}} L_l^{(0)}\left(\frac{r^2}{\sigma^2}\right) \exp\frac{-r^2}{2\sigma^2} \qquad (2.109)$$

with maximums at the center of the core, $r = 0$. The lth Laguerre polynomial has l zeros along the radial direction, which corresponds to the different orders of meridional rays. The mode with $l = 0$ is the fundamental mode:

$$E_{00}(r, \varphi) = \frac{E_0}{\sigma\sqrt{\pi}} \exp\frac{-r^2}{2\sigma^2} \qquad (2.110)$$

with the propagation constant of $\beta_{00} \approx n_1 k_0 - \sqrt{2\Delta}a^{-1}$. It is deduced that the field of modes with $v > 0$ has a zero at $r = 0$; the modes can then be regarded as the right or left helical rays corresponding to the azimuthal terms of $\exp(jv\varphi)$ and $\exp(-jv\varphi)$, which are the alternative forms of sinusoidal functions in (2.18). It is seen from the term of $(r/\sigma)^v$ that the field amplitudes around the original point decrease with v increasing, corresponding to larger r_{min}, as described by ray optics in (2.85). In addition the field expands more in radial direction as the mode index v and l increase, corresponding to larger r_{max}. For high v and l, the cladding layer will play guiding roles, no longer negligible; then the approximation of infinitive r has to be revised, and (2.107) becomes not precise enough. However, for the fundamental mode and low order modes the Laguerre–Gaussian function is a good model.

The mode characteristics are often evaluated by a beam emitting from the fiber end, because a Fourier transform relation holds between the near field and the far field. The far field pattern is described by the so-called Huygens–Fresnel diffraction [10,34]:

$$U(P_1) = \frac{1}{j\lambda z} \int U_{in}(x, y) \exp\left\{jk\left[\frac{(x_1 - x)^2}{2z} + \frac{(y_1 - y)^2}{2z}\right]\right\} \mathrm{d}x\mathrm{d}y.$$

$$(2.111)$$

In case of $z \gg ka^2/2$, which is satisfied mostly in practice, it is further approximated to Fraunhofer diffraction as

$$U(P_1) = \frac{1}{j\lambda z} \exp\left(jk\frac{r_1^2}{2z}\right) \int U_{in}(r, \varphi) \exp[j(kr_1r/z)\cos\varphi]r\,dr\,d\varphi, \qquad (2.112)$$

where the coordinate is changed to the cylindrical one, and $U_{in}(r, \varphi)$ is the near field at the end facet of the fiber. For the fundamental mode, the angle integration can be carried out by using properties of Bessel functions, resulting in

$$U(P_1) = \frac{2\pi}{j\lambda z} \exp\left(jk\frac{r_1^2}{2z}\right) \int_0^\infty U_{in}(r)J_0\left(\frac{kr_1r}{z}\right)r\,dr. \qquad (2.113)$$

The far field pattern can thus be calculated both for step-index fibers and for GRIN fibers with few complicated and lengthy expressions. Interested readers can refer to the related references [35–37].

The mode radius is one of the widely used parameters to describe the near field [38]. Some definitions for mode radius have been proposed with more or less difference. Since the Gaussian beam is a good and widely-used form in free space, it is not only suitable for parabolic index fiber mode, as analyzed above, but is also a good description for the fundamental mode of step-index fibers. References [35,36] discuss the difference between Gaussian approximation and the rigorous solution. The Gaussian beam approximation for the fundamental mode is written as

$$E_G = \frac{\sqrt{2}}{\bar{w}} \exp\frac{-r^2}{\bar{w}^2}, \qquad (2.114)$$

which is the same as (2.110) for the fundamental Laguerre–Gaussian mode with $\bar{w} = \sqrt{2}\sigma$ as used in referred literatures and with a normalization of $\int_0^\infty E_G^2 r\,dr = 1$. The matched waist width $\bar{w}$ is determined by $\partial(\Delta I)/\partial\bar{w} = 0$, where $\Delta I = \int_0^\infty (E_G - E_0)^2 r\,dr$ with the fundamental mode field E_0, resulting in an expression of [39,40]:

$$\bar{w}^2 \equiv \frac{\int_0^\infty E_G^2 r^3\,dr}{\int_0^\infty E_G^2 r\,dr} = \frac{\int_0^\infty E_0E_G r^3\,dr}{\int_0^\infty E_0E_G r\,dr}. \qquad (2.115)$$

Reference [35] gives a digital fit of beam width as a function of normalized frequency V:

$$\frac{\bar{w}}{a} = 0.65 + \frac{1.619}{V^{3/2}} + \frac{2.879}{V^6}. \qquad (2.116)$$

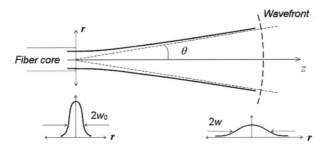

Figure 2.19   Emitting beam approximated as a Gaussian beam.

Based on the Gaussian beam theory, the emitting beam is described as

$$E = E_0 \frac{w_0}{w(z)} \exp \frac{-r^2}{w^2(z)} \exp j \left[kz + \frac{kr^2}{2R(z)} + \Gamma(z) \right], \qquad (2.117)$$

where $w_0 = \bar{w}$ stands for the beam waist of near field at $z = 0$, the beam width of far field is expressed as $w(z) = (w_0^2 + 4z^2/k^2w_0^2)^{1/2}$, and $R(z) = \sqrt{z^2 + z_0^2}$ is the curvature radius of the wavefront in far field, with the Rayleigh range of $z_0 = kw_0^2/2$; the last parameter is the Gouy phase shift: $\Gamma(z) = \tan^{-1}(2z/kw_0^2) = \tan^{-1} z/z_0$.

Figure 2.19 gives a schematic diagram of a Gaussian beam. The beam divergence for a sufficiently large propagation distance z is deduced from (2.117) to be

$$\tan \theta = \lim_{z \to \infty} \frac{w(z)}{z} = \frac{2}{kw_0} = \frac{w_0}{z_0}, \qquad (2.118)$$

which gives an expression of numerical aperture for the SMFs. The field distribution for MMFs is slightly more complicated, as analyzed in [41]. Measurements of the far field intensity distribution provide a characterization of mode field propagating inside the fiber.

The mode field analysis and simulations are important for fiber designs, especially for those with special index distribution for dispersion adjustment by the waveguide dispersion effect, such as DSF [42], NZDSF, and dispersion compensated fibers, probably with W-type, triangle-type index distributions, as shown schematically in Figure 2.20 [13].

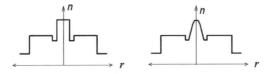

Figure 2.20   Index distributions of some special fibers.

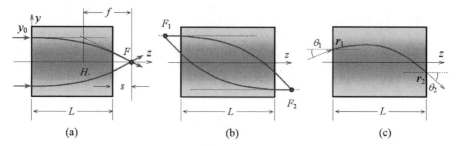

Figure 2.21   Optical ray in a GRIN lens. (a) $L \leq l_p/4$; (b) $L \leq l_p/2$; and (c) general optical path.

2.3.4   Basic Characteristics of Gradient Index Lens

The gradient index materials are also used to make optical components, such as GRIN lens, in which the refractive index has a cylindrically symmetric parabolic distribution, similar to the GRIN fiber described above. The main technical issue to be clarified is about the focal length. Figure 2.21 shows an optical ray in a section of GRIN material with the length of L and the same index distribution as (2.81). By the solution of ray equations, as shown in (2.85), the position of ray input parallel to the axis z at y_0 is written as

$$y(z) = y_0 \cos \alpha z. \tag{2.119}$$

The angle of ray to the axis is

$$\theta(z) = \frac{dy}{dz} = -\alpha y_0 \sin \alpha z. \tag{2.120}$$

At the output plane, we have $y(L) = y_0 \cos \alpha L$ and $\theta(L) = -\alpha y_0 \sin \alpha L$. The refractive angle outside the lens obeys Snell's law as $\sin \theta' = n(L) \sin \theta(L) = n(L) \sin(\alpha y_0 \sin \alpha L)$. It is approximated to be $\sin \theta' \approx n_0 \alpha y_0 \sin \alpha L$, which is just the numerical aperture of the lens if y_0 is replaced by the radius of the lens. The focal length, defined as the length of HF depicted in Figure 2.21, is deduced as

$$f = \frac{y_0}{\tan \theta'} \approx \frac{y_0}{\sin \theta'} = \frac{1}{n_0 \alpha \sin \alpha L}. \tag{2.121}$$

It is noticed that the focal length does not depend on the input position y_0 for axial beams, meaning that the GRIN lens acts similarly to an ordinary focal lens. The distance between the focal point F and the front end facet is a parameter with importance in practical applications, called working distance or back focal length, which is deduced as

$$s = \frac{y_L}{\tan \theta'} \approx \frac{y_0 \cos \alpha L}{\sin \theta'} = \frac{\cos \alpha L}{n_0 \alpha \sin \alpha L}. \tag{2.122}$$

Referring to (2.85), with position r_1 and angle θ_1 of the input optical ray used as independent variables, the output ray is expressed by the following matrix equation. The optical paths are shown in Figure 2.21(c).

$$\begin{pmatrix} r_2 \\ \theta_2 \end{pmatrix} = \begin{pmatrix} \cos \alpha L & (\sin \alpha L)/(n_0 \alpha) \\ -n_0 \alpha \sin \alpha L & \cos \alpha L \end{pmatrix} \begin{pmatrix} r_1 \\ \theta_1 \end{pmatrix}. \tag{2.123}$$

The GRIN lens can be fabricated by several technologies, including ion exchange on multielement compound glasses, diffusion of dopants [43], and CVD [44]. The GRIN lenses are useful in fiber technology because of its small size, large NA, and other advantages. They are usually categorized according to their length divided by the pitch length, $L/l_p = \alpha L/2\pi$, such as 0.23 pitch, 0.29 pitch, and 0.5 pitch. The GRIN lenses with such pitches are used for collimating the output beam from the fiber facet; for coupling between fiber and bulky optical components; for point-to-point coupling between a laser and fiber end, and so on.

2.4 SPECIAL OPTICAL FIBERS

Apart from fibers introduced in Sections 2.2 and 2.3, there are several kinds of special fibers with different characteristics and functions. Among them, rare-earth-doped fiber and DCFs, PMFs, and PCFs are used most frequently, and play important roles in fiber communications and fiber sensors.

2.4.1 Rare-Earth-Doped Fibers and Double-Cladding Fibers

2.4.1.1 Rare-Earth-Doped Fibers Rare-earth ions doped in the fiber will amplify input light when they are optically pumped to a higher energy level, just like other solid-state laser materials [45,46]. It is usually called an active fiber. Therefore, the rare-earth-doped fibers are used to make optical amplifiers and lasers, with its unique features, such as good beam quality, better heat dissipation, higher energy efficiency, and good compatibility with transmission fibers. Among them, EDF is

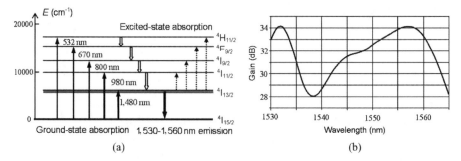

(a)   (b)

Figure 2.22   (a) Energy levels of Er-doped fiber and (b) typical gain spectrum of C-band EDFA. (Reprinted with permission from reference [12].)

the most important one; its energy levels are shown in Figure 2.22(a) [47]. When it is irradiated by pump light at its ground-state absorption band, especially around 980 or 1,480 nm, high gain can be obtained around 1,550 nm, which is the band with the lowest loss for silica fibers, with spectral bandwidth of 30 nm, as shown in Figure 2.22(b) for the conventional band (C-band) [12]. For longer wavelengths up to 1,625 nm, the long-wavelength band (L-band) erbium-doped fiber amplifier (EDFA) has also been well developed [48,49].

It is noticed that the upper level of 1,550 nm band emission is actually in the same band as that of the 1,480 nm absorption band, which causes an optical amplification with high quantum efficiency. In addition, the 1,480 nm pump is basically located in the low-loss window of silica fibers, which makes so-called remote pumping possible. However, it is found that the EDF has higher absorption at 980 nm than that at 1,480 nm pump, which helps to build an optical amplifier with a short EDF and higher amplification. Combination of two band pumps is often used to achieve optimum effectiveness [50].

The EDFA is widely recognized as one of the most important devices in fiber communications, especially as one of the keys to dense wavelength division multiplexing (DWDM) technology, because the signals carried by different wavelengths can be amplified simultaneously and separately without electro–optic–electro conversion. The erbium-doped fiber laser (EDFL) is also very useful in fiber technologies and laser technology, including fiber sensors. Table A4 in Appendix 4 gives main specifications of an EDF production.

Another rare-earth group dopant, ytterbium (Yb), is often used, especially for high-power fiber lasers. Figure 2.23(a) shows its energy levels [50], which is basically a two-level system, consisting of the $^2F_{7/2}$ ground-state and $^2F_{5/2}$ excited-state manifolds separated by about

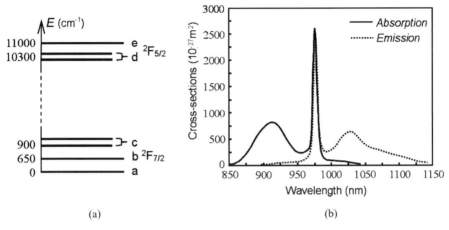

Figure 2.23 (a) Yb^{3+} energy levels and (b) its absorption (solid line) and emission (dot line) spectra. (Reprinted with permission from reference [50].)

$10,000$ cm^{-1}. Its broad absorption and emission bands, shown in Figure 2.23(b) [11], allow for high energy efficiency and wide tunability. Yb-doped silica fiber and Yb-doped YAG are two examples widely used for high-power lasers.

It is found experimentally that the doping concentration of Er^{3+} ions in silica glass is limited by the segregation effect, which consequently limits the efficiency of EDFA and power level of EDFL. There exist other drawback effects, such as excited-state absorption and lifetime quenching. The usual doping concentration is limited around $\sim 1 \times 10^{25}$ m^{-3}. Some technologies are developed to solve or to mitigate the problems. One of them is co-doping with several dopants, especially the Er–Yb co-doping technology [51]. The other is to use new host materials, especially phosphate glass [52].

2.4.1.2 Double-Cladding Fibers

Because the active volume is a thin waveguide, one of the key techniques in fiber amplifiers and fiber lasers is how to inject pump power as much as possible into the core of active fibers for absorption by ions. DCFs, emerged in the 1990s as an important progress, allowing the pump energy propagating in a larger area inner cladding, and absorbed by the active core in propagation [53–55].

The most important feature of DCF is its irregular geometry configurations. It is analyzed that the pump light injected into the inner cladding is not necessarily absorbed by the core effectively and totally [55]. From the view point of ray optics, some of the injected pump lights

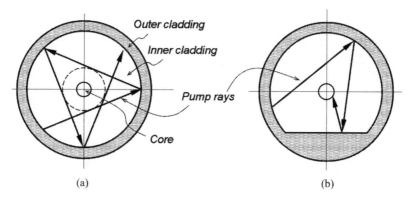

Figure 2.24 Double cladding fibers. (a) Cylindrically symmetric and (b) D-shaped DCF.

are skew rays, not passing through the core area. The larger the inner cladding, the lesser the portion of meridional rays in the pump energy. It is shown that for fibers with a rigorous cylindrical configuration, rays tend to keep their angle with the axis unchanged, that is, the skew light is hardly converted to the meridional light, as shown in Figure 2.24(a). From the view of wave optics, no coupling between the core mode and the high-order cladding modes occurs in regular symmetric fibers. A variety of designs are proposed and developed to overcome the difficulty, such as rectangular, flower-shaped, and offset core configurations [56]. It is needed to calculate the coupling efficiency of the injected pump power to the core mode. The widely used method of DCF design is the three-dimensional ray tracking [57,58]. Figure 2.24(b) shows schematically a D-shaped DCF, which is widely used for high-power fiber lasers. Other considerations in DCF design include matching with the pump beams, fabrication feasibility, and production cost.

2.4.2 Polarization Maintaining Fibers

2.4.2.1 Basic Structures and Fabrications of PMFs

As analyzed in Section 2.2, the fundamental mode in conventional fibers is a twofold degenerate mode, that is, two polarization modes have the same propagation properties with no birefringence. Actually, the fiber possesses more or less birefringence due to imperfections in fabrication [59,62]. PMFs are the fibers with high birefringence introduced intentionally by special structure design and fabrication [60,61]. The birefringence is characterized by

$$B = |\beta_x - \beta_y|/k_0 = |n_x - n_y|, \qquad (2.124)$$

where $n_x = n_{\text{eff}}^{(x)}$ and $n_y = n_{\text{eff}}^{(y)}$ are effective indexes for x- and y-polarization modes, respectively. Difference in the effective indexes for two polarizations is introduced either by unsymmetrical waveguide, such as with an elliptical core, or by index distribution caused by unsymmetrical strains in the core. The strain causes relative displacement of atoms and an accompanying change of electron state; this consequently leads to a change in the dielectric constant, termed the photoelastic effect. The index change is basically proportional to the strain, which is a six-component vector in an elastic medium, including three normal strains and three shear strains. The dielectric constant increment is also a six-component vector correspondingly. Their proportionality is expressed as a 6×6 tensor [10]. Because of the symmetry of material structure, the photoelastic effect for isotropic medium is described as

$$
\Delta \left(\frac{1}{\varepsilon} \right) = \frac{-1}{\varepsilon^2} \begin{pmatrix} \Delta \varepsilon_x \\ \Delta \varepsilon_y \\ \Delta \varepsilon_z \\ \Delta \varepsilon_{yz} \\ \Delta \varepsilon_{zx} \\ \Delta \varepsilon_{xy} \end{pmatrix} = \begin{pmatrix} p_{11} & p_{12} & p_{12} & 0 & 0 & 0 \\ p_{12} & p_{11} & p_{12} & 0 & 0 & 0 \\ p_{12} & p_{12} & p_{11} & 0 & 0 & 0 \\ 0 & 0 & 0 & p_{44} & 0 & 0 \\ 0 & 0 & 0 & 0 & p_{44} & 0 \\ 0 & 0 & 0 & 0 & 0 & p_{44} \end{pmatrix} \begin{pmatrix} e_x \\ e_y \\ e_z \\ e_{yz} \\ e_{zx} \\ e_{xy} \end{pmatrix},
$$

(2.125)

where $e_i (i = x, y, z)$ are the normal strains, and $e_{ij} (i, j = x, y, z)$ are the shear strains; and $p_{ij} (i, j = 1, 2)$ are the strain-optic coefficients with relation of $p_{44} = (p_{11} - p_{12})/2$ for isotropic materials, implying that among the 36 elements of the matrix only two are independent. Silica glass is a typical isotropic material; the photoelastic coefficients of bulk silica is measured to be $p_{11} = 0.121$ and $p_{11} = 0.270$ at a wavelength of 632.8 nm [63,64], which are used to estimate the elastic optic effect of the fibers near the infrared wave band without major errors.

Therefore, the modal birefringence can be realized if a proper strain is generated in fabrication of fiber preform and in its drawing processing. In the built-in strain PMF, two stress-applied parts (SAP) are formed in the fiber, which are usually boron (B_2O_3)-doped silica; thus the thermal expansion coefficient difference between SAP and pure silica induces thermal stress. Three kinds of PMFs—panda fiber, bow tie fiber, and elliptical cladding fiber—are now available commercially and used widely, shown schematically in Figure 2.25.

In panda fiber [65], material with thermal expansion coefficient higher than silica were buried in the two holes at the panda eye positions. The thermal stress occurs during preform sintering and fiber

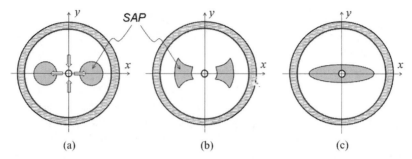

Figure 2.25 Cross section of typical PMF: (a) Panda fiber; (b) bow tie fiber; and (c) elliptical cladding fiber.

drawing processing in temperature variation from about 1,600°C to room temperature. The stress has to be relaxed by transverse deformation; whereas the axial stress is removed because the fiber is very long and thin. The transverse deformation induces strain distribution around the core, which is suppression in the up–down direction (y) and stretching in the left–right direction (x). In other words, the core suffers a positive strain in x direction and negative strain in y direction. From Equation (2.125), the index difference in this case is deduced to be

$$\Delta n_x - \Delta n_y = \frac{1}{2}n^3(p_{12} - p_{11})(e_x - e_y). \tag{2.126}$$

This means that the polarized mode parallel to x direction will propagate with slower velocity; it is said that x is the slow axis, and y is the fast axis.

Many papers on topics related to PFM have been published. Reference [66] summarizes PMF technology, including fabrications, strain analyses, characteristics measurements, and applications.

2.4.2.2 Basic Characteristics and Measurement of PMF   It is understandable that with birefringence of Equation (2.124) the PMF is regarded as a uniaxial crystal with principal axes in the x and y directions, and performs like a wave plate with very large thickness, whose effect varies with the propagation distance and can be described by a transmission matrix in a coordinate coinciding with the principal axis of PMF as [67]

$$T = \begin{pmatrix} \exp(jn_xk_0z) & 0 \\ 0 & \exp(jn_yk_0z) \end{pmatrix} = \begin{pmatrix} e^{j\delta/2} & 0 \\ 0 & e^{-j\delta/2} \end{pmatrix} \exp(j\beta z),$$

$$\tag{2.127}$$

where $\delta = Bk_0z$ ($n_x > n_y$ is assumed), and $\beta = (n_x + n_y)k_0/2$. An optical beam linearly polarized at a certain angle in laboratory coordinates can be decomposed into two beams polarized at x and y directions when it inputs into a PMF, and the two components will propagate with different phase velocities. If the input polarization coincides with one of the two principal axes of PMF, x or y, the polarization direction will be kept unchanged all the way in the fiber. This is the reason for using the term PMF. However, if the input polarization direction does not coincide with the principal axis, the composed polarization state changes periodically in the propagation, from linear polarization through elliptical polarization to circular polarization, described as

$$E = \begin{pmatrix} e^{j\delta/2} & 0 \\ 0 & e^{-j\delta/2} \end{pmatrix} \begin{pmatrix} \cos\theta \\ \sin\theta \end{pmatrix} = \begin{pmatrix} e^{j\delta/2}\cos\theta \\ e^{-j\delta/2}\sin\theta \end{pmatrix}, \qquad (2.128)$$

where θ is the direction of a normalized input linear polarized beam. The polarization direction is not necessarily kept constant, and the maintenance of polarization in PMF is not unconditional. To keep the polarization state, the input polarization direction should be adjusted to coincide with the principle axis of PMF, that is, to let $\theta = 0$ or $\theta = \pi/2$.

It is reasonable that the rate of polarization rotation depends on the birefringence of the fiber. A useful parameter to characterize PMF, called beat length L_B, is then defined as the length for the output polarization state returning to the input. Obviously, L_B depends directly on the birefringence B, expressed as

$$L_B = \lambda/|n_x - n_y|. \qquad (2.129)$$

Therefore, the birefringence B can be calculated by the measured beat length L_B. It is shown that the beat length is a direct function of wavelength, which provides a method to measure L_B by a spectrum analyzer.

Figure 2.26 shows a typical measurement setup of beat length, in which an L-long PMF is connected between the polarizer P and analyzer A. The optical field after the analyzer is expressed as

$$E_A = \begin{pmatrix} \cos^2\alpha & \cos\alpha\sin\alpha \\ \cos\alpha\sin\alpha & \sin^2\alpha \end{pmatrix} \begin{pmatrix} e^{j\delta_L/2}\cos\theta \\ e^{-j\delta_L/2}\sin\theta \end{pmatrix}, \qquad (2.130)$$

where the first 2×2 matrix in the right-hand side is for the transmission of analyzer A, placed at an angle of α in the laboratory coordinate, and $\delta_L = Bk_0L$. The optical intensity received by the optical spectrum

Figure 2.26   Beat length measurement setup. P, polarizer; A, analyzer; C, fiber connection.

analyzer is then written as

$$I_A = \frac{1}{2}[1 + \cos 2\alpha \cos 2\theta + \sin 2\alpha \sin 2\theta \cos \delta_L]. \tag{2.131}$$

The last term gives a sinusoidal oscillating spectrum, and the amplitude of the oscillation reaches the maximum when the orientations of input polarization and the analyzer are adjusted in coincidence with the principal axis of PMF. The birefringence B can be deduced from the wavelength spacing of peaks: $B = \lambda^2/L\Delta\lambda$.

The properties discussed above are for ideal PMFs and under ideal application conditions. Some practical situations have to be considered, for example, the PMF may be bent, twisted, or pressed, and placed in a variable and nonuniform temperature environment. Moreover, the input optical signal may not be an ideal polarized wave. Some of these effects will be discussed in Chapter 3.

The PMF is useful in fiber communication and fiber sensors especially when used to connect devices and components, for which a certain determined polarization direction is required; and also when used to make fiber devices, including sensors. A variety of PMFs are commercially available now. Table A4.5 of Appendix 4 shows the main specifications of Corning PANDA PMF production.

2.4.3   Photonic Crystal Fiber and Microstructure Fiber

2.4.3.1 Photonic Crystals   Photonic crystal (PC) is a medium with a spatially periodic-distributed index [68–73]. It is verified both theoretically and experimentally that the medium has a transmission spectrum with features similar to the energy band of solid state. The PCs include one-dimensional structures, such as stacks of identical multilayer materials and fiber gratings; and two-dimensional ones, such as PCF, and also three-dimensional ones, as shown in Figure 2.27.

As is known, the wave function of the electrons in solid-state crystals obeys a Schrödinger equation with periodical potentials, and the Bloch

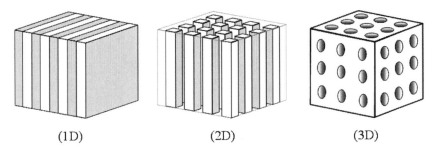

(1D) (2D) (3D)

Figure 2.27 Photonic crystals with one-dimensional (1D), two-dimensional (2D), and three-dimensional (3D) structures.

wave method is used to reduce the problem of solving the equation within a cell. The periodic index is written by means of a Bloch wave as [69]

$$n^2(\boldsymbol{r}) = n^2(\boldsymbol{r} + \boldsymbol{R}_l) = \sum_K n_K^2(\boldsymbol{r}) exp(-j\boldsymbol{K} \cdot \boldsymbol{r}), \qquad (2.132)$$

where $\boldsymbol{R}_l = l_1\boldsymbol{a}_1 + l_2\boldsymbol{a}_2 + l_3\boldsymbol{a}_3$ is the periodic position vector with integers l_1, l_2, and l_3; and $\boldsymbol{K}$ is the wave vector in reciprocal lattice space, that is, *Bloch wave vector*. Similar to the electronic energy, the optical field is expanded as a sum of Bloch waves:

$$E(\boldsymbol{r}) = \sum_K \boldsymbol{E}_K(\boldsymbol{r}) exp(-j\boldsymbol{K} \cdot \boldsymbol{r}), \qquad (2.133)$$

and the propagation constant of optical waves in the PC structure also shows similar behavior to that of an electronic wave vector in solid state. Figure 2.28 shows a schematic diagram of photonic band, where Γ stands for the original point of wave vector space, X and L are the boundary points in (100) and (111) direction of the Brillouin zone. It is shown that a bandgap exists in $\omega \sim k$ space, where the optical wave is not allowed to transmit, that is, it will be totally reflected. However, if some defects or imperfections are introduced into the structure, so that the periodicity of PC is destroyed in some degree, permitted states may appear in the bandgap. This property is the same as that of impurities in pure semiconductor crystals.

The PCs have demonstrated very attractive and unique characteristics, such as control of light emission and photon traps, corner turning of an optical beam, and a negative refractive index. By using microprocessing technology, two-dimensional photonic materials have been fabricated and used in PC devices. Three-dimensional PCs have been used in microwave systems.

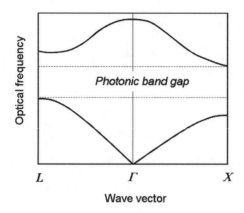

Figure 2.28   Schematic diagram of photonic band.

One-dimensional PCs, as periodical structures, exist actually before the proposal of the PC concept, such as multiple-layered dielectric films, volume gratings, and fiber gratings. The concept of PC provides a new method to understand and exploit their properties, and to expand their applications. The fiber grating will be discussed in detail in Chapter 4 of this book.

2.4.3.2 Photonic Crystal Fiber   The PCF [74–79] is a fiber with multiple periodically arranged holes extended along the fiber, which is also called the holey fiber. It is a typical two-dimensional crystal, and has found practical applications as a commercially available product. Figure 2.29 shows cross section photos of two typical PCFs. The preform of PCF is usually fabricated by a bundle of silica tubes that are melted together at high temperature.

Generally, the PCF is sorted into two kinds: index guiding and photonic bandgap (PBG) guiding, as shown in Figure 2.29(a) [76,80] and b [77] respectively. In index guiding PCF, the central air hole is filled with the same material as the background, acting as the core, surrounded by the two-dimensional PC area, which has an equivalent index lower than the core. The physical mechanism of wave guiding is the same as conventional fibers in principle.

In PBG guiding, the light wave is confined in the central hole owing to the band structure, although with lower index than the cladding. The photons of corresponding frequency are not allowed in the cladding, being forbidden by the bandgap effect of the PC, and thus propagate only in the air hole at the center.

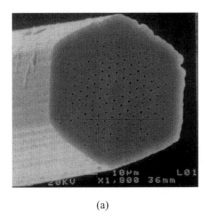 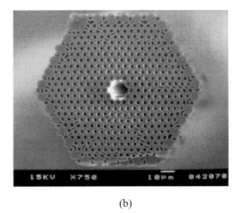

(a)   (b)

Figure 2.29 (a) Cross sections of index guiding PCF. (Reprinted with permission from reference [76].) (b) Photonic band-gap guiding PCF. (Reprinted with permission form reference [77].)

PCFs have particular properties, quite different from the conventional fibers. By proper design and fabrication PCF possesses possibilities of following unique characteristics and performances:

1. *Endless single-mode propagation* [76]: It is analyzed in Section 2.2 that in the conventional fibers higher order modes will be allowed to propagate for short wavelengths as the normalized frequency V increases. In the index-guiding PCF, the equivalent index of holey material as the cladding is an ascending function of frequency because the depth of evanescent field in the air hole decreases with the frequency increasing, making the normalized frequency decreasing. By proper design PCF keeps single-mode propagation in a much shorter wavelength range than the conventional fibers.

2. *With high nonlinear optical effect* [78]: It is known that nonlinear optic effects are enhanced by the intensity density. A much smaller core area of PCF than the ordinary fibers can be designed and fabricated, resulting in higher intensity density and higher nonlinear effects. High nonlinear fibers play important roles in some applications, such as super-continuum generation.

3. *Designable and controllable polarization characteristics* [79]: Obviously, many structure parameters can be used to adjust and control characteristics of PCF, including layout of the holes and their symmetric types, size of the holes, spacing between the holes, and shape of the hole. High birefringence is obtained by some special designs, such as laying out the holes on rectangular-like lattices,

especially in the area surrounding the core; and even single polarization propagation can be realized.

4. *Designable and controllable dispersion characteristics*: Fiber dispersion can be adjusted by the design of core index distribution, based on waveguide dispersion. The structure of PCF provides more possibilities to design its dispersion characteristics, such as large spectral range of low (zero) dispersion.

5. *High-power and high-energy transportation*: Generally, the optical absorption and thermal effects of glass materials limit the performance of energy transportation by fibers. The photonic bandgap effect provides a possibility to transport the light in an air hole with extendable area to mitigate the negative effects. The feature may also expand the spectral range to farther infrared.

The fabrication cost for PCF with complicated structures is surely higher than that of ordinary fibers. However, it is not necessary to use precisely controlled doped silica as the core of PCF, which helps to reduce the cost in mass production.

2.4.3.3 *Other Microstructure Fibers*   Apart from the PCFs introduced above, various microstructure fibers have been developed. Figure 2.30 shows an example, called "grape-fruit" fiber [81]. It has six large holes with thin walls, leaving a region at the center as the core to guide light waves. The "grape-fruit" fibers have been exploited to be used as a special sensor because the big holes allow fluid material penetration inside the fiber, making its propagation characteristics sensitive to the fluid due to the change of cladding index.

Figure 2.30   Cross section of a grape-fruit fiber. (Reprinted with permission form reference [81].)

There have been other special types, such as two-core or multicore fibers [82], and POFs [83]. Many references on these topics can be found in journals and books.

PROBLEMS

2.1 Derive the relation between group velocity and the coefficients of formulas (2.6).

2.2 dBm is used as a unit of power, which is 10 times the logarithm of power with respect to 1 mW. Calculate how many mW is the power for -30 dBm.

2.3 Calculate the output power of a 20-km-long fiber with loss coefficients of 2.1 dB/km for 800 nm, 0.33 dB/km for 1,310 nm, and 0.22 dB/km for 1,550 nm, in case of 1 mW power coupled into the input end.

2.4 At least how many percentages will be reflected backward to the fiber by the end facet without AR coating? What is the influence of the fact to loss measurement of a fiber, and how can the error coming from the influence be avoided?

2.5 Compare dispersion characteristics of an SMF and an MMF.

2.6 For a step-index fiber with core index n_1 and cladding index n_2, derive an expression of the numerical aperture (2.8). If the fiber is soaked in water, how much will the angle of output emitting beam be changed?

2.7 What are the experimental bases of Maxwell equations, what are their integral forms?

2.8 Derive Equations (2.15) from the Maxwell equation.

2.9 Explain the physical meaning of the boundary conditions. What is the difference of the boundary condition at interfaces between dielectric media from the boundary condition at interface between metal and air?

2.10 Derive the expression of normalized frequency V for the step-index fiber.

2.11 Derive the eigenvalue equation and prove the consistence of Equations (2.30a) and (2.30b).

2.12 What is the weakly guiding approximation? What is the relation and difference between hybrid modes and linear polarization modes?

2.13 Explain the characteristics of HE modes and EH modes, state their differences. Which modes are the transverse electric or magnetic modes?

2.14 What is a cutoff frequency? Derive the equations of cutoff frequency; derive the cutoff condition of the fundamental mode.

2.15 Describe the polarization characteristics of the hybrid modes and linear polarization modes.

2.16 Derive the pulse width varied with the propagation distance. Explain the mechanisms of pulse width broadening of optical pulses propagating in different fibers. Explain the three mechanisms of the optical dispersion in fiber.

2.17 Derive the sun light ray in the atmospheric layer from the ray equation. Explain the variation of the sun's visual size from noon to dusk.

2.18 What are the helical ray and meridional ray? Explain why the parabolic gradient-index fiber has the minimum mode dispersion.

2.19 Write the field distribution of the fundamental mode in Gaussian with the mode radius. Discuss the numerical aperture of the SMF.

2.20 Describe the basic characteristics of graded-index lens and its main functions.

2.21 Design the basic structure of EDFA, and depict a schematic diagram.

2.22 What are the advantages of DCF?

2.23 Describe the basic structure and characteristics of PMFs. Derive expression (2.131) by Jones matrix.

2.24 What are the birefringence and the beat length of PMF? How is the beat length measured?

2.25 What is a photonic crystal? Describe its main properties and features.

2.26 What is a PCF? Explain its waveguiding mechanism and the features that are different from ordinary fibers.

REFERENCES

1. Marcuse D. *Light Transmission Optics*. New York, Cincinnati, Toronto, Melbourne: Van Nostrand Reinhold Company, 1982.
2. Snyder AW, Love JD. *Optical Waveguide Theory*. London, New York: Chapman and Hall, 1983.
3. Yariv A. *Optical Electronics in Modern Communications*. Fifth Edition. New York: Oxford University Press, 1997.
4. Agrawal GP. *Nonlinear Fiber Optics*. Singapore: Elsevier Science, 2004.
5. Liao YB. *Fiber Optics*, Beijing: Tsinghua University Press, 2000. [In Chinese]
6. Okoshi T. *Optical Fibers*. New York, London, Paris, San Diego, San Francisco, Sao Paulo, Sydney, Tokyo, Toronto: Press, 1982.
7. Vassallo C. *Optical Waveguide Concepts*. Amsterdam, Oxford, New York, Tokyo: Elsevier, 1991.
8. Tsao C. *Optical Fiber Waveguide Analysis*. Oxford, New York, Tokyo: Oxford University Press, 1992.
9. Saleh BEA, Teich MC. *Fundamentals of Photonics*. Hoboken New Jersey: John Wiley & Sons, 2007.
10. Born M, Wolf E. *Principles of Optics*. Seventh Edition. Cambridge, UK: Cambridge University Press, 1999.
11. Kaminov IP, Koch TL. *Optical Fiber Telecommunications*. Academic Press, San Diego, London, Boston, New York, Sydney, Tokyo, Toronto, 1997.
12. Kaminow IP, Li T. *Optical Fiber Telecommunications IV*. Elsevier Inc., San Diego, London, Boston, New York, Sydney, Tokyo, Toronto, 2002.
13. Mynbaev DK, Scheiner LL. *Fiber-Optic Communications Technology*. Science Press and Pearson Education North Asia Limited, Beijing, 2002.
14. Bass M, Van Stryland EW. *Fiber Optics Handbook—Fiber, Devices and Systems for Optical Communications*. McGraw-Hill, New York, Chicago, San Francisco, Lisbon, London, Madrid, Mexico City, Milan, New Delhi, San Juan, Seoul, Singapore, Sydney, Toronto, 2002.
15. Akamatsu T, Okamura K, Ueda Y. Fabrication of long fibers by an improved chemical vapor-deposition method (HCVD Method). *Applied Physics Letters* 1977; 31: 174–176.
16. Okada M, Kawachi M, Kawana A. Improved chemical vapor-deposition method for long-length optical fiber. *Electronics Letters* 1978; 14: 89–90.
17. Miya T, Terunuma Y, Hosaka T, Miyashita T. Ultimate low-loss single-mode fiber at 1.55 μm. *Electronics Letters* 1979; 15: 106–108.
18. Imoto K, Sumi M. Modified VAD method for optical-fiber fabrication. *Electronics Letters* 1981; 17: 525–526.

19. Shibata S, Horiguchi M, Jinguji K, Mitachi S, Kanamori T, Manabe T. Prediction of loss minima in infrared optical fibers. *Electronics Letters* 1981; 17: 775–777.

20. Tanaka M, Okuno T, Omori H, Kato T, Yokoyama Y, Takaoka S, Kunitake K, Uchiyama K, Hanazuka S, Nishimura M. Water-peak-suppressed non-zero dispersion shifted fiber for full spectrum coarse WDM transmission in metro networks. *Technical Digest of Optical Fiber Communications Conference (OFC), Anaheim, 2002*, 2002, Anaheim, CA, paper WA2.

21. Malitson IH. Interspecimen comparison of refractive index of fused silica. *Journal of Optical Society America* 1965; 55: 1205–1209.

22. Cohen LG. Comparison of single-mode fiber dispersion measurement techniques. *Journal of Lightwave Technology* 1985; 3: 958–966.

23. ITU-T, G.652, Characteristics of a single-mode optical fiber and cable. http://www.itu.int.

24. Corning, InfiniCor 62.5 μm optical fiber product information sheet. www.corning.com.

25. ITU-T, G.653, Characteristics of a dispersion-shifted single-mode optical fiber and cable. http://www.itu.int.

26. ITU-T, G.655, Characteristics of a non-zero dispersion-shifted single-mode optical fiber and cable. http://www.itu.int.

27. ITU-T, G.657, Characteristics of a bending loss insensitive single-mode optical fiber and cable for the access network. http://www.itu.int.

28. Snitzer E. Cylindrical dielectric waveguide modes. *Journal of the Optical Society America* 1961; 51: 491–498.

29. Gloge D. Weakly guiding fibers. *Applied Optics* 1971; 10: 2252–2258.

30. Kim BY, Blake JN, Huang SY, Shaw HJ. Use of highly elliptic core fibers for 2-mode fiber devices. *Optics Letters* 1987; 12: 729–731.

31. Erdogan T. Cladding-mode resonances in short- and long-period fiber grating filters. *Journal of Optical Society America A* 1997; 14: 1760–1773.

32. Ankiewicz A, Pask C. Geometric optics approach to light acceptance and propagation in graded index fibers. *Optical and Quantum Electronics* 1977; 9: 87–109.

33. Pask C. Exact expressions for scalar modal eigenvalues and group delays in power-law optical fibers. *Journal of Optical Society America* 1979; 69: 1599–1603.

34. Goodman JW. *Introduction to Fourier Optics*. New York, St. Louis, Toranto, London, Sydney:McGraw-Hill, 1968.

35. Marcuse D. Loss analysis of single-mode fiber splices. *Bell System Technical Journal* 1977; 56: 703–718.

36. Marcuse D. Gaussian approximation of fundamental modes of graded-index fibers. *Journal of Optical Society America* 1978; 68: 103–109.

37. Snyder AW, Young WR. Modes of optical-waveguides. *Journal of Optical Society America* 1978; 68: 297–309.

38. Klemas AT, Shenk DS, Reed WA, Saifi MA. Analysis of mode field radius calculated from single-mode fiber far-field radiation-patterns. *Journal of Lightwave Technology* 1985; 3: 967–970.

39. Petermann K. Fundamental mode microbending loss in graded-index and W fibers. *Optical and Quantum Electronics* 1977; 9: 167–175.

40. Petermann K. Constraints for fundamental-mode spot size for broad-band dispersion- compensated single-mode fibers. *Electronics Letters* 1983; 19: 712–714.

41. Gloge D, Marcatili Ea. Multimode theory of graded core fibers. *Bell System Technical Journal* 1973; 52: 1563–1578.

42. Cohen LG, Lin C, French WG. Tailoring zero chromatic dispersion into the 1.5–1.6 μm low-loss spectral region of single-mode fibers. *Electronics Letters* 1979; 15: 334–335.

43. Ye CF, McLeod RR. GRIN lens and lens array fabrication with diffusion-driven photopolymer. *Optics Letters* 2008; 33: 2575–2577.

44. Watanabe Y, Ohmori H, Lin W, Uehara Y, Suzuki T, Morita S, Mitsuishi N, Makinouchi A. Development of aspherical gradient index (GRIN) lens fabrication system based on VCAD concept and FLID grinding, *Proceeding of International Conference on Leading Edge Manufacturing in 21st Century*, 2005, Nagoya Japan, pp. 697–702.

45. Mears RJ, Reekie L, Jauncey IM, Payne DN. Low-noise erbium-doped fiber amplifier operating at 1.54 μm. *Electronics Letters* 1987; 23: 1026–1028.

46. Desurvire E, Giles CR, Simpson JR, Zyskind JL. Efficient erbium-doped fiber amplifier at a 1.53 μm wavelength with a high output saturation power. *Optics Letters* 1989; 14: 1266–1268.

47. Wysocki PF, Digonnet MJF, Kim BY, Shaw HJ. Characteristics of erbium-doped superfluorescent fiber sources for interferometric sensor applications. *Journal of Lightwave Technology* 1994; 12: 550–567.

48. Lee J, Ryu UC, Ahn SJ, Park N. Enhancement of power conversion efficiency for an L-band EDFA with a secondary pumping effect in the unpumped EDF section. *IEEE Photonics Technology Letters* 1999; 11: 42–44.

49. Chang CL, Wang L, Chiang YJ. A dual pumped double-pass L-band EDFA with high gain and low noise. *Optics Communications* 2006; 267: 108–112.

50. Hanna DC, Percival RM, Perry IR, Smart RG, Suni PJ, Tropper AC. An Ytterbium-doped monomode fiber laser—broadly tunable operation from 1.010 μm to 1.162 μm and 3-level operation at 974 nm. *Journal of Modern Optics* 1990; 37: 517–525.

51. Kosterin A, Erwin JK, Fallahi M, Mansuripur M. Heat and temperature distribution in a cladding-pumped, Er:Yb co-doped phosphate fiber. *Review of Scientific Instruments* 2004; 75: 5166–5172.
52. Li L, Schulzgen A, Temyanko VL, Qiu T, Morrell MM, Wang Q, Mafi A, Moloney JV, Peyghambarian N. Short-length microstructured phosphate glass fiber lasers with large mode areas. *Optics Letters* 2005; 30: 1141–1143.
53. Po H, Cao JD, Laliberte BM, Minns RA, Robinson RF, Rockney BH, Tricca RR, Zhang YH. High-power neodymium-doped single transverse-mode fiber laser. *Electronics Letters* 1993; 29: 1500–1501.
54. Zellmer H, Willamowski U, Tünnermann A, Welling H, Unger S, Reichel V, Müller HR, Kirchhof J, Albers P. High-power cw neodymium-doped fiber laser operating at 9.2 W with high beam quality. *Optics Letters* 1995; 20: 578–580.
55. Liu A, Ueda K. The absorption characteristics of circular, offset, and rectangular double-clad fibers. *Optics Communications* 1996; 132: 511–518.
56. Doya V, Legrand O, Mortessagne F. Optimized absorption in a chaotic double-clad fiber amplifier. *Optics Letters* 2001; 26: 872–874.
57. Leproux P, Fevrier S, Doya V, Roy P, Pagnoux D. Modeling and optimization of double-clad fiber amplifiers using chaotic propagation of the pump. *Optical Fiber Technology* 2001; 7: 324–339.
58. Dritsas I, Sun T, Grattan KTV. Stochastic optimization of conventional and holey double-clad fibers. *Journal of Optics A — Pure and Applied Optics* 2007; 9: 405–421.
59. Papp A, Harms H. Polarization optics of index-gradient optical-waveguide fibers. *Applied Optics* 1975; 14: 2406–2411.
60. Kaminow IP, Simpson JR, Presby HM, Macchesney JB. Strain birefringence in single-polarization germanosilicate optical fibers. *Electronics Letters* 1979; 15: 677–679.
61. Okamoto K, Edahiro T, Shibata N. Polarization properties of single-polarization fibers. *Optics Letters* 1982; 7: 569–571.
62. Rashleigh SC. Origins and control of polarization effects in single-mode fibers. *Journal of Lightwave Technology* 1983; 1: 312–331.
63. Namihira Y. Opto-elastic constant in single-mode optical fibers. *Journal of Lightwave Technology* 1985; 3: 1078–1083.
64. Bertholds A, Dandliker R. Determination of the individual strain-optic coefficients in single-mode optical fibers. *Journal of Lightwave Technology* 1988; 6: 17–20.
65. Tajima K, Ohashi M, Sasaki Y. A new single-polarization optical fiber. *Journal of Lightwave Technology* 1989; 7: 1499–1503.
66. Noda J, Okamoto K, Sasaki Y. Polarization-maintaining fibers and their applications. *Journal of Lightwave Technology* 1986; 4: 1071–1089.

67. VanWiggeren GD, Roy R. Transmission of linearly polarized light through a single-mode fiber with random fluctuations of birefringence. *Applied Optics* 1999; 38: 3888–3892.

68. Yablonovitch E. Inhibited spontaneous emission in solid-state physics and electronics. *Physical Review Letters* 1987; 58: 2059–2062.

69. Zhang Z, Satpathy S. Electromagnetic-wave propagation in periodic structures—Bloch wave solution of Maxwell equations. *Physical Review Letters* 1990; 65: 2650–2653.

70. Yablonovitch E. Photonic band-gap structures. *Journal of Optical Society America B* 1993; 10: 283–295.

71. Sozuer HS, Haus JW. Photonic bands—simple-cubic lattice. *Journal of Optical Society America B* 1993; 10: 296–302.

72. Smith DR, Dalichaouch R, Kroll N, Schultz S, Mccall SL, Platzman PM. Photonic band-structure and defects in one and two dimensions. *Journal of Optical Society America B* 1993; 10: 314–321.

73. Noda S, Ogawa SP, Imada M, Yoshimoto S, Okano M. Control of light emission by 3D photonic crystals. *Science* 2004; 305: 227–229.

74. Birks TA, Roberts PJ, Russel PSJ, Atkin DM, Shepherd TJ. Full 2-D photonic band gaps in silica/air structures. *Electronics Letters* 1995; 31: 1941–1943.

75. Knight JC, Birks TA, Russell PSJ, Atkin DM. All-silica single-mode optical fiber with photonic crystal cladding. *Optics Letters* 1996; 21: 1547–1549.

76. Birks TA, Knight JC, Russell PSJ. Endlessly single-mode photonic crystal fiber. *Optics Letters* 1997; 22: 961–963.

77. Russell PSJ, Cregan RF, Mangan BJ, Knight JC, Birks TA, Roberts PJ, Allan DC. Single-mode photonic band gap guidance of light in air. *Science* 1999; 285: 1537–1539.

78. Husakou AV, Herrmann J. Supercontinuum generation of higher-order solitons by fission in photonic crystal fibers. *Physical Review Letters* 2001; 87: 203901-1–203901-4.

79. Rosa L, Poli F, Foroni M, Cucinotta A, Selleri S. Polarization splitter based on a square-lattice photonic-crystal fiber. *Optics Letters* 2006; 31: 441–443.

80. Russell PSJ. Photonic-crystal fibers. *Journal of Lightwave Technology* 2006; 24: 4729–4749.

81. Mägi EC, Nguyen HC, Eggleton BJ. Air-hole collapse and mode transitions in microstructured fiber photonic wires. *Optics Express* 2005; 13: 453–459.

82. Chiang KS. Intermodal dispersion in two-core optical fibers. *Optics Letters* 1995; 20: 997–999.

83. Eldada L, Shacklette LW. Advances in polymer integrated optics. *IEEE Journal of Selected Topics Quantum Electronics* 2000; 6: 54–68.

CHAPTER 3

FIBER SENSITIVITIES
AND FIBER DEVICES

Characteristics of light propagation in the fiber will be affected by physical conditions that are encountered by the fiber. On the other hand, the fiber's sensitivity to physical conditions is the basis of fiber sensors and most intrinsic to fiber devices. In this chapter, the sensitivities and their physical mechanisms are expounded in Section 3.1. Characteristics and basic theory of fiber couplers are stated in Section 3.2; and devices incorporated with couplers are introduced in Section 3.3. Polarization characteristics are discussed in Section 3.4; and polarization-related fiber devices are presented in the last section.

3.1 FIBER SENSITIVITIES TO PHYSICAL CONDITIONS

It is interesting to investigate the fiber sensitivities to various physical conditions and to understand their mechanisms. Such sensitivities are the basis fiber devices and fiber sensors. In this section, sensitivities of single-mode fibers (SMF) to strain, stress, and temperature are analyzed. Sensitivity of germanium-doped silica fiber to ultraviolet laser irradiation, which is the basis of fiber gratings, will be discussed in Chapter 4. Nonlinear optical (NLO) effects, which are used in distributed sensors, will be analyzed in Chapter 5.

Fundamentals of Optical Fiber Sensors, First Edition.
Zujie Fang, Ken K. Chin, Ronghui Qu, and Haiwen Cai.
© 2012 John Wiley & Sons, Inc. Published 2012 by John Wiley & Sons, Inc.

3.1.1 Sensitivity to Axial Strain

When a force is applied to a section of fiber in its axial direction, it will be stretched or compressed, that is, an axial strain occurs. According to the theory of elasticity, both strain and stress in solid-state materials have to be expressed as tensors; and they can be reduced to a six-component vector with three normal components and three for shearing directions. The fundamental of elasticity is given in Appendix 2 of the book.

In case of no shearing force, according to Hooke's law, the strain is proportional to the stress, expressed as

$$\begin{pmatrix} e_x \\ e_y \\ e_z \end{pmatrix} = \frac{1}{Y} \begin{pmatrix} 1 & -\nu & -\nu \\ -\nu & 1 & -\nu \\ -\nu & -\nu & 1 \end{pmatrix} \begin{pmatrix} \sigma_x \\ \sigma_y \\ \sigma_z \end{pmatrix}, \tag{3.1}$$

where σ_x, σ_y, and σ_z are stresses applied to the fiber, e_x, e_y, and e_z are strains caused by the stresses; Y is Young's modulus of material; ν is its Poisson's ratio, which describes transverse deformations induced by a longitudinal deformation. When the external force is in axial direction, $\sigma_x = \sigma_y = 0$, and $\sigma_z = F/A$, where F is the axial force and A is the fiber's cross-section area. Therefore, the generated strain can be written as

$$\begin{pmatrix} e_x \\ e_y \\ e_z \end{pmatrix} = \frac{F}{AY} \begin{pmatrix} 1 & -\nu & -\nu \\ -\nu & 1 & -\nu \\ -\nu & -\nu & 1 \end{pmatrix} \begin{pmatrix} 0 \\ 0 \\ 1 \end{pmatrix} = \frac{F}{AY} \begin{pmatrix} -\nu \\ -\nu \\ 1 \end{pmatrix}. \tag{3.2}$$

If the axial force is positive, $F > 0$, the fiber is stretched, and its cross section shrinks simultaneously. Inversely, it is compressed, and its cross section is expanded, as long as it can be kept straight by some support. Young's Modulus Y and Poisson's ratio ν are parameters of materials. It is measured for silica that $Y = 6.5 \times 10^{10} \ N/m^2$ and $\nu = 0.17$ [1,2].

As introduced in Section 2.4.2, strains will cause the photoelastic effect, that is, the refractive index increases proportionally with the strain. For an axially strained fiber the index change is described as:

$$\Delta\left(\frac{1}{n^2}\right) = \frac{-2}{n^3} \begin{pmatrix} \Delta n_x \\ \Delta n_y \\ \Delta n_z \end{pmatrix} = \begin{pmatrix} p_{11} & p_{12} & p_{12} \\ p_{12} & p_{11} & p_{12} \\ p_{12} & p_{12} & p_{11} \end{pmatrix} \begin{pmatrix} e_x \\ e_y \\ e_z \end{pmatrix}$$

$$= \begin{pmatrix} p_{11}e_x + p_{12}(e_y + e_z) \\ p_{11}e_y + p_{12}(e_x + e_z) \\ p_{11}e_z + p_{12}(e_x + e_y) \end{pmatrix}. \tag{3.3}$$

Substituting (3.2) into (3.3), we have

$$
\begin{pmatrix} \Delta n_x \\ \Delta n_y \\ \Delta n_z \end{pmatrix} = \frac{-n^3 e_z}{2} \begin{pmatrix} (1-v)p_{12} - vp_{11} \\ (1-v)p_{12} - vp_{11} \\ p_{11} - 2vp_{12} \end{pmatrix} = \frac{-n^3 F}{2AY} \begin{pmatrix} (1-v)p_{12} - vp_{11} \\ (1-v)p_{12} - vp_{11} \\ p_{11} - 2vp_{12} \end{pmatrix}.
$$

$$(3.4)$$

In SMFs the propagating light wave is basically a transverse mode, as analyzed in Chapter 2; thus the effective index change is approximately equal to Δn_x:

$$
\Delta n_{\text{eff}} = -n^3[(1-v)p_{12} - vp_{11}]e_z/2 = \gamma n e_z, \tag{3.5}
$$

where $\gamma = -n^2[(1-v)p_{12} - vp_{11}]/2$ is usually called the effective photoelastic coefficient. The photoelastic coefficients of bulk silica is measured to be $p_{11} = 0.113$ and $p_{12} = 0.252$ at a wavelength of 632.8 nm [3,4], which are used to estimate the photoelastic effect near the infrared wave band without major errors, resulting in $\gamma = -0.22$ for silica fibers. Combining with change of fiber length $\delta L = e_z L$, the axial strain-induced phase shift can be written as

$$
\Delta \phi = (1 + \gamma)nkLe_z \approx 0.78nkLe_z. \tag{3.6}
$$

The effect can be utilized to develop a variety of fiber devices and sensors. An electrically driven phase modulator is constructed simply by winding several turns of fiber tightly on a piezo transducer (PZT) cylinder [5]. A fiber sensor can be realized by sticking a section of fiber on the surface of the concerned part to detect deformations of a machine or a building. The strain-induced phase shift can be measured by fiber interferometers, such as a fiber Mach–Zehnder interferometer (MZI) or a fiber Michelson interferometer (MI). In fiber gratings, the strain causes a change in their reflection spectra. Detailed analyses are given in related sections of this book.

3.1.2 Sensitivity to Lateral Pressure

Conditions of lateral pressure applied to fibers can be mostly categorized into two situations: one is radial pressure and the other is diametrically unidirectional pressure, as shown in Figures 3.1(a) and (b), respectively.

Because the fiber section under pressure is usually much longer than its transverse size, the axial stress can be neglected here, and the axial

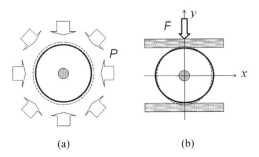

Figure 3.1 Fibers under radial pressure (a) and unidirectional pressure (b).

strain can be deduced from the transverse stress as $e_z = -v(\sigma_x + \sigma_y)/Y$. Then the stress/strain state is considered as a two-dimensional problem and treated as a plane deformation. Under this condition, stresses and stains exist only in the x–y plane; and shearing stresses with respect to z-direction vanish, $\sigma_{zx} = \sigma_{zy} = 0$. The parameters to be solved remain e_x, e_y, e_{xy}, σ_x, and σ_y. They have to meet Hooke's law and the force equilibrium equations (see Appendix 2), solved for the following two situations.

3.1.2.1 Under a Radial Pressure

The effect of radial pressure has been analyzed in reference [6]. An example of radial pressure is a section of fiber sunk in water at a certain depth. The cross section of fiber will be compressed under the pressure. In polar coordinates only radial stress σ_r exists under the radial pressure, whereas $\sigma_\varphi = 0$. Deformation occurs also only in a radial direction with a deformation vector of $u_r = u(r)$ and $u_\varphi = 0$. Strains in radial and azimuthal directions are expressed as [6] $e_r = \partial u_r/\partial r$, $e_\varphi = (\partial u_\varphi/\partial \varphi + u_r)/r = u_r/r$. The equilibrium of stresses leads to an equation of deformation vector as $\nabla(\nabla \cdot \mathbf{u}) = 0$; thus one has

$$\nabla \cdot \mathbf{u} = \frac{1}{r}\frac{d[ru(r)]}{dr} = \text{const.} \qquad (3.7)$$

It is solved to be $u(r) = ar + b/r$, where a and b are constants to be determined by the boundary conditions; therefore, the strains are obtained as

$$e_r = a - b/r^2 \text{ and } e_\varphi = a + b/r^2. \qquad (3.8)$$

By applying the boundary conditions of $\sigma_r(R) = P$, where P is the pressure applied on the outer surface of $r = R$, and $u(0) = 0$ at the

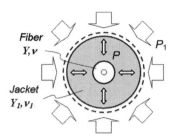

Figure 3.2 Jacketed fiber in uniform environment pressure.

center of the fiber, constants a and b can be determined as $a = (1 + v)(1 - 2v)P/Y$, and $b = 0$. The strains are then deduced to be

$$e_x = e_y = e_r = (1 + v)(1 - 2v)P/Y. \tag{3.9}$$

It is noticed that the strains are uniform inside the fiber, and independent of the fiber radius.

For a practical fiber, the nude silica fiber is coated with a jacket, as shown in Figure 3.2, which surely affects the pressure transfer. In this case equation (3.7) and its solution (3.8) hold as well for radial pressure, but a two-layer model must be considered with different Young's Modulus, Y and Y_1, and different Poisson's ratios, v and v_1 [7]. In the fiber, formula (3.9) still holds. At the boundary between cladding and jacket, the radial stresses and azimuthal strains should satisfy the continuity condition, expressed as

$$\sigma_{r1}(R) = Y_1 \frac{a - b(1 - 2v_1)/R^2}{(1 + v_1)(1 - 2v_1)} = \sigma_r(R) = P, \tag{3.10a}$$

$$e_{\varphi 1}(R) = a + b/R^2 = e_\varphi(R) = (1 + v)(1 - 2v)P/Y. \tag{3.10b}$$

At the outer boundary of the jacket, the radial stress is just the pressure outside:

$$\sigma_{r1}(R_1) = Y_1 \frac{a - b(1 - 2v_1)/R_1^2}{(1 + v_1)(1 - 2v_1)} = P_1. \tag{3.10c}$$

Thus the ratio of pressure inside the fiber over the outer pressure is deduced as

$$\frac{P_1}{P} = \frac{1}{2}\left[\frac{Y_1}{Y}\frac{(1 + v)(1 - 2v)}{1 - v_1^2}\left(1 - \frac{R^2/R_1^2}{1 - 2v_1}\right) + \frac{1 + R^2/R_1^2}{1 - v_1}\right]. \tag{3.11}$$

The dependence of strain e_r inside the fiber on the environmental pressure P_1 can then be deduced by formula (3.9). The pressure ratio is simplified in case of $R_1 \gg R$ as

$$\frac{P_1}{P} \approx \frac{1}{2(1 - v_1)} \left[\frac{Y_1}{Y} \frac{(1 + v)(1 - 2v)}{1 + v_1} + 1 \right]. \qquad (3.11a)$$

Substituting the deduced strain into the photoelastic equation, one has

$$\Delta \left(\frac{1}{n^2} \right) = \frac{(1 + v)(1 - 2v)P}{Y} \begin{pmatrix} p_{11} & p_{12} & p_{12} \\ p_{12} & p_{11} & p_{12} \\ p_{12} & p_{12} & p_{11} \end{pmatrix} \begin{pmatrix} 1 \\ 1 \\ -v \end{pmatrix}. \qquad (3.12)$$

Index increments are written as

$$\Delta n_x = \Delta n_y = -\frac{n_0^3}{2Y}(1 + v)(1 - 2v)(p_{11} + p_{12})P. \qquad (3.13)$$

The phase shift in the fiber section with length L is obtained to be

$$\Delta \phi = -\frac{n_0^3}{2Y}(1 + v)(1 - 2v)(p_{11} + p_{12})PkL. \qquad (3.14)$$

Analysis of the radial pressure effect is needed in sensor technology, such as in hydrophones [8] and in case of deep underwater applications of the fiber.

3.1.2.2 Under a Diametrically Unidirectional Pressure

The strain state of a fiber under unidirectional pressure is also considered as a two-dimensional problem of plane deformation. The stress and strain distributions are analyzed in circular disk model, compressed by two equal and opposite pressures P applied at the ends of a diameter, at Point A $(d/2, -\pi/2)$ and Point B $(d/2, \pi/2)$ in the polar coordinates, as shown in Figure 3.3, where d is fiber diameter. References [6,7] give a complete discussion of the problem.

Let us observe the stress at point $C(r, \varphi)$, as shown in the figure. The external forces P in y direction applied at points A and B give stresses in A–C and B–C directions, which are proportional to the cosine of

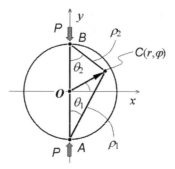

Figure 3.3 Stress distribution under diametrical pressure.

the angles to the y-axis, and inversely proportional to their distances, expressed as

$$\sigma_{\rho_1\rho_1}^{(1)} = \frac{-2P}{\pi\rho_1}\cos\theta_1, \tag{3.15a}$$

$$\sigma_{\rho_2\rho_2}^{(2)} = \frac{-2P}{\pi\rho_2}\cos\theta_2. \tag{3.15b}$$

In the directions perpendicular to A–C and B–C no normal stress and tangential stress exist, that is, $\sigma_{\theta_1\theta_1}^{(1)} = \sigma_{\rho_1\theta_1}^{(1)} = \sigma_{\theta_2\theta_2}^{(2)} = \sigma_{\rho_2\theta_2}^{(2)} = 0$. The stress analysis requires that no normal and azimuthal stress exist at the edge circle of the disk, except for points A and B: $\sigma_\varphi|_A = \sigma_\varphi|_B = 0$, $\sigma_r|_A = -\sigma_r|_B = 2P/\pi d$. To meet the boundary condition, the third stress component is needed:

$$\sigma_{ik}^{(3)} = (2P/\pi d)\delta_{ik}, \quad (i,k = \rho,\theta) \tag{3.15c}$$

which describes the uniform extension.

Referring to the tensor conversion between the polar coordinates and Cartesian coordinates (see Appendix 2), the stresses are obtained as [7]

$$\sigma_x = \frac{2P}{\pi}\left(\frac{\sin^2\theta_1\cos\theta_1}{\rho_1} + \frac{\sin^2\theta_2\cos\theta_2}{\rho_2}\right) - \frac{2P}{\pi d}, \tag{3.16a}$$

$$\sigma_y = \frac{2P}{\pi}\left(\frac{\cos^3\theta_1}{\rho_1} + \frac{\cos^3\theta_2}{\rho_2}\right) - \frac{2P}{\pi d}, \tag{3.16b}$$

$$\sigma_{xy} = -\frac{2P}{\pi}\left(\frac{\sin\theta_1\cos^2\theta_1}{\rho_1} - \frac{\sin\theta_2\cos^2\theta_2}{\rho_2}\right). \tag{3.16c}$$

For the single mode fiber the core radius is much smaller than its outer radius d/2, it is widely considered that the "center strain approximation" holds; that is, the stresses in the fiber core are uniform and equal to those at the center of $\rho_1 = \rho_2 \approx d/2$, and $\theta_1 = \theta_2 = 0$. In the core, we have [9]

$$\sigma_x = -\frac{2P}{\pi d}, \quad \sigma_y = \frac{6P}{\pi d}, \quad \sigma_{xy} = 0, \tag{3.17}$$

and

$$e_x = \frac{1}{Y}(\sigma_x - v\sigma_y - v\sigma_z) = -(1 + 3v)\frac{2P}{\pi Y d}, \tag{3.18a}$$

$$e_y = \frac{1}{Y}(\sigma_y - v\sigma_x - v\sigma_z) = (3 + v)\frac{2P}{\pi Y d}, \tag{3.18b}$$

$$e_x - e_y = -\frac{8(1 + v)}{\pi Y d}P. \tag{3.18c}$$

Substituting them into the photoelastic effect formula, a birefringence occurs in the fiber:

$$B = \Delta n_y - \Delta n_x = \frac{4n_0^3}{\pi Y d}(P_{12} - P_{11})(1 + v)P. \tag{3.19}$$

This effect is used to make a fiber polarization controller (PC), and is also used in sensors [10], as will be discussed in Section 3.5. By using typical parameters of silica fiber, a typical value of B is obtained as $B = 4.51 * 10^{-6} P$ with the pressure P in N/cm.

3.1.3 Bending-Induced Birefringence

Fibers in practical devices and systems are always bent in some way and some degree. It was found early in the 1970s that fiber bending would cause birefringence, which was explained theoretically in the literature. References [11–13] analyze the stress and strain states in bent fibers and in tension-coiled SMFs, deduce the related formulas by photoelastic effect, and explain experimental phenomena. Detailed discussions are given in references [14,15].

A bent fiber can be considered a bent cylindrical silica rod. Figure 3.4 shows a section of fiber, which is not only bent but also stretched, because bending and axial deformations occur often simultaneously in practice. In a tension-coiled fiber, a pressure from a cylindrical support must be applied on the fiber at its inner side of bending to balance the axial stretching force. A neutral surface exists inside the rod, at $x = -x_0$ in the figure, on which there is no axial extension, nor does compression occur; that is, the fiber length on the

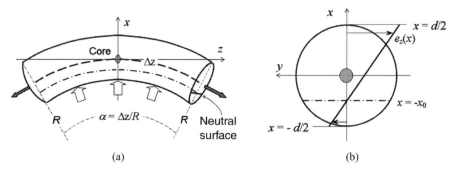

Figure 3.4 (a) Bent and stretched fiber and (b) its strain distribution.

neutral surface is equal to its original length under no bending and no stretching. In case of pure bending, without axial stretching, the neutral surface passes through the fiber axis, $x_0 = 0$. It is shown that the part above the neutral surface suffers stretching axially, whereas the part beneath is compressed. The axial strain is expressed approximately as a linear function of x, measured from the neutral surface, with a coefficient proportional to the bending curvature $1/R$:

$$e_z = \frac{x + x_0}{R}. \tag{3.20}$$

Under the first-order approximation, the averaged axial strain caused by an axial stress is

$$\bar{\sigma}_z = \frac{4Y}{\pi d^2} \int e_z \mathrm{d}x \mathrm{d}y = \frac{Y}{R} x_0 = Y \bar{e}_z, \tag{3.21}$$

where d is the diameter of the fiber. The formula gives the dependence of neutral surface position on the axial stress.

The basic effect found in bent fiber experimentally is birefringence. Obviously, mere axial strain can not explain the phenomenon. It is necessary to analyze the transverse deformation of bent fiber. It is conjectured that the round rod is deformed to be an elliptical rod, just like what happens when a rub rod is bent; and asymmetric strain occurs around the center. To reveal the mechanism, we must understand the transverse distribution of stress and strain inside the fiber.

To solve the elastic state of bent fiber some approximations are taken. The above described axial strain and stress distribution is the first-order one:

$$\sigma_z^{(1)} = Y e_z = \frac{Y}{R}(x + x_0). \tag{3.22}$$

Second, it is supposed that any component of strain and stress is not a function of axial position z, in case uniform curvature bending is considered. Then the three-dimensional problem is reduced to a two-dimensional plane deformation problem. In the second-order approximation, the shearing force σ_{xz} and σ_{yz}, induced by bending, is regarded as an equivalent body force in the transverse cross-sectional plane, which has only one component in x direction, expressed as

$$F_x = \frac{\partial \sigma_x^{(1)}}{\partial x} = \frac{\sigma_z^{(1)}}{R+x} \approx \frac{\sigma_z^{(1)}}{R}. \tag{3.23}$$

The force equilibrium equations are then written as [6,7]

$$\frac{\partial \sigma_x}{\partial x} + \frac{\partial \sigma_{xy}}{\partial y} + \frac{\partial \sigma_{xz}}{\partial z} - F_x = \frac{\partial \sigma_x}{\partial x} + \frac{\partial \sigma_{xy}}{\partial y} - \frac{\sigma_z^{(1)}}{R} = 0, \tag{3.24a}$$

$$\frac{\partial \sigma_{yx}}{\partial x} + \frac{\partial \sigma_y}{\partial y} + \frac{\partial \sigma_{yz}}{\partial z} = \frac{\partial \sigma_{yx}}{\partial x} + \frac{\partial \sigma_y}{\partial y} = 0, \tag{3.24b}$$

$$\frac{\partial \sigma_{zx}}{\partial x} + \frac{\partial \sigma_{zy}}{\partial y} + \frac{\partial \sigma_z}{\partial z} = \frac{\partial \sigma_{zx}}{\partial x} + \frac{\partial \sigma_{zy}}{\partial y} = 0. \tag{3.24c}$$

Due to the continuity relation of deformation and Hooke's law in the isotropic body, a basic equation for stresses $\sigma_x(x, y)$ and $\sigma_y(x, y)$ is obtained as

$$\nabla^2(\sigma_x + \sigma_y) = \frac{1}{(1-v)R} \frac{\partial \sigma_z^{(1)}}{\partial x} = \frac{Y}{(1-v)R^2}, \tag{3.25}$$

where differential operator ∇ is in x–y-plane. To solve (3.25), a stress potential U and a body force potential V are introduced with the following relations:

$$\sigma_x = \frac{\partial^2 U}{\partial y^2} + V, \tag{3.26a}$$

$$\sigma_y = \frac{\partial^2 U}{\partial x^2} + V, \tag{3.26b}$$

$$\sigma_{xy} = \frac{\partial^2 U}{\partial x \partial y}, \tag{3.26c}$$

$$F_x = \frac{\partial V}{\partial x} = \frac{\sigma_z^{(1)}}{R} = \frac{Y(x + x_0)}{R^2}, \tag{3.26d}$$

$$F_y = \frac{\partial V}{\partial y} = 0. \tag{3.26e}$$

The body force potential is obtained to be

$$V = -\frac{Y}{2R^2}(x^2 - d^2/4) - \frac{Y}{R^2}x_0(x - d/2). \tag{3.27}$$

Substituting the variables into (3.25), a biharmonic equation for the stress potential is deduced:

$$\nabla^4 U = -\frac{1 - 2v}{1 - v}\frac{Y}{R^2}. \tag{3.28}$$

Its solution should meet the following conditions: (1) the stress and strain are symmetric to the x-axis. (2) At the outer circle of the fiber, $x^2 + y^2 = d^2/4$, stress σ_y should vanish; and stress σ_x at point of $x = r_0$ should vanish as well. (3) In case of moderate bending, deformation of the neutral surface is neglected. It keeps a flat plane at $x = -x_0$, and no shear strain exists in the neutral surface. (4) At point $x = -d/2$ a certain stress σ_x exists to balance with the reaction force from the cylindrical support. Under these considerations, a stress potential is assumed in the form of

$$U = a(x^2 + y^2)^2 + b(x^2 + y^2) + c(x^4 - y^4) + d(x^2 - y^2) + exy^2 + fx^3. \tag{3.29}$$

Stresses are then deduced by applying boundary and symmetry conditions:

$$\sigma_x = \frac{Y}{2R^2}\left[\frac{7 - 6v}{8(1 - v)}\left(x^2 + y^2 - \frac{d^2}{4}\right)\right.$$
$$\left. - \frac{1 - 2v}{2(1 - v)}y^2 + \frac{7 - 6v}{4(1 - v)}x_0\left(x - \frac{d}{2}\right)\right] \tag{3.30a}$$

$$\sigma_y = \frac{1 - 2v}{16(1 - v)}\frac{Y}{R^2}\left(x^2 + y^2 - \frac{d^2}{4}\right), \tag{3.30b}$$

$$\sigma_{xy} = \frac{1 - 2v}{8(1 - v)}\frac{Y}{R^2}(x + x_0)y. \tag{3.30c}$$

At the center of fiber, we have

$$\sigma_{x0} = -\frac{Y}{R^2}\frac{7 - 6v}{16(1 - v)}\left(\frac{d^2}{4} + x_0 d\right), \tag{3.31a}$$

$$\sigma_{y0} = \frac{Y}{R^2}\frac{1 - 2v}{16(1 - v)}\frac{d^2}{4}. \tag{3.31b}$$

By the "center strain approximation," the birefringence is obtained to be

$$
\begin{aligned}
B &= -\frac{n^3}{16}(p_{11} - p_{12})(1 + v)\left[\frac{d^2}{R^2} - \frac{7 - 6v}{2(1 - v)}\frac{x_0 d}{R^2}\right] \\
&= -\frac{n^3}{16}(p_{11} - p_{12})(1 + v)\left[\frac{d^2}{R^2} - \frac{7 - 6v}{2(1 - v)}\frac{\bar{e}_z d}{R}\right].
\end{aligned}
\tag{3.32}
$$

As an example, the birefringence of a conventional SMF wound on a cylinder with a diameter of 3 cm without stretching is calculated to be about 2×10^{-6}.

Some other effects are considered in the literature. The fiber bending and the lateral pressure also cause a geometric deformation of the fiber core, from a circular area to a slightly elliptical one. According to electromagnetic theory, it leads to a difference of effective indexes between the two axes. Reference [13] gives a formula of retardation per meter as $\delta_s = 0.125(e^2/a)(2\Delta)^{3/2}$, where e is the core ellipticity, a is the core radius, and Δ is the relative index difference of core and cladding; but the factor is much smaller than that of the photoelastic effect.

The bending-induced birefringence is shown to depend strongly on the bending curvature $(1/R)$. It is observed experimentally that the birefringence becomes quite notable when the curvature radius reaches less than a few centimeters. The effect is utilized to develop fiber devices, such as a fiber polarization controller (PC), which will be introduced in Section 3.5.

3.1.4 Torsion-Induced Polarization Mode Cross-Coupling

Fiber twisting occurs very often but has almost no influence on usual applications of the SMF for optical pulse signal transportation. However, twists certainly cause changes in the polarization state of the propagating light wave, which are important in some situations, such as in fiber PCs. Reference [13] presents experimental results on bending and torsion of fibers; reference [16] gives detailed analyses on polarization state evolution in twisted fibers based on coupled mode theory (CMT). References [17] and [18] analyze torsion-induced biaxial refraction properties.

The twisted fiber is regarded as a twisted elastic rod with its axis kept straight, as shown in Figure 3.5. The torsion is characterized by the twist rate, namely twisted angle per unit length of fiber: $\tau = d\varphi/dz = $

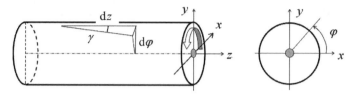

Figure 3.5 Twisted fiber and its cross section.

$\gamma/(d/2)$; the displacement line appears as spirals. In case of $\tau d \ll 1$, the relative displacement of adjoining parts of the rod is small, and the cross section keeps a circular plate without deformation in the x–y-plane. With relations of $x = r \cos\varphi$ and $y = r \sin\varphi$, shear strains in z-direction are deduced to be

$$e_{zx} = -\tau y \quad \text{and} \quad e_{zy} = \tau x, \tag{3.33}$$

whereas other strain components can be neglected in case of small torsion, that is,

$$e_x = e_y = e_z = e_{xy} = 0. \tag{3.34}$$

The photoelastic effect induced refractive index increment is then obtained to be:

$$\Delta\left(\frac{1}{\varepsilon}\right) = \frac{-1}{\varepsilon^2}\begin{pmatrix} \varepsilon_{xz} \\ \varepsilon_{yz} \\ \varepsilon_{xy} \end{pmatrix} = \begin{pmatrix} p_{44} & 0 & 0 \\ 0 & p_{44} & 0 \\ 0 & 0 & p_{44} \end{pmatrix}\begin{pmatrix} \tau x \\ -\tau y \\ 0 \end{pmatrix} = \tau p_{44}\begin{pmatrix} x \\ -y \\ 0 \end{pmatrix}. \tag{3.35}$$

With such an index change the original isotropic medium becomes anisotropic, and the index ellipsoid (the optical indicatrix) is expressed as

$$\frac{s_x^2}{n^2} + \frac{s_y^2}{n^2} + \frac{s_z^2}{n^2} - y p_{44}\tau s_x s_z + x p_{44}\tau s_y s_z = 1, \tag{3.36}$$

where s_x, s_y and s_z are the coordinates of the index ellipsoid. By a proper coordinate conversion from (s_x, s_y, s_z) to (q_x, q_y, q_z), the ellipsoid equation is orthogonalized as

$$\frac{1}{n^2}q_1^2 + \left(\frac{1}{n^2} + \frac{p_{44}\tau r}{2}\right)q_2^2 + \left(\frac{1}{n^2} - \frac{p_{44}\tau r}{2}\right)q_3^2 = 1. \tag{3.37}$$

This means that the medium has a property similar to a biaxial crystal, as pointed out by reference [18].

To investigate its effect on polarization characteristics we have to start from the basic equation. For anisotropic media, the Maxwell equation is written as

$$\nabla^2 E + k_0^2 \varepsilon \cdot E - \nabla(\nabla \cdot E) = 0, \qquad (3.38)$$

with a dielectric constant tensor of

$$\varepsilon = \bar{\varepsilon} + \tau p_{44} \bar{\varepsilon}^2 \begin{pmatrix} 0 & 0 & y \\ 0 & 0 & -x \\ y & -x & 0 \end{pmatrix} = \bar{\varepsilon} + \tilde{\varepsilon}, \qquad (3.39)$$

where $\bar{\varepsilon}$ is the dielectric constant of material without shear strain. By using the equation of $\nabla \cdot D = 0 = \nabla \cdot (\varepsilon \cdot E) = \bar{\varepsilon} \nabla \cdot E + \nabla \cdot (\tilde{\varepsilon} \cdot E)$ for a medium with no free electric charge, the Maxwell equation is rewritten as

$$\nabla^2 E + k_0^2 \bar{\varepsilon} E + k_0^2 \tilde{\varepsilon} E + \frac{1}{\bar{\varepsilon}} \nabla[\nabla \cdot (\tilde{\varepsilon} E)] = 0. \qquad (3.40)$$

According to the CMT, solutions of (3.40) with $\tilde{\varepsilon} = 0$ are taken as the zeroth-order results, that is, the two degenerate fundamental modes LP_{01x} and LP_{01y} for the single mode fiber, described by their Cartesian components for the core area of

$$E_1 = \begin{pmatrix} J_0(\beta_t r) \\ 0 \\ jq \cos \varphi J_1(\beta_t r) \end{pmatrix} e^{j\beta z} \quad \text{and} \quad E_2 = \begin{pmatrix} 0 \\ J_0(\beta_t r) \\ jq \sin \varphi J_1(\beta_t r) \end{pmatrix} e^{j\beta z},$$

$$(3.40a)$$

where $q = \beta_t/\beta$; and by similar expressions for the cladding.

The terms with $\tilde{\varepsilon}$ in the equation are taken as perturbations. The optical field should now be written as a sum of two eigenmodes with their amplitudes as functions of propagating distance: $E = [a_1(z)E_1^t(x, y) + a_2(z)E_2^t(x, y)]e^{j\beta z}$, where the superscript t stands for the transverse distribution. By substituting it into Equation (3.40), multiplying by each of the eigenmodes, and integrating the equation in x-y plane, the equation is transformed to a coupled mode equation (CME) of

$$a_1' = j\langle E_1^* \hat{\Phi} E \rangle / 2\beta, \qquad (3.41a)$$

$$a_2' = j\langle E_2^* \hat{\Phi} E \rangle / 2\beta, \qquad (3.41b)$$

where $\langle E_i \hat{\Phi} E \rangle = \langle E_i \{k^2 \tilde{\varepsilon} E + \nabla[\nabla \cdot (\tilde{\varepsilon} E)]/\tilde{\varepsilon}\}\rangle$ is the integral of the perturbation. The second-order differentials in the equation are omitted for the slowly varying amplitude.

The weakly guiding approximation of conventional fibers gives $q \ll 1$ and $E_z \ll E_{x,y}$. Neglecting the z-components results in $\tilde{\varepsilon} E_1^t \propto \left(0 \; 0 \; y J_0\right)^T$, $\tilde{\varepsilon} E_2^t \propto \left(0 \; 0 \; -x J_0\right)^T$, and $\int E_i^* \cdot (\tilde{\varepsilon} \cdot E_j) dS = 0$, both for $i = j$ and $i \neq j$.

For the second perturbation term, it is obtained that $\nabla \cdot (\tilde{\varepsilon} E) = j\beta(\tilde{\varepsilon} E)_z$ and

$$\nabla[\nabla \cdot (\tilde{\varepsilon} E_1)] = j\beta p_{44} \tau \tilde{\varepsilon} \left[\vec{i} \frac{xy}{r} \frac{\partial J_0}{\partial r} + \vec{j} \left(J_0 + \frac{y^2}{r} \frac{\partial J_0}{\partial r} \right) \right] e^{j\beta z}, \qquad (3.42a)$$

$$\nabla[\nabla \cdot (\tilde{\varepsilon} E_2)] = -j\beta p_{44} \tau \tilde{\varepsilon} \left[\vec{i} \left(J_0 + \frac{x^2}{r} \frac{\partial J_0}{\partial r} \right) + \vec{j} \frac{xy}{r} \frac{\partial J_0}{\partial r} \right] e^{j\beta z}. \qquad (3.42b)$$

The integral are then obtained to be $\int E_i \cdot \nabla[\nabla \cdot (\tilde{\varepsilon} E_i)] dS = 0$ due to the parity of function xy; and the cross-integral are obtained to be

$$\int E_2^* \cdot \nabla[\nabla \cdot (\tilde{\varepsilon} E_1)] dS = -\int E_1^* \cdot \nabla[\nabla \cdot (\tilde{\varepsilon} E_2)] dS = \beta \tilde{\varepsilon} p_{44} \tau. \qquad (3.43)$$

The coupled mode equation is then derived as

$$a_1' = \kappa a_2,$$
$$a_2' = -\kappa a_1, \qquad (3.44)$$

where $\kappa = n_{\text{eff}}^2 p_{44} \tau / 2$. The coupled modes are solved to be

$$a_1(z) = a_0 \cos \kappa z,$$
$$a_2(z) = -a_0 \sin \kappa z. \qquad (3.45)$$

Or in the form of a Jones vector, $E \propto a_0 \begin{pmatrix} \cos \kappa z \\ -\sin \kappa z \end{pmatrix} e^{j\beta z}$. It is also written as a sum of right and left circularly polarized light waves:

$$E = \frac{1}{2} \begin{pmatrix} 1 \\ j \end{pmatrix} \exp(j\beta_R z) + \frac{1}{2} \begin{pmatrix} 1 \\ -j \end{pmatrix} \exp(j\beta_L z), \qquad (3.46)$$

where $\beta_R = \beta + \kappa$ and $\beta_L = \beta - \kappa$. NB: Conventional definitions of right-handed and left-handed circular polarizations are used here [19,20].

It is deduced that the polarization rotates proportionally to the twisted angle as

$$\alpha = (\beta_L - \beta_R)z/2 = -\kappa z = -n_{\text{eff}}^2 p_{44} \tau z/2. \qquad (3.47)$$

This means that fiber twisting causes an optical activity. It is worth noting that the polarization rotation does not depend on the wavelength of propagating light, different from that in the birefringence phase retard of $\delta = Bk_0 z$. By using the data of silica in the literature, $p_{44} = -0.07$, the torsion-induced polarization rotation is about $\alpha \approx 0.074 \Delta \varphi$.

The complete deduction, taking the z-component of the eigenmodes into account, indicates that the coupling coefficients will contain terms proportional to $\gamma = \beta_t / \beta$ and γ^2, which give minor contributions. The torsion induced characteristics for birefringent fibers [e.g., polarization maintaining fiber (PMF)] will be discussed later in Section 3.4.

3.1.5 Bending Loss

Fiber bending not only induces birefringence but also causes propagation loss. Fiber loss is one of the key performances people are concerned with first, and also one of the main characteristics in fiber sensors. Research on bending loss have been carried out since the initial stage of fiber invention and development, as well as on bending loss of dielectric waveguides [21,22].

It is geometrically imaged that the bending will cause the peak of the mode field to move from fiber center toward the outer circle, and some energy of the guided mode may radiate out, as shown in Figure 3.6. To understand its mechanism, a Helmholtz equation under a toroidal boundary should be solved, where a constant bending curvature is considered. Although no rigorous analytic solution is obtained in

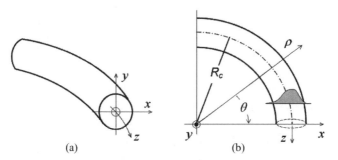

Figure 3.6 (a) Bent fiber and (b) toroidal coordinate.

toroidal coordinates, quite a number of approximations have been presented [23,24].

Considering the fact that the curvature radius of bending, R_c, is usually much larger than fiber diameter, a commonly used approximation is to convert the curved fiber into a straight waveguide with an equivalent index, which is a function of position x, instead of the original uniform step index. The concept is expounded as follows. Because the bending curvature keeps constant in the fiber section under investigation, variable separation can be used in cylindrical coordinates (ρ, θ, y), where $\rho = R_c + x$, and direction y in Figure 3.6(b) is perpendicular to the paper, leading to an azimuthal function equivalent of the wave in z-direction:

$$\exp(jq\theta) = \exp(j\beta z) = \exp(j\beta R_c \theta), \qquad (3.48)$$

where $q = \beta R_c$ is a positive number. By substituting it into a Maxwell equation of $\nabla \times \nabla \times E = k^2 n^2 E$ and $\nabla \cdot E = 0$, three equations are derived for field components [14]:

$$\left(\frac{\partial^2}{\partial \rho^2} + \frac{1}{\rho}\frac{\partial}{\partial \rho} + \frac{\partial^2}{\partial y^2} - \frac{q^2}{\rho^2} + k^2 n^2\right)\left(\frac{\rho}{R_c}E_\rho\right) = -\frac{2}{R_c}\frac{\partial E_y}{\partial y}, \qquad (3.49a)$$

$$\left(\frac{\partial^2}{\partial \rho^2} + \frac{1}{\rho}\frac{\partial}{\partial \rho} + \frac{\partial^2}{\partial y^2} - \frac{q^2}{\rho^2} + k^2 n^2\right)E_y = 0, \qquad (3.49b)$$

$$\left(\frac{1}{\rho} + \frac{\partial}{\partial \rho}\right)E_\rho + \frac{\partial E_y}{\partial y} + j\frac{q}{\rho}E_\theta = 0. \qquad (3.49c)$$

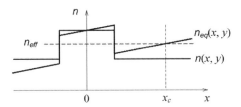

Figure 3.7 Schematic diagram of equivalent index for bent fiber.

As R_c in practice is so large that the term in the right-hand side of (3.49a) can be neglected, equations for E_y and ρE_ρ have the same forms, but coupled with each other. Introducing new functions of $\psi_\rho(x, y) = (\rho/R_c)^{3/2} E_\rho(\rho, y)$ and $\psi_y(x, y) = (\rho/R_c)^{1/2} E_y(\rho, y)$, and converting the coordinate back to a Cartesian one originated at the center of the fiber, the equations are in the form of

$$\left[\frac{\partial^2}{\partial x^2} + \frac{\partial^2}{\partial y^2} + n_{eq}^2(x)k^2\right]\psi_{\rho,y}(x, y) = 0, \qquad (3.50)$$

where an equivalent index is introduced as

$$n_{eq}^2 k^2 = n^2 k^2 + \frac{1 - 4\beta^2 R_c^2}{4(R_c + x)^2}$$

$$\approx (n^2 - n_{eff}^2)k^2 + \beta^2 \frac{2x}{R_c} = \beta_t^2 + \beta^2 \frac{2x}{R_c}. \qquad (3.51)$$

Figure 3.7 gives a schematic diagram of an equivalent index in x-direction. It is shown that there is a critical position of x_c beyond which the equivalent index is larger than the effective index of the straight fiber; and the optical wave in the region becomes a radiating wave, resulting in fiber bending loss.

By variable conversion of $\chi = -(2\beta^2 R^2)^{1/3}[\beta_t^2/(2\beta^2) + x/R]$, and $\bar\psi(\chi) = \int_{-\infty}^{\infty} \psi(x, y)dy/a$, equation (3.50) can be transformed to

$$\frac{\partial^2 \bar\psi(\chi)}{\partial \chi^2} - \chi\bar\psi(\chi) = 0. \qquad (3.52)$$

It is an Airy equation with a solution of Airy functions. For a large argument, the solution is approximated as $\bar\psi(\xi) \propto |\xi|^{-1/4}$

$[A \sin(2|\xi|^{3/2}/3) + B \cos(2|\xi|^{3/2}/3)]$, showing features of radiation. Some simplifications and approximations are then used to calculate the radiation losses, and an imaginary propagation constant is deduced [14] as

$$\beta_{i,0} = \frac{\beta_{t1}^2}{4V^2} \sqrt{\frac{\pi}{R_c \beta_{t2}^3}} \frac{1}{K_1^2(\beta_{t2}a)} \exp\left(-\frac{2\beta_{t2}^3}{3\beta^3}\beta R_c\right) \qquad (3.53a)$$

for the fundamental mode LP_{01}, and

$$\beta_{i,m} = \frac{\beta_{t1}^2}{2V^2} \sqrt{\frac{\pi}{R_c \beta_{t2}^3}} \frac{1}{K_{m+1}(\beta_{t2}a)K_{m-1}(\beta_{t2}a)} \exp\left(-\frac{2\beta_{t2}^3}{3\beta^3}\beta R_c\right) \quad (3.53b)$$

for LP_{ml} modes, where K_m is the mth-order modified Bessel functions of the second kind. There are also other deductions for bending losses based on asymptotic forms of the Hankel functions that give similar results [23,24].

Bending loss should be minimized in designing and fabricating fiber devices. It is noticed in (3.53a) and (3.53b) that the bending loss is proportional to an exponential term of $R_c^{-1/2} \exp(-C R_c)$, which has been shown in experiments [25,26]; and the formulas give a critical bending curvature to be taken into account in fiber pavement and fiber device building.

As seen in Figure 3.7 the bending moves the mode distribution toward the outward circle of the bending, which results in mode mismatching at points between sections of fiber with different bending curvatures, especially between a straight section and a bent section. This is another mechanism of bending loss, called transition loss. The mode displacement parameter p, defined as $\phi(x, y) = \phi_0(x - p, y)$ for mode distributions after and before bending, is introduced and deduced to be $p \simeq k^2 n^2 w_d^4/4R_c$ for a Gaussian mode approximation with mode width of w_d. The transmission due to the mode transition is then written as [14]

$$T = \frac{(\int \phi_0 \phi_1 dS)^2}{\int \phi_0^2 dS \cdot \int \phi_1^2 dS} \propto 1 - \frac{p^2}{w_d^2} \approx 1 - \frac{k^4 n^4 w_d^6}{16R_c^2}. \qquad (3.54)$$

Microbending loss, which is attributed to random bends, is one of the important specifications for judging the quality of fiber fabrication and cabling. Numerous papers are devoted to the topic [25–30], especially in the developing stage of optical fiber and cable.

It is noticed that the lost energy caused by bending is converted to radiation mode, which can be detected outside the fiber. Such a mechanism has been utilized to develop a special device, the fiber taper for monitoring and inspecting the optical signals propagating in the fiber without interruption and intrusion. The bending loss is utilized to develop fiber sensors, such as intrusion sensors.

It is surely attractive to develop low bending loss fibers, because more and more fibers are paved in houses and in apparatus with sharper corners. It is noted that bending loss depends on the fiber structure, and low bending loss fibers can be designed and fabricated [31]. This has been standardized as ITU-T G. 657 fiber.

3.1.6 Vibration and Mechanical Waves in Fiber

The above analyses give static sensitivities of fiber. More generally, dynamic physical influences should be considered in practice, such as mechanical vibrations. When a vibration in an axial direction occurs somewhere, the deformation will propagate along the fiber to form a longitudinal mechanical wave in its axial direction. Because the fiber is very thin, and its transverse deformation can occur freely in case of no external force around the circle, the propagation equation is written as [6]

$$\rho \frac{\partial^2 u_z}{\partial t^2} = Y \frac{\partial^2 u_z}{\partial z^2},$$
(3.55)

where ρ is the density of fiber (e.g., silica). The solution is a wave in the form of $\exp[j(Kz - \Omega t)]$ with wavelength $\Lambda = 2\pi/K$ and vibration frequency Ω; the propagation velocity is $V_a = \partial\Omega/\partial K = \sqrt{Y/\rho}$, which does not depend on frequency Ω.

More often, fiber will suffer transverse forces and transverse vibrations. A dynamic transverse deformation is then propagated along the fiber. Taking x-direction bending into consideration, the transverse movement obeys the equation of

$$\rho \frac{\partial^2 u_x}{\partial t^2} = Y \frac{I}{S} \frac{\partial^4 u_x}{\partial z^4},$$
(3.56)

where $I = \iint x^2 dS = \pi d^4/64$ is the bending moment of fiber, $S = \pi d^2/4$ is its cross-sectional area. The wave vector is deduced to be

$K = (\Omega^2 \rho S/YI)^{1/4} = 2(\Omega^2 \rho/Yd^2)^{1/4}$, and propagation velocity is $V_a = \partial\Omega/\partial K = \sqrt{Y/\rho}Kd/4$, giving a dispersive relation on vibration frequency [6]. The situation will be much more complicated when the jacket and cabling materials are considered. Generally waves of $u_x(x, t)$ and $u_y(y, t)$ occur simultaneously; ideally they obey their respective independent equations without mutual coupling. The composite wave is their linear combination.

Torsion can also propagate along the fiber, though this is relatively seldom in fiber sensor applications. Interested readers can find its basic theory in reference [6].

On the basis of the photoelastic effect, the mechanical waves will lead to changes in polarization states of the light wave propagating along the fiber. The transverse vibration wave means a bending wave along the fiber, and a dynamic distribution of the curvature:

$$\frac{1}{R_c} \propto u_0 K^2 \cos(Kz - \Omega t) = \frac{4u_0\Omega}{d}\sqrt{\frac{\rho}{Y}}\cos(Kz - \Omega t), \qquad (3.57)$$

where u_0 is the vibration amplitude. Consequently a dynamic birefringence occurs:

$$B \propto \frac{(p_{11} - p_{12})(1 + v)}{Y}n^3 u_0^2 \Omega^2 \rho \cos^2(Kz - \Omega t). \qquad (3.58)$$

The section of fiber can be regarded as a vibrating wave plate, which makes the polarization state of propagating light wave vibrate, with a dynamic phase retard of $\delta \propto \cos^2(Kl - \Omega t)$ and consequent polarization rotation. The dynamic effect of the fiber has been utilized in developing fiber sensors, such as hydrophones.

The above analysis is based on the elasticity of fiber. It is necessary in some cases to consider stress relaxation process, that is, the creep of material, especially in high-precision and high-stability sensors. Detailed analysis on the influence of creep is beyond the scope of this book.

3.1.7 Sensitivity to Temperature

Effects of temperature changes on fiber characteristics include thermal expansion and thermal optic effect. The former occurs in almost all materials and is described as

$$\Delta L = \alpha L \Delta T, \qquad (3.59)$$

where L is the length of fiber being observed, and α is called the thermal expansion coefficient. For silica fibers, $\alpha = 0.55 \times 10^{-6}$°C^{-1} [1,2].

The second is the effect of temperature on the fiber effective index, described as

$$\Delta n_{\text{eff}} = \xi n_{\text{eff}} \Delta T, \tag{3.60}$$

which is attributed mainly to the thermal optic effect of silica material, measured as $dn/dT \sim 1 \times 10^{-5}$ K^{-1}, with a negligible effect of the waveguide. The composite coefficient is about $\xi \approx 7 \times 10^{-6}$ K^{-1}, which is one order larger than the thermal expansion effect. Thus, the phase of $\phi = n_{\text{eff}}kL$ will be modulated by temperature change as

$$\Delta \phi = (\alpha + \xi) n_{\text{eff}} kL \Delta T. \tag{3.61}$$

Temperature variation will also bring about thermal stress, which comes from the different thermal expansion coefficients between fiber and its packaging structures, including its jackets and protective materials in cabling, and intentionally designed mount structures for fiber sensors. Such thermal stresses have to be analyzed individually for different situations. In practice, temporal variation of temperature and related thermal conduction phenomena induce the dynamic thermal effect, usually in low speed, and cause drifts of signal, which must be dealt with seriously, especially for high-precision and high-stability sensors.

3.2 FIBER COUPLERS

One of the important properties of optical waves in comparison with electric currents in conductors is that different optical beams can propagate in a common space without mutual disturbances. Now that optical waveguides, including fibers, provide functions of one-dimensional in-line propagation, without the limitation of straight propagation, it is of great interest to know how different optical beams can be combined to propagate in a single waveguide. Quite a lot of devices have been invented for this purpose; among them directional couplers are the most often used, which utilize lateral coupling between two tightly arranged waveguides to make the optical beam propagating in one waveguide transfer into the adjacent waveguide. In this section, the basic structure and fabrications of couplers are described; the basic characteristics and theory of mode coupling are analyzed with emphasis on 2×2 and 3×3 couplers; mode coupling in an axial direction is also discussed.

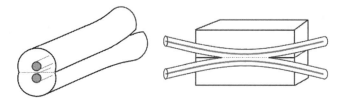

Figure 3.8   Structure of D-shaped fiber coupler.

3.2.1   Structures and Fabrications of 2×2 Couplers

Different fiber couplers have been developed, including 2×2, 3×3, $N \times N$, $N \times 1$, and $1 \times N$ couplers by the port numbers, among which the most basic is the 2×2 coupler. At earlier stages it is fabricated by two D-shaped fibers, which have been polished laterally near the fiber core and then placed together in an up-and-down fashion, as shown in Figure 3.8 [32,33]. Another configuration is the fused conical coupler, which is fabricated by torching two parallel and closely laid fibers, and pulling axially at the same time, so that the heated fiber sections are fused together and get thinner, as shown in Figure 3.9 [34,35].

The principle of 2×2 couplers is based on evanescent field coupling, that is, if two waveguides are placed so close that the evanescent field of one waveguide penetrates partly into the other, the optical energy in one waveguide is then transferred to the other. In a fused conical coupler, shown in Figure 3.9, not only is the spacing between two cores greatly reduced by fusing, the cores also become thinner, making the mode expand transversely. More energy propagates in the cladding, and the depth of the evanescent field is enlarged, resulting in enhanced coupling.

The directional couplers can be realized in planar waveguide structures. Figure 3.10 shows an X-type 2×2 coupler and a Y-type 2×1 (or 1×2) coupler, fabricated mostly by photolithography and ion-exchange technology on electro-optic crystals, or by epitaxial growth on semiconductors. Their functions are the same as fiber couplers with the merit of a smaller volume than fiber couplers; however, input and output fiber

Figure 3.9   Structure of fused conical coupler.

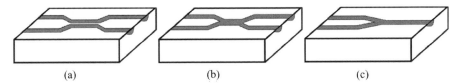

Figure 3.10   (a), (b) X-type 2×2 waveguide coupler; (c) Y-type 2×1 coupler.

pigtails have to be coupled to the waveguide facets in fiber technology applications.

3.2.2 Basic Characteristics and Theoretical Analyses of the Coupler

3.2.2.1 Coupler Mode Theory for the Doubled Waveguide   The CMT is widely used in analyzing and designing waveguides and waveguide devices [36–38]. In analyzing the coupling between two waveguides, the theory takes the eigenmode of each solitary waveguide with the other one omitted as the zeroth-order approximation, and uses the eigenmodes to calculate the coupling coefficient, as the first-order perturbation. A simplified model with two single-mode waveguides closely placed and parallel to each other is shown in Figure 3.11, where β_A and β_B are the propagation constants of the respective waveguides when they exist solitarily.

The perturbation of the evanescent field from the next waveguide makes the mode amplitude vary with the propagation distance z. Denoting the mode amplitudes by $A(z)$ and $B(z)$, they obey the CMEs as follows:

$$\frac{dA}{dz} = j\beta_A A + j\kappa_{BA} B, \qquad (3.62a)$$

$$\frac{dB}{dz} = j\beta_B B + j\kappa_{AB} A. \qquad (3.62b)$$

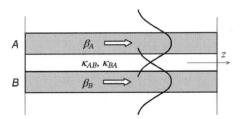

Figure 3.11   Coupling between two parallel waveguides.

κ_{AB} and κ_{BA} are the coupling coefficients between the two waveguides, expressed as [24]

$$\kappa_{AB} = \frac{k_0 \iint (n^2 - n_B^2)\Phi_A \Phi_B \, dS}{2(\iint |\Phi_A|^2 \, dS \cdot \iint |\Phi_B|^2 \, dS)^{1/2}}, \qquad (3.63a)$$

$$\kappa_{AB} = \frac{k_0 \iint (n^2 - n_A^2)\Phi_A \Phi_B \, dS}{2(\iint |\Phi_A|^2 \, dS \cdot \iint |\Phi_B|^2 \, dS)^{1/2}}, \qquad (3.63b)$$

where Φ_A and Φ_B are the mode fields of the respective waveguides, which are supposed to exist independently, and their indexes are described by $n_A(x, y)$ and $n_B(x, y)$; $n(x, y)$ is the index distribution in the whole space of twin waveguides. The integration is carried in the whole x–y space, in which the cladding-air bound is supposed too far away to be considered for simplified analysis.

The optical loss in the waveguides is assumed negligibly low, and the energy conservation requires

$$\frac{d}{dz}(AA^* + BB^*) = 0. \qquad (3.64)$$

It is then deduced that the coupling coefficients meet a relation of

$$\kappa_{AB} = \kappa_{BA}^*. \qquad (3.65)$$

Combining Equations (3.62a) and (3.62b), we obtain

$$\frac{d^2 A}{dz^2} - i(\beta_A + \beta_B)\frac{dA}{dz} - (\beta_A \beta_B - \kappa^2)A = 0, \qquad (3.66)$$

with $\kappa^2 = \kappa_{AB}\kappa_{BA}$, and a similar equation for B. The general solution of Equation (3.66) is written as

$$A(z) = a_1 \exp j\gamma_+ z + a_2 \exp j\gamma_- z, \qquad (3.67a)$$

$$B(z) = b_1 \exp j\gamma_+ z + b_2 \exp j\gamma_- z, \qquad (3.67b)$$

where $\gamma_\pm = (\beta_A + \beta_B)/2 \pm \sqrt{(\beta_A - \beta_B)^2/4 + \kappa^2}$, denoted as $\gamma_\pm = \bar{\beta} \pm \sqrt{\tilde{\beta}^2 + \kappa^2} = \bar{\beta} \pm \gamma$ below. By boundary conditions of $A(0) = A_0$ and

$B(0) = B_0$, the amplitudes are obtained to be

$$A(z) = \left[A_0 \left(\cos \gamma z + j \frac{\tilde{\beta}}{\gamma} \sin \gamma_1 z \right) + j B_0 \frac{\kappa_{BA}}{\gamma} \sin \gamma z \right] e^{j\bar{\beta}z}, \qquad (3.68a)$$

$$B(z) = \left[B_0 \left(\cos \gamma z - j \frac{\tilde{\beta}}{\gamma} \sin \gamma z \right) + j A_0 \frac{\kappa_{AB}}{\gamma} \sin \gamma z \right] e^{j\bar{\beta}z}, \qquad (3.68b)$$

or in a matrix form

$$\begin{pmatrix} A(z) \\ B(z) \end{pmatrix} = \begin{pmatrix} T_{11} & T_{12} \\ T_{21} & T_{22} \end{pmatrix} \begin{pmatrix} A_0 \\ B_0 \end{pmatrix} \exp(j\bar{\beta}z), \qquad (3.68c)$$

with $T_{11} = \cos \gamma z + j(\tilde{\beta}/\gamma) \sin \gamma z = T_{22}^*$ and $T_{12} = j(\kappa_{BA}/\gamma) \sin \gamma z = -T_{21}^*$, which meet relations of $T_{11}T_{11}^* + T_{12}T_{12}^* = 1$, $T_{22}T_{22}^* + T_{21}T_{21}^* = 1$, $T_{11}T_{12}^* + T_{21}T_{22}^* = 0$, and $\det(T) = 1$. This means that the transmission matrix is a unitary matrix obeying the reversibility of the optical path.

For a symmetric 2×2 coupler composed by two identical waveguides with equal parameters of $\beta_A = \beta_B = \beta$, $\kappa_{AB} = \kappa_{BA} = \kappa$, and consequently $\gamma = \kappa$, the mode fields are simplified as

$$\begin{pmatrix} A(z) \\ B(z) \end{pmatrix} = \begin{pmatrix} \cos \kappa z & j \sin \kappa z \\ j \sin \kappa z & \cos \kappa z \end{pmatrix} \begin{pmatrix} A_0 \\ B_0 \end{pmatrix} e^{j\bar{\beta}z}. \qquad (3.69)$$

If the coupling between two waveguides, or two fibers, is stopped at distance $z = l$, the mode $A(l)$ and $B(l)$ will propagate continuously in the respective fibers as the output ports, and thus a 2×2 coupler is realized.

The evolutions of intensities I_A and I_B are obtained from (3.68a–c) to be

$$I_A(z) = A_0^2 \cos^2 \gamma z + [g^2 A_0^2 + |h|^2 B_0^2 + (h + h^*)g A_0 B_0] \sin^2 \gamma z$$
$$+ i(h - h^*)A_0 B_0 \sin \gamma z \cos \gamma z, \qquad (3.70a)$$

$$I_B(z) = B_0^2 \cos^2 \gamma z + [g^2 B_0^2 + |h|^2 A_0^2 - (h + h^*)g A_0 B_0] \sin^2 \gamma z$$
$$+ i(h^* - h)A_0 B_0 \sin \gamma z \cos \gamma z, \qquad (3.70b)$$

where $g = \tilde{\beta}/\gamma$ and $h = \kappa_{BA}/\gamma$ are denoted for simplicity, with relation of $g^2 + hh^* = 1$. The sum of two intensities keeps constant, $I_A(z) + I_B(z) = A_0^2 + B_0^2$, as the energy conservation law requires. The coupling coefficients can be considered real numbers without loss of generalization; (3.70a and 3.70b) are then rewritten as

$$I_A(z) = A_0^2 \cos^2 \gamma z + (g A_0 + h B_0)^2 \sin^2 \gamma z,$$
$$I_B(z) = B_0^2 \cos^2 \gamma z + (g B_0 - h A_0)^2 \sin^2 \gamma z. \tag{3.71}$$

For a symmetric coupler, it is deduced from (3.69) that

$$\begin{pmatrix} I_A(z) \\ I_B(z) \end{pmatrix} = \begin{pmatrix} \cos^2 \kappa z & \sin^2 \kappa z \\ \sin^2 \kappa z & \cos^2 \kappa z \end{pmatrix} \begin{pmatrix} I_A(0) \\ I_B(0) \end{pmatrix}. \tag{3.72}$$

The analysis above can also help to discuss the cross-talk between two waveguides, which are supposed to transport optical signals independently, but placed so close laterally that the undesired mutual coupling between the two waveguides occurs.

3.2.2.2 *Applications of 2 × 2 Couplers* 2×2 couplers are widely used in a variety of applications with related functions. The three most important devices are described here: the *beam splitter*, the *beam combiner*, and the *wavelength division multiplexer* (WDM).

The symmetric coupler is mostly manufactured in practice due to its low cost and market demand. It is seen that the optical energy is transferred from one fiber core to the other, back and forth. When an optical signal inputs fiber A, that is, $A(0) = A_0$ and $B(0) = 0$, and the coupling stops at $z = l$, the mode amplitudes and intensities in the respective fibers after l are written as

$$A(l) = A_0 \cos(\kappa l) e^{j\beta l},$$
$$B(l) = j A_0 \sin(\kappa l) e^{j\beta l}, \tag{3.73}$$

$$I_A(l) = I_0 \cos^2 \kappa l,$$
$$I_B(l) = I_0 \sin^2 \kappa l. \tag{3.74}$$

The 2×2 fiber coupler acts as a *beam splitter*; the split ratio is determined by the phase factor κl: $\cos^2 \kappa l / \sin^2 \kappa l$. It is important to notice that besides the complementary amplitude evolution, a phase difference of $\pi/2$ exists between $A(l)$ and $B(l)$. This property is important in designing coupler-incorporated devices. The behavior is similar to the phase shift between a reflected beam and transmitted beam at the surface of two dielectric media for an input beam from the lower index medium to the higher.

If the phase factor is controlled to meet $\kappa l = (m + 1/4)\pi$, two equal outputs are obtained, $I_A = I_B = 0.5I_0$, which is the so-called 3 dB beam splitter, the most often used splitter. The transferring matrix of the 3 dB coupler is then written as

$$\begin{pmatrix} A_1 \\ B_1 \end{pmatrix} = \frac{\sqrt{2}}{2} \begin{pmatrix} 1 & j \\ j & 1 \end{pmatrix} \begin{pmatrix} A_0 \\ B_0 \end{pmatrix}. \tag{3.75}$$

It is obvious that the device can also be regarded as *a beam combiner*, if optical signals from both input fibers are required to output in one fiber port. It is necessary to note in formula (3.72) that it is impossible to combine two input energies in one output fiber; at least 3 dB loss is inevitable. This is a physically theoretical limitation, similar to the impossibility of increasing brightness by any lens system. This physical limitation holds in coherent summation, that is, under such a condition that both fibers (or waveguides as well) are SMF, and the wavelengths of the two inputs are the same.

It is noted that the parameter κ is inversely proportional to the wavelength. For a fabricated coupler with fixed coupling length l, the output amplitudes are periodic functions of wavelength [39]. Figure 3.12 shows a typical output spectrum [40].

In fabrication processing, both the coupling coefficient κ and coupling length l are changing with fusing and drawing, so that the power received from the output fiber pigtails vary as a result. Figure 3.13

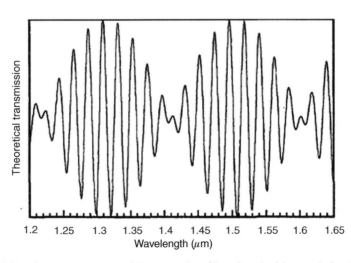

Figure 3.12 Output spectrum of fiber coupler. (Reprinted with permission from reference [40].)

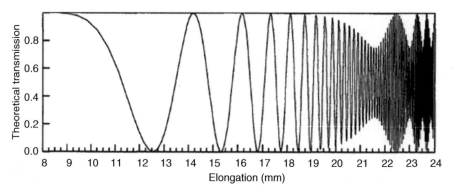

Figure 3.13   Variations of monitored output with the elongation. (Reprinted with permission from reference [40].)

shows typical curves of power received from one of the fiber pigtails that vary with elongation, which are generally used for process monitoring and control [40]. Inside the coupler, the optical energy goes back and forth between the two fiber cores in sinusoidal functions with a beat length of $L_b = 2\pi/\gamma \simeq 2\pi/\kappa$. The apparent beat length can be estimated by the transmission spectrum.

The 2×2 fiber coupler is also used as a wavelength *multiplexer* or a *demultiplexer*. If the wavelengths of two inputs are not the same, the 3 dB loss limitation is broken through; that is, the whole energy of two beams with different wavelengths can be combined together into one of the two output ports. In this case, the 2×2 fiber coupler should be treated as an asymmetric doubled waveguide, and the analysis is based on formulae (3.68a–c) and (3.70a and 3.70b).

The boundary condition is now written as: $A(0) = A_{\lambda 1}$, and $B(0) = B_{\lambda 2}$. Formulae (3.70a and 3.70b) should be rewritten separately for the two wavelengths λ_1 and λ_2, respectively:

$$I_A(l, \lambda_1) = A_{\lambda 1}^2 [g_{\lambda 1}^2 + (1 - g_{\lambda 1}^2)\cos^2 \gamma_{\lambda 1} l], \tag{3.76a}$$

$$I_B(l, \lambda_1) = A_{\lambda 1}^2 |h_{\lambda 1}|^2 \sin^2 \gamma_{\lambda 1} l, \tag{3.76b}$$

$$I_A(l, \lambda_2) = B_{\lambda 2}^2 |h_{\lambda 2}|^2 \sin^2 \gamma_{\lambda 2} l, \tag{3.76c}$$

$$I_B(l, \lambda_2) = B_{\lambda 2}^2 [g_{\lambda 2}^2 + (1 - g_{\lambda 2}^2)\cos^2 \gamma_{\lambda 2} l], \tag{3.76d}$$

where dependences of two propagation constants and two coupling coefficients on wavelength are taken into account, and denoted by their subscripts.

By carefully designing the structure and controlling the parameters of materials and the fabrication processing, the phase factors for two wavelengths are adjusted to meet conditions of $\gamma_{\lambda 1} l = (m + 1/2)\pi$ and $\gamma_{\lambda 2} l = m\pi$ with $m = 0, 1, \dots$, thus a particular output state is realized as follows:

$$I_A(\lambda_1) = A_{\lambda 1}^2 g_{\lambda 1}^2, \tag{3.77a}$$

$$I_B(\lambda_1) = A_{\lambda 1}^2 |h_{\lambda 1}|^2, \tag{3.77b}$$

$$I_A(\lambda_2) = 0, \tag{3.77c}$$

$$I_B(\lambda_2) = B_{\lambda 2}^2. \tag{3.77d}$$

The parameters can be controlled with $g_{\lambda 1}^2 \ll |h_{\lambda 1}|^2$ and $|h_{\lambda 1}|^2 \approx 1$, thus the two inputs are mostly combined into waveguide B, $I_B \approx A_{\lambda 1}^2 + B_{\lambda 2}^2$, and a little portion of beam λ_1 is left in waveguide A. A wavelength division multiplexer (WDM) is then realized.

On the basis of the reversibility of linear optics, two beams with different wavelengths of λ_1 and λ_2 propagating in one fiber can be divided into two output fibers accordingly by the coupler, which is termed a demultiplexer. WDM is very useful in fiber technologies. In fiber amplifiers, for example, the pump light and the input optical signals to be amplified are multiplexed into one fiber and connected to a rare-earth-doped active fiber, for example, at 980 nm and at 1550 nm band for an erbium-doped silica fiber amplifier (EDFA). In wavelength division multiplexing technology of fiber communications, two optical signals at different wavelengths can be injected into one fiber by a WDM; and vice versa to separate two different wavelength signals input from one fiber.

3.2.2.3 *Mode Evolution in the Fiber Coupler* As analyzed above, the coupling is caused by the mutual penetration of the evanescent fields between the two parallel waveguides, and the coefficients are calculated by the mode overlap integral, as expressed by Equation (3.63). This means that they depend on the spacing between the two waveguides and also on the transverse extension of the modes. The mode extension is determined by the normalized frequency V and the mode parameter u (or w) of both the solitary fibers, as shown in Section 2.2. Based on the mode theory of step-index fibers, the coupling coefficient between two identical fiber cores in the couplers is deduced as [23]

$$\kappa = \left(\frac{\pi \Delta}{w d_c a} \right)^{1/2} \frac{u^2}{V^3} \frac{\exp(-w d_c / a)}{K_1^2(w)}, \tag{3.78}$$

where d_c is the spacing between two core centers, which is supposed to be large enough so that the mode field can be described as a solitary cylindrical waveguide; a is the core radius. For the thinned cores in the coupler, approximation of $u \approx V$ and $K_1(w) \approx 1/w$ holds, and we have

$$
\kappa \approx \left(\frac{\pi \Delta w}{d_c a} \right)^{1/2} \frac{1}{V} \exp\left(-\frac{w d_c}{a} \right)
$$

$$
= \left[\frac{\pi \Delta \sqrt{n_{\mathrm{eff}}^2 - n_2^2}}{(n_1^2 - n_2^2) k d_c} \right]^{1/2} \frac{1}{a} \exp\left(-\sqrt{n_{\mathrm{eff}}^2 - n_2^2} k d_c \right). \qquad (3.79)
$$

It is indicated that the coupling coefficient increases with a decreasing of d_c and a.

In fiber communication applications the so-called return loss is one of the most important specifications, because any small reflection will cause serious impairment of optical signals. Such reflections are introduced by imperfections of processing and geometrical size controlling. Among them, a key issue is the smooth transition from the solitary fiber to coupled fibers at the beginning and ending sections. That is, the coupling should arise from nil smoothly to a certain amount at the start point, and vice versa at the end; otherwise harmful reflections would occur at the sharp discontinuity.

Therefore, the transverse distance between two cores must be gradually changing at the input and output parts, and the coupling coefficients are not constants, but functions of z. In such cases, the coupling equations of (3.62) may not be solved analytically. There are many digital methods for simulations; among them the transfer matrix method (TMM) is commonly used. In TMM, the region, where the equation is defined, is divided into many short segments; each of them is regarded as a regular waveguide, with no dependence on position z, and can be solved as a propagation matrix; the solution is obtained by multiplying the concatenated matrices.

For symmetric couplers, if the coupling coefficient variation meets the condition of $d\kappa/dz \ll \kappa^2$, the general solutions can be expressed approximately as [23]

$$
A^{\pm}(z) \approx a_0^{\pm} \cdot \exp\left\{ j \int_0^z [\beta(z) \pm \kappa(z)] \mathrm{d}z \right\}, \qquad (3.80a)
$$

$$
B^{\pm}(z) \approx b_0^{\pm} \cdot \exp\left\{ j \int_0^z [\beta(z) \pm \kappa(z)] \mathrm{d}z \right\}. \qquad (3.80b)
$$

The formulas listed above for constant β and κ can then be used with modified parameters of $\bar{\beta} = \int_0^l \beta(z) \mathrm{d}z/l$ and $\bar{\kappa} = \int_0^l \kappa(z) \mathrm{d}z/l$.

As stated above, the CMT takes the coupling as a perturbation term. In practical fiber couplers the coupling is so strong that it cannot be treated as a perturbation, the merged and thinned region is in fact a multimode waveguide, supporting many cladding modes. Among them the symmetric fundamental mode (LP_{01}) and the asymmetric first-order mode (LP_{11}) are the most intensive modes. In such cases, the CMT is not precise enough for quantitative analyses and processing control; and the so-called super-mode method is then introduced. The mode excitation is described in forms of [41]

$$\psi_1 \leftrightarrow \eta_1 \psi_e - \eta_2 \psi_o,$$
$$\psi_2 \leftrightarrow \eta_2 \psi_e + \eta_1 \psi_o, \qquad (3.81)$$

where ψ_1 and ψ_2 stand for the modes in the individual fibers at the entrance and exit, ψ_e and ψ_o stand for the even mode LP_{01} and the odd mode LP_{11} of the composite waveguide in the waist, respectively. The ratio of modal power is expressed by the coupling coefficients as η_2^2/η_1^2. The two modes have different propagation constants β_e and β_o, resulting in a coupling phase of

$$\phi = (\beta_o - \beta_e)L. \qquad (3.82)$$

The outputs are obtained to be $P_2 = M \cos^2(\phi/2)$ and $P_1 = 1 - P_2$, similar to the results of CMT, but with a maximum coupled power of $M = 4\eta_1^2\eta_2^2/(\eta_1^2 + \eta_2^2)^2$, which may be less than 1. In an ideal case with two identical fibers, the two modes can be excited equally, we have $\eta_2^2 = \eta_1^2$ and $M = 1$.

A special fiber coupler, called null coupler, is developed to compose functional fiber optic devices for applications, in which the modal power ratio η_2^2/η_1^2 and the maximum coupled power M reach zero, meaning that no power is coupled from the input fiber to the other [41,42]. It is a broadband zero splitting, different from the case with phase factors of $\kappa l = m\pi$ or $(m + 1/2)\pi$ in formula (3.74). It is actually a highly mismatched coupler made of two fibers with very different propagation constants, such as with highly different diameters. Qualitatively, it can be understood by CMT with formula (3.71) under conditions of $\tilde{\beta} \gg \kappa$ and $h \ll g \approx 1$.

Although a null coupler behaves as a pair of fibers without interaction, the light waves interpenetrate and overlap at the waist. Combined with the acousto-optic effect or other mechanisms, the null-couplers are used to build a mode converter, a polarizer and other devices [43,44].

Figure 3.14 Fabrication of a polarization-independent fiber coupler.

3.2.2.4 *Polarization Behaviors of the Fiber Couplers* When the coupling between two waveguides is expressed by Equations (3.62a and 3.62b) and coefficients (3.63a and 3.63b), an assumption is implied that the modes have the same polarization. It is expected that a polarization-dependent coupling occurs in parallel waveguides, because the configuration is obviously anisotropic in x–y plane; and moreover the cross section of two cores may be deformed from circular to noncircular during the fusing and drawing process of the fused conical fiber couplers. As a result, the coupling coefficients for the two polarizations will differ from each other. References [45–47] offer analyses and discussion on the issue.

Such polarization dependence does not meet the requirement of most practical applications. To fabricate a polarization-independent coupler one of the schemes is shown in Figure 3.14, that is, to twist the fiber pair when it is fused and drawn in processing; the coupling coefficient is then averaged over two polarizations.

On the other hand, polarization dependence is welcome in some applications, where it is required to divide two polarizations by a single device, called polarization beam splitter (PBS), and to combine two orthogonal polarized beams to one fiber (polarization beam combiner, PBC). It was proposed to utilize form birefringence, which is caused by the dumbbell form of the merged fibers [48,49]. The polarization dependences of D-shaped fiber coupler and coupling between two PMFs are analyzed in references [50] and [51]. Figure 3.15 shows fiber couplers made of PMF: structure (a) is a PBS, by which a linearly polarized beam is divided into two polarized beams according to the axis of the PMF; and (b) is a polarization beam combiner (PBC), by which two orthogonal beams polarized at the axis of PMF are combined into one of the output ports. The qualitative understanding is similar to the 2×2 WDM fiber coupler, as expressed by Equations (3.76a–3.76d), in which the parameters of two wavelengths are replaced by those of two polarizations.

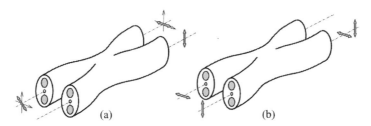

Figure 3.15 Structure of (a) polarization beam splitter and (b) polarization beam combiner.

3.2.2.5 *Sensitivities of Fiber Couplers* The stability of fiber couplers concerns most users of fiber technology applications. On the other hand, their sensitivity to various conditions is interesting for the exploitation of different sensors.

The diameter of a coupler's waist is very small, typically 30–40 µm, causing it to be very fragile. Even if not broken it is very easy to bend, which would change its characteristics greatly. Therefore, practical fused fiber couplers must be packaged in a silica tube with the fused section kept straight and surrounded by air, and protected by the tube from the airflow.

The coupler experiences a large temperature decrease from around 1,000°C to room temperature during its fabrication; therefore, the coupling coefficient and the performances evolve consequently due to the thermal expansion effect and the thermal effect of the refractive index. The evolution of power and its spectrum measured at the fiber pigtail is usually used to monitor the specifications of the coupler during processing; and it is necessary to take its temperature dependence into consideration, often by using computer-aided controllers.

Such temperature dependence remains in the ambient temperature range. For a WDM coupler, temperature sensitivity is one of the limitations to be used in the dense WDM (DWDM) technology with wavelength spacing of ≤0.8 nm, because it is difficult to control and stabilize the wavelengths at precision of tenth nanometer for a conical tapered fiber coupler. However, fiber couplers have been utilized to develop devices for the coarse WDM (CWDM) technology, in which the wavelength spacing is typically around 10–30 nm. Nevertheless, precise designing and processing are necessary for high-quality devices, including material composition design and processing control.

On the other hand, the temperature dependence of the coupler's characteristics can be used as a temperature sensor in the ambient temperature range with merits of low cost, immunity of electric shock, and so on. Because the fiber coupler is very thin in the fused and drawn parts, its transmission characteristics are sensitive to the index of the surrounding medium, which is neglected approximately in the discussion of mode coupling by formulas (3.62a and 3.62b) and (3.63a and 3.63b) for simplicity. As mentioned above, the thin waist is actually a multimode waveguide; its surrounding medium will surely affect the mode conversion and coupling. To be used as a sensor, the coupler can be soaked in a liquid to be tested; or is coated by a special film on its surface, which has high sensitivity to some environmental parameters, including chemical specimens absorbed by the film. Vibrations and acoustic waves can be introduced on to the thin coupling region, and the induced effect can be utilized to develop all-fiber acousto-optic devices, which will be discussed in Chapter 4.

The NLO effect, mainly Kerr effect, occurs also in the fiber coupler, with a sensitivity to the intensity of the propagating light beam. This book will not cover this topic as quite a number of papers have presented analyses and demonstrations [52,53].

3.2.3 $N \times N$ and $1 \times N$ Fiber Star Couplers

3.2.3.1 N×N and 1×N Couplers and Their Applications Most fiber devices can find their counterpart in optics built by bulky components. The counterpart of a 2×2 fiber coupler corresponds to a beam splitter made of a plate with a coating of required reflection and transmission. However, the $N \times N$ star coupler ($N > 2$) may be an exception, which divides input optical signal into N path outputs by a single component. A complicated group of bulky components would be needed to realize such a function in conventional optics. Figure 3.16 shows a schematic structure of an $N \times N$ fused fiber coupler, in which a two-dimensional waveguide array is composed of the fibers, and every fiber may be coupled to other fibers, with the strongest coupling to the adjacent ones [38,39,54]. To analyze the characteristics of an $N \times N$ star coupler theoretically, the number of CMEs increases to N, and the number of coupling terms increases also to the number of adjacent fibers.

Figure 3.16   Schematic diagram of an $N \times N$ star coupler.

References [23,24,35] present the basic method of analyzing the arrayed waveguide.

In principle the $1 \times N$ and $N \times 1$ couplers have the same configuration as an $N \times N$ coupler, in which one of the N input fiber pigtails is taken as the input port, while N fibers on its other side are used as output ports, or vice versa. The $N \times M$ coupler is useful in fiber communications and fiber sensor systems as well. For example, the $1 \times N$ star coupler is a necessary component in PON (passive optical network) technology, which divides input signals into N output and sends them downstream to N users. In multiple path sensor systems the star coupler transports the source light to N sensor heads at different places, and picks up the reflected light with sensed signals. On the other hand, $N \times 1$ couplers are used to combine signals from multiple input channels, where a multiplex scheme should be adopted, such as wavelength division multiplexing (WDM), or time division multiplexing (TDM).

3.2.3.2 Characteristics of 3×3 Fiber Couplers

Among various $N \times N$ couplers, the 3×3 coupler shows unique features and is used in some special devices [55–59]. To understand its general properties it is necessary to start from the coupled mode equations, similar to (3.62a and 3.62b):

$$\frac{\mathrm{d}A}{\mathrm{d}z} = j\beta_1 A + j\kappa_{21} B + j\kappa_{31} C, \tag{3.83a}$$

$$\frac{\mathrm{d}B}{\mathrm{d}z} = j\beta_2 B + j\kappa_{32} C + j\kappa_{12} A, \tag{3.83b}$$

$$\frac{\mathrm{d}C}{\mathrm{d}z} = j\beta_3 C + j\kappa_{13} A + j\kappa_{23} B, \tag{3.83c}$$

where A, B, and C are the optical field amplitudes with their respective propagation constant β_i; κ_{ij} are the coupling coefficients. In case of lossless waveguides, the energy conservation, $AA^* + BB^* + CC^* = $ Const., leads to relations of $\kappa_{12} = \kappa_{21}^*$, $\kappa_{23} = \kappa_{32}^*$, $\kappa_{31} = \kappa_{13}^*$. Most practical 3×3 couplers are made of three identical fibers with the same effective index; the coupling equations are simplified as

$$\frac{\mathrm{d}}{\mathrm{d}z} \begin{pmatrix} A \\ B \\ C \end{pmatrix} = j \begin{pmatrix} \beta & \kappa_{12}^* & \kappa_{31} \\ \kappa_{12} & \beta & \kappa_{23}^* \\ \kappa_{31}^* & \kappa_{23} & \beta \end{pmatrix} \begin{pmatrix} A \\ B \\ C \end{pmatrix}. \tag{3.84}$$

The equation can be solved by a trial solution in the form of $\propto \exp(j\gamma z)$ with a composite propagation constant γ, determined by the eigenvalue equation of:

$$\begin{pmatrix} \beta - \gamma & \kappa_{12}^* & \kappa_{31} \\ \kappa_{12} & \beta - \gamma & \kappa_{23}^* \\ \kappa_{31}^* & \kappa_{23} & \beta - \gamma \end{pmatrix} \begin{pmatrix} A \\ B \\ C \end{pmatrix} = 0. \tag{3.85}$$

To ensure a nonzero solution its determinant must be equal to zero, resulting in an equation:

$$x^3 - 3p^2 x - 2q^3 = 0, \tag{3.86}$$

where $x = \gamma - \beta$, $3p^2 = |\kappa_{12}|^2 + |\kappa_{23}|^2 + |\kappa_{31}|^2$, and $2q^3 = \kappa_{12}^* \kappa_{23}^* \kappa_{31}^* + \kappa_{12} \kappa_{23} \kappa_{31}$ are denoted for simplicity. The last term can be written as $2q^3 = 2|\kappa_{12} \kappa_{23} \kappa_{31}| \cos(\phi_1 + \phi_2 + \phi_3)$ with the phases ϕ_i of the respective complex coupling coefficients. The solution is obtained to be

$$\gamma_n = \beta - 2p \cos\left[\frac{2(n-1)\pi}{3} + \frac{\theta}{3}\right] \qquad (n = 1, 2, 3), \tag{3.87}$$

with $\theta = \cos^{-1}(-q^3/p^3)$. The transmission characteristics are then given if the coupling coefficients are known.

The 3×3 coupler is more complicated than the 2×2 coupler since there are more geometrical and physical parameters to influence its characteristics, and to be adjusted also. Due to the feasibility of fabrication, two typical structures are used mostly, as shown schematically in Figure 3.17: triangle configuration (Type-1), and planar configuration (Type-2). For the triangle coupler, it is reasonable to suppose that the coupling coefficients have equal magnitudes, $|\kappa_{12}| = |\kappa_{23}| = |\kappa_{31}| = \kappa$, and equal phases, $\phi_1 = \phi_2 = \phi_3 = \phi$. The eigenvalue is simplified as $\gamma_n = \beta - 2\kappa \cos[2(n-1)\pi/3 + \phi]$. Substituting it into (3.84), the transmission of symmetric 3×3 coupler is obtained to be

$$\begin{pmatrix} A \\ B \\ C \end{pmatrix} = \begin{pmatrix} \cos\vartheta - j(1/3)\sin\vartheta & j(2/3)\sin\vartheta & j(2/3)\sin\vartheta \\ j(2/3)\sin\vartheta & \cos\vartheta - j(1/3)\sin\vartheta & j(2/3)\sin\vartheta \\ j(2/3)\sin\vartheta & j(2/3)\sin\vartheta & \cos\vartheta - j(1/3)\sin\vartheta \end{pmatrix} \begin{pmatrix} A_0 \\ B_0 \\ C_0 \end{pmatrix}$$

with $\vartheta = 3\kappa z/2$, which can be simplified as

$$\begin{pmatrix} A \\ B \\ C \end{pmatrix} = j\frac{2}{3}\sin\vartheta \begin{pmatrix} s & 1 & 1 \\ 1 & s & 1 \\ 1 & 1 & s \end{pmatrix} \begin{pmatrix} A_0 \\ B_0 \\ C_0 \end{pmatrix}, \tag{3.88}$$

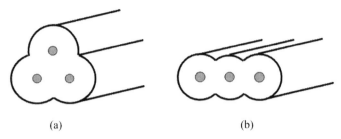

(a)   (b)

Figure 3.17 Two typical configurations of 3×3 fiber couplers: (a) triangle and (b) planar.

where $s = -(1 + j3 \cot \vartheta)/2$ is denoted, and $\phi = 0$ is assumed without loss of generality; a common phase factor of $\exp[j(\beta + \kappa/2)z]$ is omitted for simplicity. In most practical applications a beam split ratio of $1 : 1 : 1$ is required; then the coupling length l and the effective index have to be controlled to get $\vartheta = 3\kappa l/2 = m\pi \pm \pi/3$, and (3.88) is rewritten as

$$\begin{pmatrix} A \\ B \\ C \end{pmatrix} = \frac{j}{\sqrt{3}} \begin{pmatrix} s & 1 & 1 \\ 1 & s & 1 \\ 1 & 1 & s \end{pmatrix} \begin{pmatrix} A_0 \\ B_0 \\ C_0 \end{pmatrix}, \tag{3.89}$$

with $s = -\exp(\pm j\pi/3)$. It is noticed that when an optical signal inputs into one port of the coupler, a directly coupled output and two cross-coupled outputs are detected with equal intensities, but with a phase shift between the direct and cross-outputs. Here, the direct and cross-couplings have only relative meanings, they are exchangeable symbols without any physical difference.

For the planar 3×3 coupler, it is assumed that no coupling exists between the two side waveguides; the CME is now written as

$$\frac{d}{dz} \begin{pmatrix} A \\ B \\ C \end{pmatrix} = j \begin{pmatrix} \beta & \kappa_1^* & 0 \\ \kappa_1 & \beta & \kappa_2^* \\ 0 & \kappa_2 & \beta \end{pmatrix} \begin{pmatrix} A \\ B \\ C \end{pmatrix}. \tag{3.90}$$

With a trial solution in the form of $\propto \exp(j\gamma z)$ an eigenvalue equation is obtained:

$$x^3 - (|\kappa_1|^2 + |\kappa_2|^2)x = 0. \tag{3.91}$$

The composite propagation constants are solved to be

$$\begin{aligned} \gamma_1 &= \beta, \\ \gamma_{2,3} &= \beta \pm \kappa, \end{aligned} \tag{3.92}$$

where $\kappa = \sqrt{|\kappa_1|^2 + |\kappa_2|^2}$. Substituting the eigenvalues into (3.90) the transmission is obtained to be

$$
\begin{pmatrix} A \\ B \\ C \end{pmatrix} = \frac{1}{2} \begin{pmatrix} \cos \kappa z + 1 & j\sqrt{2}\sin \kappa z & \cos \kappa z - 1 \\ j\sqrt{2}\sin \kappa z & \sqrt{2}\cos \kappa z & j\sqrt{2}\sin \kappa z \\ \cos \kappa z - 1 & j\sqrt{2}\sin \kappa z & \cos \kappa z + 1 \end{pmatrix} \begin{pmatrix} A_0 \\ B_0 \\ C_0 \end{pmatrix}. \qquad (3.93)
$$

A special case is the coupling phase factor meeting $\cos \kappa l = 1$, making the transmission a 3×3 identity matrix. Another useful planar 3×3 coupler is a device in which the coupling length l and the coupling coefficient are controlled to meet $\cos \kappa l = 0$, and the transmission matrix is written as

$$
\begin{pmatrix} A \\ B \\ C \end{pmatrix} = \frac{1}{2} \begin{pmatrix} 1 & j\sqrt{2} & -1 \\ j\sqrt{2} & 0 & j\sqrt{2} \\ -1 & j\sqrt{2} & 1 \end{pmatrix} \begin{pmatrix} A_0 \\ B_0 \\ C_0 \end{pmatrix}. \qquad (3.94)
$$

It is shown that when an optical signal inputs into the middle port of the coupler, no signal is detected at the middle output port, whereas two equal and in-phase signals are obtained from the side fiber pigtails.

It is worth noting that the symmetric couplers, whether it is a 2×2 coupler or 3×3 coupler, are bidirectional, that is, the characteristics are described by the same transmission matrix for the input and output ports exchanged.

The 3×3 fiber couplers are useful in many applications, such as fiber optical gyroscopes, and fiber interferometers, and other fiber optic devices incorporating 3×3 couplers, which will be introduced in the next section. Formulas (3.88) and (3.93), describing characteristics for the two typical 3×3 couplers, are helpful in analyzing their sensitivities and stabilities, and improving their performance. Reference [60] describes a Mach–Zehnder interferometer, which is composed of planar and triangle 3×3 couplers, showing good performance as a multiplexer.

3.2.4 Coupling in Axial Direction and Tapered Fiber

Apart from the lateral coupling between waveguides, the longitudinal coupling plays important roles, including internal couplings between propagating modes at any irregular point in the waveguide and external couplings at the fiber facet. The longitudinal coupling involves a number of mechanisms and technical issues, which are beyond the

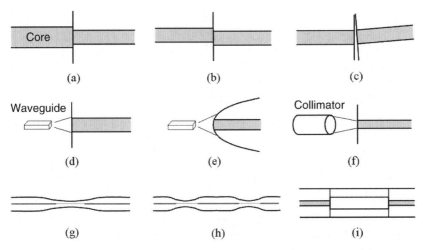

Figure 3.18 Typical axial couplings: (a) between two different fibers; (b) with transverse displacement; (c) with tilted displacement; (d) between laser diode and fiber; (e) with lensed fiber; (f) between collimator and fiber; (g) tapered fiber; (h) concatenated fiber tapers; and (i) single–multi–single-mode structure.

scope of this chapter. Some basic principles for typical cases are introduced here.

Figure 3.18 shows typical axial couplings with structural deviations from the regular waveguide, including coupling between two fibers with discontinuity and errors, coupling between fiber and other optic devices, and tapered fibers. SMFs are mostly used in practical applications; multimode fiber or few-mode fibers are also used for some special devices.

Figure 3.19 gives a conceptional diagram of longitudinal coupling at a discontinuity, which causes conversion between modes and also radiation losses [14]. At the discontinuity the input beam is divided into three parts, a forward transmitted beam, a reflected beam, and scattered radiations. Energy conservation must be obeyed in the conversion, and general optical field continuity conditions have to be considered. The optical field of the input beam can be expanded as a sum of forward

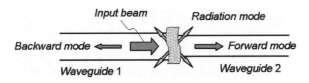

Figure 3.19 Conceptional diagram of longitudinal coupling.

mode, backward mode, and radiation mode:

$$E_{\text{in}} = E_{\text{tr}} + E_{\text{re}} + E_{\text{rad}} = \eta E_{\text{in}} + r E_{\text{in}} + E_{\text{rad}}. \tag{3.95}$$

The theoretical analysis indicates that the coupling coefficient is determined by transverse overlaps of input mode and transmitted mode, and the coefficient for power's coupling is expressed as [14]

$$\eta = \eta_{\text{opt}} \frac{\left| \iint \Phi_{\text{in}}^* \Phi_{\text{out}} \, \mathrm{d}s \right|^2}{\iint |\Phi_{\text{in}}|^2 \, \mathrm{d}s \cdot \iint |\Phi_{\text{out}}|^2 \, \mathrm{d}s}, \tag{3.96}$$

where Φ_{in} and Φ_{out} are the input mode and output mode in the respective waveguides. The integral on the numerator is a mode matching factor between the two waveguides, which sets a theoretical limitation on the coupling coefficient; η_{opt} is the optimum coupling coefficient, taking into account the losses other than mode field mismatching.

Irregular factors also exist between sections of a same waveguide, such as internal defects, bending, twisting, and other deformations. End coupling between two fibers or between fiber and other optical components occurs very often, in which various factors are involved, such as surface Fresnel reflection, beam divergence in free space, and scattering loss due-to imperfections. The fiber grating is one of the most interesting subjects, involving a periodical perturbation, which will be discussed in Chapter 4.

One of the important cases is the tapered fiber, as shown in Figure 3.20(a). The tapered fiber itself has attractive features to be used for some devices, such as the mode converter and sensors. For the fused fiber coupler, loss at the taper must be reduced as much as possible, and

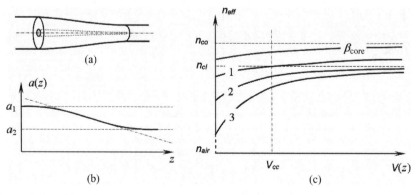

Figure 3.20   (a) Geometrical illustration of tapped fiber; (b) core radius variation; and (c) effective index of LP_{0m} modes of the fiber with a finite cladding.

its adiabaticity is one of the important parameters to be investigated; whereas for sensors the loss is the main parameter to be measured.

The fiber tapers have been fully investigated [60–64]. It is reasonable to consider that mode evolutions occur at the taper, including change of mode field distribution and coupling between modes. In the frame of optical modes in cylindrical waveguides, the variation of core radius means variation of the normalized frequency:

$$V(z) = \frac{2\pi}{\lambda}\sqrt{n_{co}^2 - n_{cl}^2}a(z) = V(0)a(z)/a(0), \qquad (3.97)$$

and consequently variation of the effective index $n_{eff} = n_{eff}(V)$.

The optical field of the input wave now expands transversely as the core radius reduces, and tends to couple with the cladding modes. Figure 3.20(c) shows the effective indices varied with $V(z)$ schematically [63]. The forward fundamental mode is now expressed with a varied propagation constant $\beta(z)$ as

$$E_0(z) = E_0(0) \exp\left[j \int_0^z \beta_0(z')dz' \right]. \qquad (3.98)$$

The cladding mode is then exited by the coupling, expressed as

$$E_1(z) = \eta E_0(0) \exp\left[j \int_0^z \beta_1(z')dz' \right]. \qquad (3.99)$$

The coupling coefficient η is expressed as [61]

$$\eta \propto \int_0^z \frac{a'(u)}{a(u)}\langle\Phi_0\Phi_1\rangle \exp\left\{ j \int_0^u [\beta_0(z') - \beta_1(z')]dz' \right\} du, \qquad (3.100)$$

where $\langle\Phi_0\Phi_1\rangle$ is the overlap integral between the two modes, and $a' = da/dz$ is the taper slope. $\langle\Phi_0\Phi_1\rangle$ and a' are the main factors determining the coefficient. It is noticed that a sinusoidal function is contained in (3.100), resulting in an oscillating behavior in the spectral domain.

To set an adiabaticity criterion for minimizing the energy loss during transmission at the taper, a beat length is defined as $z_b = 2\pi/(\beta_0 - \beta_1)$, and the taper is approximated to be linear as shown in Figure 3.20(b) with taper length $z_t \approx (a_1 - a_2)/a'$. The criteria for the fundamental mode propagating along the taper adiabatically is obtained approximately by $z_t \gg z_b$, that is,

$$a' \ll \frac{(a_1 - a_2)(\beta_0 - \beta_1)}{2\pi}. \tag{3.101}$$

The taper can make couplings from the fundamental core mode to cladding modes; on the other hand, reversed coupling occurs also in a taper with expanded core radius. Detailed discussion and analyses on the fiber taper can be found in references [60–65]. The fiber taper by itself can be used as a sensor, and an optical filter as well [66].

3.3 FIBER LOOP DEVICES INCORPORATED WITH COUPLERS

Fiber devices incorporated with fiber couplers have been studied and developed since the early 1960s up to recent years, and their applications are exploited continuously. Limited by the volume of this book, we can not expound all the variety of fiber devices, but just give some introduction of several fundamental configurations; among them basic properties of fiber Sagnac loops, fiber rings, fiber Mach–Zehnder interferometers (MZI), and fiber Michelson interferometers are discussed in Sections 3.3.1–3.3.3, respectively. Section 3.3.4 introduces loops incorporated with 3×3 fiber couplers, which have unique features.

3.3.1 Fiber Sagnac Loops

3.3.1.1 Basic Properties of Fiber Sagnac Loops   The Fiber Sagnac loop is one of the simplest fiber devices. The loop is constructed with a 2×2 fiber coupler, as shown in Figure 3.21; and the remaining two legs are taken as input (#0) and output (#1) ports. It is equivalent to the optical system in which Sagnac discovered the effect after his name [67].

The input optical beam is divided at the fiber coupler into two beams: one goes clockwise (CW) in the loop from point a to point b,

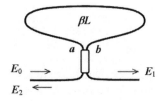

Figure 3.21   Basic configuration of a fiber Sagnac loop.

and the other goes in counterclockwise (CCW) way from b to a. The transmission matrix of the 2×2 coupler is expressed as

$$T_c = \begin{pmatrix} t & jr \\ jr & t \end{pmatrix}, \tag{3.102}$$

which is just a simplified form of formula (3.69) with $t = \cos kl$ and $r = \sqrt{1-t^2} = \sin \kappa l$, implying that a lossless coupler is considered here. When the phase shifts of CW and CCW propagations are equal, the waves transmitted and reflected from the loop are deduced as

$$\begin{aligned} E_1 &= (t^2 - r^2) \exp(j\beta L) E_0, \\ E_2 &= j2rt \exp(j\beta L) E_0. \end{aligned} \tag{3.103}$$

Their intensities are expressed as

$$\begin{aligned} I_1 &= (t^2 - r^2)^2 I_0, \\ I_2 &= 4r^2 t^2 I_0. \end{aligned} \tag{3.104}$$

A special result occurs if a 3 dB coupler with $t^2 = r^2 = 1/2$ is used: $I_1 = 0$ and $I_2 = I_0$, indicating that the beam is totally reflected back to the input port, whereas no optical power outputs from port #1. This result does not depend on the loop length L, and on the wavelength neither, within the spectral band of 3 dB split ratio of the coupler. The device is thus also called a *fiber loop mirror* (FLM), and used in fiber systems where a highly reflective mirror is needed.

It is noticed that the phase factor disappears in (3.104), which means that the output does not depend on the effective index n_{eff} and its spatial variation along the fiber loop. It is because the CW and CCW propagating beams experience identical phase shifts, which can be canceled out in interference at the coupler. Therefore, the fiber loop has a unique feature of eliminating external condition disturbances.

However, if a nonreciprocal element exists in the loop, or some nonreciprocal effects occur there, such as a magneto-optic element, Faraday effect in the fiber, and rotation of the loop in the inertial coordinate, which make the CW and CCW beams suffer different phases, their difference will be manifested in the interfered results.

If the loop rotates in the inertial coordinate, as shown in Figure 3.22, the time delays for CW and CCW propagations in vacuum are deduced as

$$\begin{aligned} t_{\text{CW}} &= (2\pi R + R\Omega t_{\text{CW}})/c = 2\pi R/(c - R\Omega), \\ t_{\text{CCW}} &= (2\pi R - R\Omega t_{\text{CCW}})/c = 2\pi R/(c + R\Omega), \end{aligned} \tag{3.105}$$

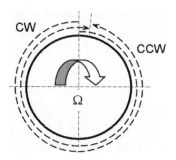

Figure 3.22   Phase shift in a rotating Sagnac loop.

where Ω is the angular velocity and R is the radius of the rotating loop. The phase shift between the two paths is deduced as

$$\Delta\phi = \omega(t_{CW} - t_{CCW}) = 4\omega A \cdot \Omega/c^2, \qquad (3.106)$$

where A is the area of the loop with its normal projected on the angular velocity vector Ω by the scalar product. If the fiber in the Sagnac loop is wound N turns, the phase shift is multiplied:

$$\Delta\phi = 4\omega N A \cdot \Omega/c^2 = 4\pi L R\Omega/c\lambda, \qquad (3.107)$$

where $L = 2\pi RN$ is the total length of the N-turn fiber loop; and codirection of A and Ω is assumed. It is proved that the Sagnac effect is a pure spatial effect, independent of medium properties, because of the Fresnel–Fizeau drag effect in a moving medium [67].

Considering the phase shift, the outputs of the loop are expressed as

$$\begin{aligned} E_1 &= (t^2 e^{j\phi_{CW}} - r^2 e^{j\phi_{CCW}})E_0, \\ E_2 &= jrt(e^{j\phi_{CW}} + e^{j\phi_{CCW}})E_0. \end{aligned} \qquad (3.108)$$

With $\Delta\phi = \phi_{CW} - \phi_{CCW}$, their intensities are expressed as

$$\begin{aligned} I_1 &= [1 - 4r^2 t^2 \cos^2(\Delta\phi/2)]I_0, \\ I_2 &= 4r^2 t^2 \cos^2(\Delta\phi/2)I_0. \end{aligned} \qquad (3.109)$$

They are simplified further as $I_1 = I_0 \sin^2(\Delta\phi/2)$ and $I_2 = I_0 \cos^2(\Delta\phi/2)$, if a 3 dB coupler is used. The above formulas describe the basic effect for a fiber gyroscope; however, lots of related mechanisms and technical issues are to be considered and solved for a practical gyro. This will be discussed in further detail in Section 6.1.

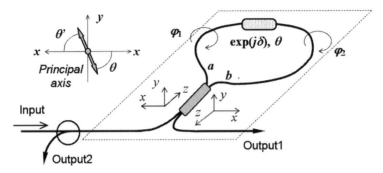

Figure 3.23 Polarization variation of forward and backward beams.

3.3.1.2 *Polarization Characteristics of the Fiber Sagnac Loop*
The birefringence in the fiber loop is an influential effect with theoretical and practical importance. Even if no polarization-dependent element is inserted in the loop and the coupler is polarization independent, fiber bending and twisting are inevitable; thus the bending-induced birefringence and torsion-induced polarization rotation must be taken into consideration. Some components are often inserted to the loop intentionally to make the device a sensor or a measurement setup.

Figure 3.23 shows a fiber loop with a phase retarder inserted, which can be a real birefringence component or a fiber section pressed by some lateral pressure, for example. It is necessary to set a coordinate to describe the principal axis of the phase retarder [68–70]. In the figure a Cartesian coordinate is depicted with y-direction perpendicular to the coupler plane, which is defined by the two parallel fiber pigtails a and b. The fiber loop may take varying forms in the three-dimensional space, the direction of polarization is judged on the fixed coupler plane at the starting and end points of propagation.

Generally, bending and twisting occur simultaneously along the fiber; the PMF may also be used to compose the loop for special functions [71]. For explicit understanding, models based on lumped parameters are usually accepted, which regard the fiber as a concatenation of wave plates [72,73]. On the basis of the model, a multiplication of 2×2 Jones matrices for phase retard and polarization rotation is used to describe the polarization evolution along the fiber. In Figure 3.23, it is assumed that a phase retard element, sandwiched by two rotators, exists in the loop, as a simplified model, and the polarization rotations are depicted by angles φ_1 and φ_2.

The different roundtrip transmissions of CW and CCW ways are denoted by two 2×2 matrices T_{CW} and T_{CCW}. It is noted that the principal axis azimuth θ of the retard has opposite signs for CW and CCW beams, that is $\theta' = -\theta$, as shown in the figure. Their transmission matrices are written as $T_{CW} = R_{\varphi 2} R_{\theta} T_{\delta} R_{-\theta} R_{\varphi 1}$ and $T_{CCW} = R_{-\varphi 1} R_{-\theta} T_{\delta} R_{\theta} R_{-\varphi 2}$, where R is the related rotation matrix and T_{δ} is the matrix of phase retard, written as

$$T_{\delta} = \begin{pmatrix} e^{-j\delta/2} & 0 \\ 0 & e^{j\delta/2} \end{pmatrix}, \tag{3.110}$$

where $\delta = (\beta_2 - \beta_1)L$ is the phase retardation. The composite CW and CCW transmission matrices are deduced to be

$$T_{CW} = \begin{pmatrix} \cos \gamma_1 \cos \dfrac{\delta}{2} - j \cos \gamma_2 \sin \dfrac{\delta}{2} & \sin \gamma_1 \cos \dfrac{\delta}{2} + j \sin \gamma_2 \sin \dfrac{\delta}{2} \\ -\sin \gamma_1 \cos \dfrac{\delta}{2} + j \sin \gamma_2 \sin \dfrac{\delta}{2} & \cos \gamma_1 \cos \dfrac{\delta}{2} + j \cos \gamma_2 \sin \dfrac{\delta}{2} \end{pmatrix}, \tag{3.111a}$$

$$T_{CCW} = \begin{pmatrix} \cos \gamma_1 \cos \dfrac{\delta}{2} - j \cos \gamma_3 \sin \dfrac{\delta}{2} & -\sin \gamma_1 \cos \dfrac{\delta}{2} + j \sin \gamma_3 \sin \dfrac{\delta}{2} \\ \sin \gamma_1 \cos \dfrac{\delta}{2} + j \sin \gamma_3 \sin \dfrac{\delta}{2} & \cos \gamma_1 \cos \dfrac{\delta}{2} + j \sin \gamma_3 \sin \dfrac{\delta}{2} \end{pmatrix}, \tag{3.111b}$$

where $\gamma_1 = \varphi_1 + \varphi_2$, $\gamma_2 = \varphi_2 - \varphi_1 - 2\theta$, and $\gamma_3 = \varphi_2 - \varphi_1 + 2\theta$. It is shown that the matrix is a unitary matrix, whose elements satisfy relations of $m_{22} = m_{11}^*$, $m_{21} = -m_{12}^*$, and $|m_{11}|^2 + |m_{12}|^2 = 1$. This property is the requirement of energy conservation in the propagation. The model can be extended to the case of multiple wave-plate concatenation, because the multiplication of unitary matrices is still unitary, as proved in [73].

The input beam, described by a Jones vector $\boldsymbol{E}_0 = (E_x \quad E_y)^T$, is divided into two beams with the same ratio for its x- and y-components by the polarization-independent coupler. After the roundtrip propagation, the two beams are combined at the coupler, resulting in two outputs:

$$\boldsymbol{E}_1 = (t^2 T_{CW} - r^2 T_{CCW}) \begin{pmatrix} E_x \\ E_y \end{pmatrix}, \tag{3.112a}$$

$$\boldsymbol{E}_2 = jtr(T_{CW} + T_{CCW}) \begin{pmatrix} E_x \\ E_y \end{pmatrix}. \tag{3.112b}$$

The output intensities are obtained to be

$$I_1 = \left[(1 - 4t^2 r^2 \cos^2 \gamma_1) \cos^2 \frac{\delta}{2} + (1 - 4t^2 r^2 \cos^2 2\theta) \sin^2 \frac{\delta}{2} \right] I_0,$$

(3.113a)

$$I_2 = 4t^2 r^2 \left(\cos^2 \gamma_1 \cos^2 \frac{\delta}{2} + \cos^2 2\theta \sin^2 \frac{\delta}{2} \right) I_0.$$ (3.113b)

It is shown that the transmitted and reflected outputs of the loop are complementary, if both the fiber and the coupler are lossless. When the split ratio of the coupler is $t^2 = r^2 = 1/2$, the outputs are written as

$$I_1 = \left(\sin^2 \gamma_1 \cos^2 \frac{\delta}{2} + \sin^2 2\theta \sin^2 \frac{\delta}{2} \right) I_0,$$ (3.114a)

$$I_2 = \left(\cos^2 \gamma_1 \cos^2 \frac{\delta}{2} + \cos^2 2\theta \sin^2 \frac{\delta}{2} \right) I_0.$$ (3.114b)

For the case without polarization rotation, that is, $\varphi_1 = \varphi_2 = 0$, the outputs are written as

$$I_1 = \left[(t^2 - r^2)^2 + 4t^2 r^2 \sin^2 2\theta \sin^2 \frac{\delta}{2} \right] I_0,$$ (3.115a)

$$I_2 = 4t^2 r^2 \left(1 - \sin^2 2\theta \sin^2 \frac{\delta}{2} \right) I_0.$$ (3.115b)

This means that the loop is now no longer a totally reflective mirror, even if a 3 dB coupler is used: $I_1/I_0 = \sin^2 2\theta \sin^2(\delta/2)$, unless the principal axis coincides with the loop plane ($\theta = 0, \pi/2$). On the other hand, the reflection is the most sensitive to the phase retard when the axis is set at $\theta = \pi/4$.

For the case of no birefringence effect, that is, $\delta = 0$, the outputs are written as

$$I_1 = (1 - 4t^2 r^2 \cos^2 \gamma_1) I_0,$$ (3.116a)

$$I_2 = (4t^2 r^2 \cos^2 \gamma_1) I_0.$$ (3.116b)

In case a 3 dB coupler is used, we have $I_1 = I_0 \sin^2 \gamma_1$ and $I_2 = I_0 \cos^2 \gamma_1$, meaning that the reflectivity of the loop mirror depends on polarization rotation in the loop. Therefore, attention must be paid to avoid fiber twisting in splicing the fiber pigtails, to reduce the background of the mirror. On the other hand, when the loop has been built, fiber twisting will introduce opposite torsions along the fiber, giving a cancelled effect on the output. In this sense the built loop is a twisting-independent unit [16].

It is seen that the output of the loop is sensitive to any externally induced birefringence, and to its orientation. In practical applications an orientation-insensitive device is desired. It is found that a half wavelength plate (HWP) inserted in the loop is helpful for the purpose [74]. The Jones matrices of HWP with azimuth of ψ for CW and CCW waves are written as

$$T_{CW}^{HWP} = \begin{pmatrix} -j\cos 2\psi & -j\sin 2\psi \\ -j\sin 2\psi & j\cos 2\psi \end{pmatrix}, \tag{3.117a}$$

$$T_{CCW}^{HWP} = \begin{pmatrix} -j\cos 2\psi & j\sin 2\psi \\ j\sin 2\psi & j\cos 2\psi \end{pmatrix}. \tag{3.117b}$$

The matrices for the round-trip propagations are deduced as

$$T_{CW} = \begin{pmatrix} -m_{11} - jm_{12} & m_{21} - jm_{22} \\ -m_{21} - jm_{22} & -m_{11} + jm_{12} \end{pmatrix}, \tag{3.118a}$$

$$T_{CCW} = \begin{pmatrix} -m_{11} - jm_{12} & m_{21} + jm_{22} \\ -m_{21} + jm_{22} & -m_{11} + jm_{12} \end{pmatrix}, \tag{3.118b}$$

where $m_{11} = \cos 2(\theta - \psi)\sin(\delta/2)$, $m_{12} = \cos 2\psi \cos(\delta/2)$, $m_{21} = \sin 2$ $(\theta - \psi)\sin(\delta/2)$, and $m_{22} = \sin 2\psi \cos(\delta/2)$ are denoted for simplicity. According to (3.112a and 3.112b) the transmitted field and intensity for a 3 dB coupler are obtained to be

$$E_1 = -j\sin 2\psi \cos(\delta/2) \begin{pmatrix} 0 & 1 \\ 1 & 0 \end{pmatrix} \begin{pmatrix} E_x \\ E_y \end{pmatrix}, \tag{3.119}$$

$$I_1 = \sin^2 2\psi \cos^2(\delta/2). \tag{3.120}$$

It is noted that the output does not depend on the principal axis azimuth θ, but is a function of phase retardence δ. It is obvious that the HWP should be rotated to $\psi = \pi/4$ to get the highest sensitivity. Because the retardation of a HWP is π, this device is called π-shift Sagnac interferometer. In a practical setup, a PC is used for the purpose [75].

The birefringence effects and polarization rotations exist often along the fiber in a randomly distributed way and are varying temporally with external conditions, such as temperature. The polarization dependence of the coupler has to be considered also, especially for highly sensitive sensors, such as fiber gyroscopes. Detailed analysis will be given in Section 3.4 and Chapter 6.

3.3.1.3 Nonlinear Optical Fiber Loop Mirror

Nonlinear optical (NLO) effects may also induce nonreciprocal effects. Among the various NLO effects, Kerr effect occurs often in case of pulsed laser beams, which induces an index increment proportional to the optical intensity:

$$n = n_0 + n_{II}I, \qquad (3.121)$$

where n_{II} is the Kerr coefficient, and the nonlinear phase is induced by self-phase modulation (SPM). A special device, called nonlinear optical fiber loop mirror (NOLM), is then proposed [76], in which the output depends on the intensity of the input beam, expressed as

$$I_1 = I_0\{1 - 2r^2t^2 + 2r^2t^2 \cos[(1 - 2r^2)n_{II}I_0kL]\}. \qquad (3.122)$$

The effect is also used to develop an optical switch, as shown in Figure 3.24. In the device, a controlling pulse I_C is injected through an additional coupler with intensity high enough for the Kerr effect to occur in fiber section L_2. Thus, the CW and CCW propagating pulses acquire different phase shifts, expressed as

$$\phi_{CW} = \beta_1 L_1 + \beta_2(I_C, \Delta t_{CW})L_2, \qquad (3.123a)$$

$$\phi_{CCW} = \beta_1 L_1 + \beta_2(I_C, \Delta t_{CCW})L_2, \qquad (3.123b)$$

where Δt_{CW} and Δt_{CCW} stand for the periods of CW and CCW pulses meeting the controlling pulse I_C. The index modulation by Kerr effect is $\Delta n = n_{II}|E_{CW}(t) + E_{CCW}(t) + E_C(t)|^2$. The difference between Δt_{CW} and Δt_{CCW} results in the phase shifts not cancelling out each other, invalidating the condition of total reflection. By properly controlling the injecting moment and duration of I_C, certain pulses can be selected from the input pulse train and output from port 1. The switch is used in

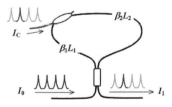

Figure 3.24   Schematic diagram of nonlinear optical fiber loop mirror.

ultra-short optical pulse processing technologies, such as optical time division multiplexing technology as a terahertz optical asymmetric demultiplexer [77] in the high-speed fiber communication system.

The NOLM is a temporal nonreciprocal effect observed in short laser pulse systems. Temporal nonreciprocity exists in a long loop with wider pulse-width optical pulses, and is used as a sensor, which will be discussed in Chapter 5.

3.3.2 Fiber Rings

3.3.2.1 Basic Properties and Characteristics of Fiber Rings   A fiber ring resonator is composed simply of a 2×2 coupler, as a beam splitter, with its legs connected in such a way as shown in Figure 3.25, where L is the length of the fiber ring from a to b [78].

Light propagates in the ring by way of recurring series: an input light wave is divided at the coupler into the output port and port a, and later returns to port b after a roundtrip in the ring and inputs the coupler again, and so on and so forth. If loss in the coupler and in the fiber section is low enough to be neglected, by using the transmission matrix of the coupler and the phase delay in the fiber section, $E_b = E_a \exp(j\beta L)$, the output is deduced to be

$$E_1 = \frac{t - e^{j\beta L}}{1 - t e^{j\beta L}} E_0 = E_0 \exp(j\varphi), \tag{3.124}$$

where the composite phase delay is expressed as

$$\varphi = \tan^{-1} \frac{(1 - t^2) \sin \beta L}{(1 + t^2) \cos \beta L - 2t}. \tag{3.125}$$

Several general properties are concluded from the formulas:

1. Optical beam passes without attenuation, $E_1^2 = E_0^2$, due to the assumption of lossless coupler and lossless fiber. Therefore, the ring is called an *all-pass filter*.
2. The composite phase delay φ depends not only on phase delay in the fiber ring, but also on the beam split ratio of the coupler.

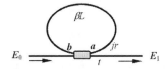

Figure 3.25   Basic configuration of a fiber ring.

3. The phase delay is a periodic function of optical frequency with resonances at $\beta L = 2m\pi$, that is the reason for calling it a *ring resonator*.

4. It is deduced that the ring possesses periodic dispersion characteristics, so that it can be used as a phase modulator, and a dispersion compensator as well.

The group delay τ_g and the dispersion parameter D_{ring} is deduced from (3.125):

$$\tau_g = \frac{\partial \varphi}{\partial \omega} = \frac{nL}{c} \frac{1 - t^2}{(1 + t^2 - 2t \cos \beta L)^2}, \tag{3.126}$$

$$D = \frac{\partial \tau_g}{\partial \lambda} = \frac{8\pi t(1 - t^4)n^2 L^2}{c\lambda^2} \frac{\sin \beta L}{(1 + t^2 - 2t \cos \beta L)^3}. \tag{3.127}$$

The fiber dispersions of orders higher than β_1 ($= \partial \beta / \partial \omega$) are neglected in (3.126) and (3.127). It is shown that the dispersion is a periodic function of wavelength, and covers a region from negative value to positive value, determined by the split ratio of coupler and fiber length. It is noticed that the basic properties and formulas hold also for waveguide rings.

It is seen that the fiber ring resembles the Gires–Tournois interferometer (GTI) [79], which consists of two parallel planar reflective surfaces: one of them has low reflectivity, and is used as the input and output surface, and the rear one is a mirror with 100% reflectivity. Similarly, the coupler can be regarded as a dielectric surface, as analyzed in Section 3.2.1, whereas the connected ring is equivalent to a totally reflective mirror. Figure 3.26 shows a calculated dispersion of a GTI varied with the wavelength, which is composed of a 1.2 mm glass plate with input surface coated to $r = 0.4$ and $r = 0.6$. The GTI is a typical all-pass filter, used as a phase shifter or as a dispersion compensation element in fiber communications and laser systems [80].

The phase delay of formula (3.125) can be expanded to a series by Fourier transform as

$$\cos \varphi = \left(\frac{1 + t^2}{1 - t^2} \cos \beta L - \frac{2t}{1 - t^2} \right) \left(1 + 2 \sum_{m=1}^{\infty} t^m \cos m\beta L \right), \tag{3.128a}$$

$$\sin \varphi = \sin \beta L \left(1 + 2 \sum_{m=1}^{\infty} t^m \cos m\beta L \right). \tag{3.128b}$$

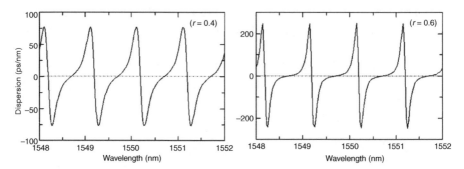

Figure 3.26   Dispersion spectrum of fiber ring.

The infinitive summations in the expansions describe the concatenated recurrence. This expression of phase delay is helpful for understanding and analyzing characteristics of fiber ring applications [81].

If optical loss in the fiber (or in the waveguide) cannot be neglected, and the coupler is not lossless, formula (3.124) has to be revised. Denoting the roundtrip loss as $\gamma = \exp(-\alpha L/2)$, and $r^2 + t^2 = \gamma_c < 1$, the transmission of the ring is then deduced as

$$\frac{E_1}{E_0} = \frac{t - \gamma_c \gamma e^{j\beta L}}{1 - t\gamma e^{j\beta L}}. \tag{3.129}$$

The power transmission is obtained to be

$$\frac{I_1}{I_0} = \frac{t^2 + \gamma_c^2 \gamma^2 - 2t\gamma_c\gamma \cos \beta L}{1 + t^2\gamma^2 - 2t\gamma \cos \beta L}, \tag{3.130}$$

which means that the ring is no longer an all-pass filter. Its transmitted power is a spectrally periodic function with the maximum of $(t + \gamma_c\gamma)^2/(1 + \gamma t)^2$ at resonance of $\cos \beta L = -1$ and the minimum of $(t - \gamma_c\gamma)^2/(1 - \gamma t)^2$ at $\cos \beta L = 1$.

It is noticed that when the beam split ratio is controlled to meet condition of $t = \gamma_c\gamma$, the transmission minimum reaches zero, indicating that all the input energy is lost in the ring due to multiple circulations. This phenomenon is called *critical coupling* [78,82]. In such cases, the fiber ring is very sensitive to any small loss change in the ring. Figure 3.27(a) shows the transmission spectrum with periodic critical couplings; Figure 3.27(b) gives the output varying with the intra-cavity loss $\alpha = \gamma_c\gamma$ for two different split ratios.

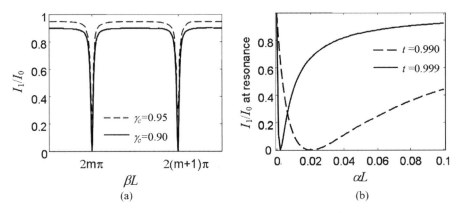

Figure 3.27 (a) Transmission spectrum of fiber ring and (b) transmission power versus loss.

Another noticeable feature is that the optical intensity inside the ring I_b at point b changes with the wavelength, expressed as

$$\frac{I_b}{I_0} = \frac{r^2\gamma^2}{1 + t^2 - 2t\gamma\cos\beta L}. \qquad (3.131)$$

At the resonances of $\cos\beta L = 1$ it is obtained that $(I_b/I_0)_{max} = r^2\gamma^2/(1 - t\gamma)^2 \rightarrow (1 + t)/(1 - t)$, where the last expression is for the lossless case ($\gamma = \gamma_c = 1$). The optical intensity inside the ring at resonances is higher than the input, and when the split ratio tends toward $1 : 0$, that is, $t \rightarrow 1$, the intensity inside the ring would tend to the infinitive, though only a small percentage of input energy enters the ring. In other words, the ring can store optical energy at the resonances. This property is utilized to develop devices with NLO effects, such as Brillouin fiber lasers [83].

The rings can be combined to build devices with useful functions for applications. Figure 3.28 is a proposed optical add-drop multiplexer (OADM) with a fiber ring consisting of two couplers [84]. It is deduced

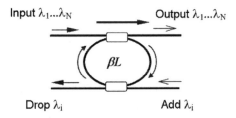

Figure 3.28 Waveguide ring add-drop multiplexer.

that when the length of ring meets the condition of $\beta_i L = 2m\pi$ the optical signal at corresponding wavelength of λ_i can be dropped from the input wavelength division multiplexed (WDM) data, and added to the downward data stream. The wavelength spacing $\Delta\lambda$ of the WDM is determined by the length of ring, expressed as $\Delta\lambda = \lambda^2/2n_{\mathrm{eff}}L$. Because the fiber ring is usually fabricated with length not short enough to match the wavelength spacing specified in practical fiber communication systems, the ring OADM is usually fabricated by planar waveguides.

3.3.2.2 Polarization Characteristics of Fiber Rings

If no polarization-dependent effect exists in the fiber section L or in the coupler, the ring is a polarization-independent device. However, the fiber in the ring often suffers bending and twisting, resulting in polarization dependence. It is necessary to analyze the polarization evolution in fiber ring. The distributed birefringence and polarization rotations lead to couplings between the two polarization modes; it can be regarded as a loss/gain factor for each individual mode, and causes its transmission fluctuation. At the same time, the resonant peaks of the two modes differ with each other by the phase shifts of $\Delta\beta L$. These mechanisms make the ring polarization dependent. For simplicity, the polarization-dependent factors are regarded as lumped elements; by the same expressions of (3.111b) the optical field is written as

$$E_b = T_{\mathrm{CCW}} E_a \exp(j\beta L) = T_{\mathrm{CCW}} E_a \mathrm{e}^{j\phi(\lambda)}. \qquad (3.132)$$

The optical field at port **a** is written as

$$E_a = jr E_0 + t E_b = (jr E_0 + t\mathrm{e}^{j\phi} T_{\mathrm{CCW}} E_a), \qquad (3.133)$$

and the output $E_1 = t E_0 + jr E_b$ is expressed as

$$E_1 = [t - r^2 \mathrm{e}^{j\phi} T_{\mathrm{CCW}} (\hat{I} - t\mathrm{e}^{j\phi} T_{\mathrm{CCW}})^{-1}] E_0, \qquad (3.134)$$

where $\hat{I}$ is the unit matrix; the superscript (-1) stands for the inversion of the matrix. Then the transmission of the fiber ring is deduced to be [85]

$$E_1 = \frac{1}{D} \left[D_1 \hat{I} + \mathrm{e}^{j\phi} r^2 \begin{pmatrix} m_{11}^* & m_{21}^* \\ -m_{21} & m_{11} \end{pmatrix} \right] E_0, \qquad (3.135)$$

where $D = 1 - t\mathrm{e}^{j\phi}(m_{11} + m_{11}^*) + t^2 \mathrm{e}^{j2\phi}$, $D_1 = (D - r^2)/t$, $m_{11} = \cos\gamma_1 \cos\frac{\delta}{2} - j\cos\gamma_3 \sin\frac{\delta}{2}$, and $m_{21} = \sin\gamma_1 \cos\frac{\delta}{2} + j\sin\gamma_3 \sin\frac{\delta}{2}$.

The polarization dependence for given birefringence and polarization rotations can then be calculated using the formula. It is concluded that the dependence is a periodic function of a wavelength with resonances, and the transmission polarization is sensitive to the external disturbances caused often by fiber bending and twisting. The sensitivity makes the device unstable; on the other hand, it provides the possibility of being developed for some sensors. The polarization dependence of the fiber ring has been utilized to develop a depolarizer [85] as discussed in Section 3.5, where the coherence of the input wave is taken into account.

3.3.3 Fiber Mach–Zehnder Interferometers and Michelson Interferometers

MZI and Michelson interferometers are two of the basic and most useful interferometers. Their principles are expounded in many textbooks. Here a brief introduction to fiber interferometers is given.

3.3.3.1 Fiber Mach–Zehnder Interferometer and Its Applications
The basic configuration of a fiber MZI is depicted in Figure 3.29. Where losses of the couplers and fibers are negligible, its transmissions are obtained to be

$$E_1 = [t_1 t_2 \exp(j\beta_1 L_1) - r_1 r_2 \exp(-j\beta_2 L_2)]E_0, \qquad (3.136a)$$

$$E_2 = j[t_1 r_2 \exp(j\beta_1 L_1) + r_1 t_2 \exp(-j\beta_2 L_2)]E_0, \qquad (3.136b)$$

and their intensities to be

$$I_1 = [(t_1 t_2 - r_1 r_2)^2 + 4t_1 t_2 r_1 r_2 \sin^2 \Delta\phi]I_0, \qquad (3.137a)$$

$$I_2 = [(t_1 r_2 + r_1 t_2)^2 - 4t_1 t_2 r_1 r_2 \sin^2 \Delta\phi]I_0, \qquad (3.137b)$$

where $\Delta\phi = \beta_2 L_2 - \beta_1 L_1$. It is seen that both outputs are sinusoidal curves but with a phase shift of π. If both couplers have a split ratio of 1:1, the intensities are simplified as $I_1 = I_0 \sin^2 \Delta\phi$, and $I_2 = I_0 \cos^2 \Delta\phi$.

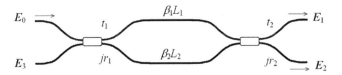

Figure 3.29   Schematic configuration of a fiber MZI.

It is shown that any change in its two beam lengths, L_1 or L_2, or in indexes ($= \beta/k_0$), will cause changes in the outputs. Therefore, the fiber MZI is widely used as a sensor, and as a measurement setup. The initial bias of the phase difference is usually set at $\pi/4$, called the quadrature point, to get a linear detection of sensed parameters with the highest sensitivity. At the point the detected phase is $\Delta\phi = \pi/4 + \delta\phi$, and the optical signal is $\delta I \propto \sin \delta\phi \approx \delta\phi$, where $\delta\phi$ is the phase variation to be detected.

On the other hand, any change of the input wavelength will also cause variation in the output intensity, because $\Delta\phi = 2\pi(n_{\text{eff1}}L_1 - n_{\text{eff2}}L_2)/\lambda$ varies with the wavelength. This function is used as one of the interrogation methods for optical signals with information carried by its wavelength, such as fiber grating sensors. It is also utilized to develop a wavelength division device, in which the beam lengths are designed to meet a condition of $\Delta\phi(\lambda_1) - \Delta\phi(\lambda_2) = \pi$, where λ_1 and λ_2 are wavelengths to be divided; thus the difference in lengths has to be adjusted to meet

$$2n_{\text{eff}}(L_1 - L_2) = \frac{\lambda_1\lambda_2}{\lambda_1 - \lambda_2} \approx \frac{\lambda^2}{\Delta\lambda}. \tag{3.138}$$

Actually the number of wavelengths to be divided in a WDM system is often more than two, such as 8 or 16 channels. The division can be realized by cascaded MZIs with $\Delta L_1 = 2\Delta L_2 = \lambda^2/\Delta\lambda$ as shown in Figure 3.30 for two stages, which allows for four WDM signals output from four different ports. In practical fiber communication applications, more specifications are required, and more technological problems have to be solved.

The fiber MZI is also used for optical data processing in a time domain. Figure 3.31(a) shows an MZI for pulse repetition rate doubling,

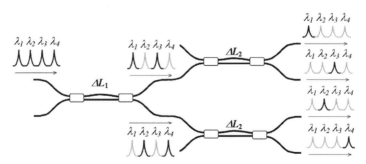

Figure 3.30 Cascaded MZI as a WDM device.

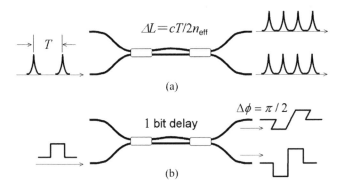

Figure 3.31   (a) Pulse repetition rate multiplier and (b) RZ-DPSK format generation.

where the optical path difference (OPD) between the two beams is designed to make the delay equal to a half of the period of input pulse train [86]. If the delay is designed to be a bit in the optical communications, the MZI is named the delayed interferometer, which is useful in optical signal generation and detection [87]. Figure 3.31(b) shows a format conversion from an intensity modulated signal to a return-to-zero (RZ) differential phase shift keying (DPSK) signal. The OPD of MZI beams is controlled to generate a bit delay, resulting in a phase difference of $\pi/2$ between the two outputs, as depicted by a slanted waveform.

Figure 3.32(a) shows a setup for measuring polarization characteristics of a component, or a section of fiber, inserted in one of the beams of MZI, denoted by phase retardation of $\delta = (\beta_{//} - \beta_{\perp})k_0l$; two analyzers are inserted at the output ports to give the dependence of polarization on concerned conditions, such as wavelength, temperature, strain, and others, of the components to be tested.

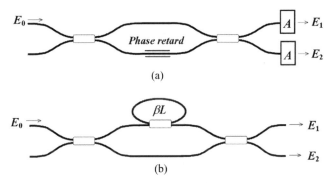

Figure 3.32   (a) MZI for polarization characteristics measurement and (b) MZI incorporated with a ring.

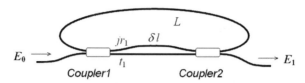

Figure 3.33   Combined structure of ring and MZI.

A combination of MZI and ring, shown in Figure 3.32(b), gives some attractive characteristics. Under certain conditions its transmission spectrum shows sharp peaks, different from the sinusoidal curves. When the ring contains some sensitive elements, the MZI plays the role of phase detection; and the ring can consist of a long fiber extending to the area to be sensed [88].

Another combination of MZI and ring is shown in Figure 3.33, which is regarded as a ring with an MZI playing the role of composite coupler. The split ratio can then be adjusted by the OPD of $n_{\text{eff}}\delta l$, even if the split ratios of coupler #1 and coupler #2 are already fixed. It is an all-pass filter basically, expressed by a transmission with only a phase shift:

$$E_1 = \frac{t_1 t_2 - r_1 r_2 \mathrm{e}^{j\beta\delta l} - \mathrm{e}^{j\beta\delta l}\mathrm{e}^{j\beta L}}{1 + (r_1 r_2 - t_1 t_2 \mathrm{e}^{j\beta\delta l})\mathrm{e}^{j\beta L}} E_0 = \mathrm{e}^{j\Phi} E_0. \qquad (3.139)$$

Such a ring structure can be developed as a dispersion compensator with more adjustable parameters [89]; it is better to be fabricated with a planar waveguide to get suitable channel spacing coincident with the DWDM ITU standards.

3.3.3.2 Fiber Michelson Interferometer and Its Applications
Fiber Michelson interferometers are widely used in measurement set-ups and sensor technologies. Compared with conventional Michelson interferometers by bulk components, the fiber Michelson interferometers have their attractive features. It is compatible to fiber systems and suitable for constructing set-ups to measure characteristics of fiber optic materials and devices. It provides flexibility in building various structures to meet different requirements.

Figure 3.34 shows a so-called all-fiber Michelson interferometer incorporated with an FLM, which provides stable, and broad band reflections for both beams of the interferometer [90]. In the setup, as an example, the sensor head is inserted in one of the interferometer beams, while a phase modulator driven by PZT is inserted in the other beam, providing an intentionally modulated phase ϕ_1 as a reference,

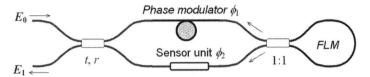

Figure 3.34 Michelson interferometer incorporated with a fiber loop mirror.

which is helpful to suppress noise and drift in signal interrogation. The output is expressed as

$$E_1 = jrt(e^{j2\phi_1} + e^{j2\phi_2})E_0, \qquad (3.140a)$$

$$I_1 = 4r^2t^2 \cos^2(\phi_1 - \phi_2)I_0. \qquad (3.140b)$$

The highest extinction ratio is obtained when the split ratio of the first coupler is 3 dB, resulting in the output of $I_1 = \cos^2(\phi_1 - \phi_2)I_0$.

The fiber Michelson interferometers are used widely in various applications, especially in metrology and sensing technology. Many technical issues have to be taken into consideration in practice.

3.3.4 Fiber Loops Incorporated with 3×3 Couplers

The 3×3 fiber coupler has unique features as described in Section 3.2.3, which is often used to construct fiber loops with attractive and special functions. In this subsection, two examples are introduced: one is a modified Mach–Zehnder interferometer applied as a sensor or a measurement setup; the other is a modified Sagnac loop in optical fiber gyros.

3.3.4.1 Basic Characteristics of MZI Incorporated with a 3×3 Coupler   The fiber MZI is very useful in sensor and metrology fields with remarkable merits, such as high sensitivity, wide applicability, simplicity and low cost. However some substantial shortcomings have been noticed in a MZI composed of two 2×2 couplers. Because the sensed signal appears in forms of sinusoidal function, two signals with phase difference of $2m\pi$ cannot be distinguished and the sensing scale is limited within phase difference of 2π. In addition, dead regions exist where the sensitivity goes to zero, when the bias of MZI is shifted to the maximum or minimum points of sinusoidal function. Due to the symmetrical property of sinusoidal functions at its extreme positions the interferometer cannot tell the direction of signal changes around the positions.

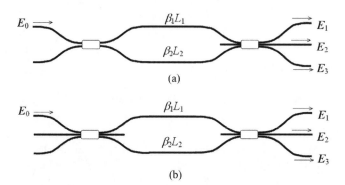

Figure 3.35 Modified MZI with (a) one and (b) two 3×3 fiber couplers.

If 3×3 fiber couplers are used to build the interferometer, replacing one of the two 2×2 couplers, or both of them, as shown in Figures 3.35(a) and (b), such shortcomings can be overcome [55–59,91–93]. The two beams of the interferometer with phase difference of $\delta\theta$ and the 2×2 coupler in Figure 3.35(a) are expressed by expanded matrices to match with the 3×3 matrix. If a 3 dB 2×2 coupler and a 1:1:1 symmetric 3×3 coupler are used, by using matrix (3.89) with $s = -e^{j\pi/3}$, the transmission matrices are now written as

$$M_{2+3} = T_{3\times3}T_{\text{arm}}T_{2\times2} = \frac{j}{\sqrt{6}} \begin{pmatrix} s & 1 & 1 \\ 1 & s & 1 \\ 1 & 1 & s \end{pmatrix} \begin{pmatrix} 1 & 0 & 0 \\ 0 & e^{j\delta\theta} & 0 \\ 0 & 0 & 0 \end{pmatrix} \begin{pmatrix} 1 & j & 0 \\ j & 1 & 0 \\ 0 & 0 & 0 \end{pmatrix}$$

$$= \frac{j}{\sqrt{6}} \begin{pmatrix} j(e^{j\delta\theta} - e^{j\pi/3}) & e^{j\delta\theta} + e^{j\pi/3} & 0 \\ 1 + e^{j(\delta\theta+\pi/3)} & j[1 - e^{j(\delta\theta+\pi/3)}] & 0 \\ 1 - je^{j\delta\theta} & j - e^{j\delta\theta} & 0 \end{pmatrix}.$$

(3.141)

And for the structure in Figure 3.35(b),

$$M_{3+3} = T_{3\times3}T_{\text{arm}}T_{3\times3}$$

$$= \frac{-1}{\sqrt{9}} \begin{pmatrix} e^{j\delta\theta} + e^{j\pi/3} & -e^{j\pi/6}(1 + e^{j\delta\theta}) & e^{j\delta\theta} - e^{j\pi/6} \\ -e^{j\pi/6}(1 + e^{j\delta\theta}) & 1 + e^{j(\delta\theta+\pi/3)} & 1 - e^{j(\delta\theta+\pi/6)} \\ e^{j\delta\theta} - e^{j\pi/6} & 1 - e^{j(\delta\theta+\pi/6)} & 1 + e^{j\delta\theta} \end{pmatrix}.$$

(3.142)

If an optical beam injects into port #0 as a light source for the interferometer, three outputs for structure (a) in Figure 3.35 are obtained

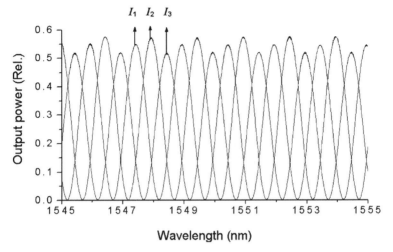

Figure 3.36 Measured outputs of a modified MZI with a 3×3 fiber coupler.

with intensities written as

$$I_1 = \frac{1}{3}\left[1 + \sin\left(\delta\theta - \frac{\pi}{3}\right)\right]I_0, \tag{3.143a}$$

$$I_2 = \frac{1}{3}\left[1 + \sin\left(\delta\theta + \frac{\pi}{3}\right)\right]I_0, \tag{3.143b}$$

$$I_3 = \frac{1}{3}(1 - \sin\delta\theta)I_0. \tag{3.143c}$$

The results of structure (b) are obtained similarly. The three outputs can be depicted as functions of phase shift $\delta\theta$. Figure 3.36 gives experimentally measured curves of the outputs varying with the source wavelength [92], which is equivalent to the phase variations to be sensed. It is shown that the wavelength spacing between the adjacent curves remains equal, corresponding to a phase shift of $\pi/3$. When one of three outputs reaches its exame value with a zero slope (zero sensitivity), the other two outputs are in a region with high sensitivities; and moreover, their derivatives have opposite signs. This property diminishes the dead region, and gives a capability of judging the direction of phase change. This improvement no longer limits the measurement range within the period 2π of sinusoidal functions. A data processing method, called the phase unwrapping algorithm, extends greatly the range of measurement so long as all the measured data are stored in a computer. The curves in Figure 3.36 do not have equal amplitude, as expected by the theoretical analysis; this is attributed to the differences in optical and electrical components used in the experiment.

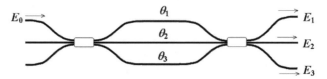

Figure 3.37   Three beam MZI as a multiple parameter sensor.

The 3×3 couplers can be used to build a MZI for multiple parameter sensors, as shown in Figure 3.37, where the three arms are regarded as sensor heads for phase changes θ_1, θ_2, and θ_3, including changes of their lengths and/or their indexes. By using matrix (3.89) the transmission of the MZI is expressed as

$$
M_{3+3+3} = \frac{-1}{3}
\begin{pmatrix} s & 1 & 1 \\ 1 & s & 1 \\ 1 & 1 & s \end{pmatrix}
\begin{pmatrix} e^{j\theta_1} & 0 & 0 \\ 0 & e^{j\theta_2} & 0 \\ 0 & 0 & e^{j\theta_3} \end{pmatrix}
\begin{pmatrix} s & 1 & 1 \\ 1 & s & 1 \\ 1 & 1 & s \end{pmatrix}. \quad (3.144)
$$

Three outputs are obtained to be

$$
I_1 = \frac{1}{9}\left[3 + 2\cos\Delta\theta_{23} + 2\cos\left(\frac{2\pi}{3} + \Delta\theta_{12}\right) + 2\cos\left(\frac{2\pi}{3} + \Delta\theta_{13}\right)\right],
$$
$$(3.145a)$$

$$
I_2 = \frac{1}{9}\left[3 + 2\cos\Delta\theta_{12} - 2\cos\left(\frac{\pi}{3} + \Delta\theta_{23}\right) - 2\cos\left(\frac{\pi}{3} + \Delta\theta_{13}\right)\right],
$$
$$(3.145b)$$

$$
I_3 = \frac{1}{9}\left[3 + 2\cos\Delta\theta_{13} - 2\cos\left(\frac{\pi}{3} + \Delta\theta_{12}\right) - 2\cos\left(\Delta\theta_{23} - \frac{\pi}{3}\right)\right],
$$
$$(3.145c)$$

where $\Delta\theta_{ij} = \theta_i - \theta_j$. The information of phase changes can then be retrieved from the three outputs [91]. Reference [94] describes a Mach–Zehnder interferometer, which is composed of a planar coupler and a triangle 3×3 coupler, that shows good performance as a multiplexer.

3.3.4.2 Basic Characteristics of a Sagnac Loop Incorporated with a 3×3 Coupler   Similar to the conventional MZI, there exists the "dead region" of low sensitivity in Sagnac loop interferometers [58,59]. The difficulty can be overcome also by a 3×3 coupler, usually with 1:1:1 symmetric split ratio, in the configuration shown in Figure 3.38. The

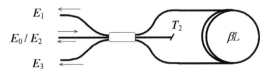

Figure 3.38 Sagnac loop incorporated with a 3×3 coupler.

source light is injected into one of three ports and the other two are used to detect reflected and interfered light waves; and a backward reflected light is generated at the input port, denoted by E_2. In the structure, any reflected light from port T_2 has to be minimized, such as by an absorbing component.

By using matrix (3.89) with $s = -e^{j\pi/3}$, the outputs are deduced as

$$
\begin{pmatrix} E_1 \\ E_2 \\ E_3 \end{pmatrix} = \frac{-1}{3} \begin{pmatrix} s & 1 & 1 \\ 1 & s & 1 \\ 1 & 1 & s \end{pmatrix} \begin{pmatrix} 0 & 0 & e^{j\varphi_{ccw}} \\ 0 & 0 & 0 \\ e^{j\varphi_{cw}} & 0 & 0 \end{pmatrix} \begin{pmatrix} s & 1 & 1 \\ 1 & s & 1 \\ 1 & 1 & s \end{pmatrix} \begin{pmatrix} 0 \\ E_0 \\ 0 \end{pmatrix}
$$

$$
= \frac{-E_0}{3} \begin{pmatrix} e^{j\varphi_{cw}} - e^{j(\varphi_{ccw}+\pi/3)} \\ e^{j\varphi_{cw}} + e^{j\varphi_{ccw}} \\ e^{j\varphi_{ccw}} - e^{j(\varphi_{cw}+\pi/3)} \end{pmatrix}. \tag{3.146}
$$

The output powers are obtained to be

$$
\begin{pmatrix} P_1 \\ P_2 \\ P_3 \end{pmatrix} = \frac{2}{9} \begin{pmatrix} 1 - \cos(\Delta\varphi + \pi/3) \\ 1 + \cos\Delta\varphi \\ 1 - \cos(\Delta\varphi - \pi/3) \end{pmatrix} P_0, \tag{3.147}
$$

where $\Delta\varphi = \varphi_{CW} - \varphi_{CCW}$ stands for the non reciprocal phase shifts in the loop. Similar to (3.143a–3.143c), the "dead region" of low sensitivity can then be eliminated.

If a planar 3×3 coupler is used in the structure of Figure 3.38, with transmission matrix (3.94) and denotation of $s = j\sqrt{2}$, the output is obtained to be

$$
\begin{pmatrix} E_1 \\ E_2 \\ E_3 \end{pmatrix} = \frac{1}{4} \begin{pmatrix} 1 & s & -1 \\ s & 1 & s \\ -1 & s & 1 \end{pmatrix} \begin{pmatrix} 0 & 0 & e^{j\varphi_{ccw}} \\ 0 & 0 & 0 \\ e^{j\varphi_{cw}} & 0 & 0 \end{pmatrix} \begin{pmatrix} 1 & s & -1 \\ s & 1 & s \\ -1 & s & 1 \end{pmatrix} \begin{pmatrix} 0 \\ E_0 \\ 0 \end{pmatrix}
$$

$$
= \frac{E_0}{4} \begin{pmatrix} j\sqrt{2}(e^{j\Delta\varphi} - 1) \\ -2(e^{j\Delta\varphi} + 1) \\ j\sqrt{2}(1 - e^{j\Delta\varphi}) \end{pmatrix} e^{j\varphi_{ccw}}. \tag{3.148}
$$

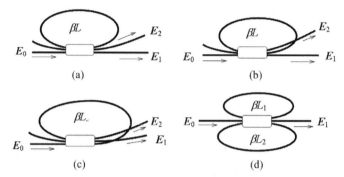

Figure 3.39   Fiber ring with a 3×3 coupler: (a)–(c) single rings with different connections and (d) double ring.

If the loop is reciprocal, $\varphi_{CW} = \varphi_{CCW}$, results of $E_1 = E_3 = 0$ and $E_2 = -E_0$ are derived, meaning that the loop is again a totally reflective mirror. However, if there exists nonreciprocal elements in the loop, two equal outputs from port 1 and from port 3 are acquired as the measurement of nonreprocity; but without the merits of (3.147), the same as a Sagnac loop with a 2×2 coupler.

The 3×3 coupler can be used to build various ring structures [95–97]. Figure 3.39 shows two typical examples: where (a), (b), and (c) are for a single ring with two outputs, and (d) for a double ring. A transmission equation for structure (a) is written as

$$
\begin{pmatrix} A \\ E_1 \\ E_2 \end{pmatrix} = M_{3\times3} \begin{pmatrix} Ae^{j\beta L} \\ 0 \\ E_0 \end{pmatrix} = \begin{pmatrix} m_{11}Ae^{j\beta L} + m_{13}E_0 \\ m_{21}Ae^{j\beta L} + m_{23}E_0 \\ m_{31}Ae^{j\beta L} + m_{33}E_0 \end{pmatrix},
\qquad (3.149)
$$

where m_{ij} are the elements of matrix $M_{3\times3}$.

The two outputs are deduced to be

$$
\frac{E_1}{E_0} = \frac{m_{23} + (m_{13}m_{21} - m_{11}m_{23})e^{j\beta L}}{1 - m_{11}e^{j\beta L}},
\qquad (3.150a)
$$

$$
\frac{E_2}{E_0} = \frac{m_{33} + (m_{13}m_{31} - m_{11}m_{33})e^{j\beta L}}{1 - m_{11}e^{j\beta L}}.
\qquad (3.150b)
$$

The transmissions are functions of the split ratio of the coupler. If a triangularly symmetric 3×3 fiber coupler is incorporated in the ring, the output intensities are deduced to be

$$\frac{I_1}{I_0} = \frac{2 - \sqrt{3}\cos\beta L + 2\sin\beta L}{4 - \sqrt{3}\cos\beta L + 3\sin\beta L}, \tag{3.151a}$$

$$\frac{I_2}{I_0} = \frac{2(1 + \sin\beta L)}{4 - \sqrt{3}\cos\beta L + 3\sin\beta L}. \tag{3.151b}$$

It is noted by the sum of two outputs that the ring is still an all-pass filter. Both outputs show resonant spectra with the same peak positions, but with different phase shifts. Figures 3.39(b) and (c) show structures with connecting ways other than that in Figure 3.39(a). The transmission behaviors are similar to that of Figure 3.39(a) with minor differences.

Double ring structures can be built by using a 3×3 coupler; an example is shown in Figure 3.39(d). Other connections by exchanging port numbers can generate another double ring with similar characteristics. Its transmission is expressed by matrix equation as

$$\begin{pmatrix} A \\ E_1 \\ B \end{pmatrix} = M \begin{pmatrix} Ae^{j\beta L_1} \\ E_0 \\ Be^{j\beta L_2} \end{pmatrix} = \begin{pmatrix} m_{11}Ae^{j\beta L_1} + m_{12}E_0 + m_{13}Be^{j\beta L_2} \\ m_{21}Ae^{j\beta L_1} + m_{22}E_0 + m_{23}Be^{j\beta L_2} \\ m_{31}Ae^{j\beta L_1} + m_{32}E_0 + m_{33}Be^{j\beta L_2} \end{pmatrix}, \tag{3.152}$$

with the solution of

$$\frac{E_1}{E_0} = \frac{m_{22} + M_{12}e^{j\beta L_1} + M_{23}e^{j\beta L_2} + (\det M)e^{j\beta(L_1+L_2)}}{1 - m_{11}e^{j\beta L_1} - m_{33}e^{j\beta L_2} - M_{13}e^{j\beta(L_1+L_2)}}, \tag{3.153}$$

with $M_{12} = m_{12}m_{21} - m_{11}m_{22}$, $M_{23} = m_{23}m_{32} - m_{22}m_{33}$, and $M_{13} = m_{13}m_{31} - m_{11}m_{33}$,

The characteristics are functions of the coupler's beam split ratios. For a planar symmetric coupler, described in (3.94), the transmission is deduced to be

$$\frac{E_1}{E_0} = -\frac{e^{j\beta L_1} + e^{j\beta L_2} - 2e^{j\beta(L_1+L_2)}}{2 - e^{j\beta L_1} - e^{j\beta L_2}}, \tag{3.154}$$

which can be simplified, by denoting $\theta = \beta(L_1 + L_2)/2$ and $\delta = \beta(L_1 - L_2)/2$, as

$$\frac{E_1}{E_0} = e^{j\theta} \frac{e^{j\theta} - \cos\delta}{1 - e^{j\theta}\cos\delta} = e^{j(\theta+\varphi)}, \qquad (3.154a)$$

with a phase factor of

$$\varphi = \tan^{-1} \frac{\sin\theta(1 - \cos^2\delta)}{\cos\theta(1 + \cos^2\delta) - 2\cos\delta}. \qquad (3.155)$$

It is shown that the ring is an all-pass filter with the phase shift different from the single ring, whose characteristics depend not only on the split ratio but also on the length difference between the two rings. The incorporation of 3×3 couplers provides more possibilities to develop various fiber loop devices, and more flexibility to adjust their specifications.

3.4 POLARIZATION CHARACTERISTICS OF FIBERS

The polarization characteristics of a fiber involve a number of physical mechanisms, and relate to various applications. Four topics are presented in this section. The polarization state evolution in fibers is discussed in Section 3.4.1, with emphasis on the mode coupling theory in twisted birefringent fibers. Basic characteristics of polarization mode dispersion (PMD) are given in Section 3.4.2. Spun fiber and circular birefringence fiber are introduced in 3.4.3. In the last subsection, magneto-optic effect and Faraday rotation are described briefly.

3.4.1 Polarization State Evolution in Fibers

It is found that the two degenerate fundamental modes LP_{01x} and LP_{01y} propagate independently without coupling, unless the fiber is twisted [13,98]. However, many inevitable factors in practice make the fiber deviate from the regular cylindrical waveguide. Intrinsically, the core may not be exactly circular; residue stress and scattering defects may exist after preform sintering and fiber drawing process; and so on. Extrinsically, in fiber cabling and field paving the fiber suffers bending, twisting, and various pressures. All these factors remove the degeneracy; resulting in coupling between the two modes and harming the performance of fiber communication systems and fiber sensors. The PMFs are thus developed. The effects of external disturbances on SMFs and PMFs are investigated widely, especially for the composite effects of various factors with time variations and spatial distribution [98–104].

The CMT is used widely in analyzing polarization characteristics [105,106]. It is also a basic method for twisted birefringent fibers [107–109], taking both the polarization rotation and the phase retard into consideration. The typical CME used to be expressed as

$$\frac{d}{dz}\begin{pmatrix} a_1 \\ a_2 \end{pmatrix} = \begin{pmatrix} j\Delta\beta/2 & \kappa_\tau \\ -\kappa_\tau & -j\Delta\beta/2 \end{pmatrix}\begin{pmatrix} a_1 \\ a_2 \end{pmatrix}, \qquad (3.156)$$

where a_1 and a_2 are the amplitudes of the two polarization modes, $\Delta\beta = \beta_2 - \beta_1$ is the difference between their propagation constants, $\kappa_\tau = n_{\text{eff}}^2 p_{44}\tau/2$ is the rate of polarization rotation induced by torsions. The CME is solved analytically, resulting in a propagation equation in the form of

$$\begin{pmatrix} a_1(z) \\ a_2(z) \end{pmatrix} = \begin{pmatrix} p & q \\ -q & p^* \end{pmatrix}\begin{pmatrix} a_1(0) \\ a_2(0) \end{pmatrix}, \qquad (3.157)$$

where the matrix components are

$$p = \cos\rho z + (j\Delta\beta/2\rho)\sin\rho z, \qquad (3.158a)$$

$$q = (\kappa/\rho)\sin\rho z, \qquad (3.158b)$$

with $\rho = \sqrt{(\Delta\beta/2)^2 + \kappa_\tau^2}$.

The polarization modes in the equations are defined on the principal axis; therefore, it is suitable for a twisted fiber under unidirectional pressure, as shown in Figure 3.40(a). However, for a twisted birefringent fiber the principal axis is rotating along with the fiber, as shown in Figure 3.40(b). Therefore, equation (3.157) does not hold for the twisted birefringent fiber.

To take the principal axis rotation into account, a direct revision adopted was to rewrite κ_τ by $(\kappa_\tau - \tau)$; and the CME was modified as [110,111]

$$\begin{pmatrix} a_1' \\ a_2' \end{pmatrix} = \begin{pmatrix} -j\Delta\beta/2 & \kappa_\tau - \tau \\ \tau - \kappa_\tau & j\Delta\beta/2 \end{pmatrix}\begin{pmatrix} a_1 \\ a_2 \end{pmatrix}. \qquad (3.159)$$

But it gives a result greatly different from (3.44) deduced in Section 3.1.4 for the case of negligible birefringence.

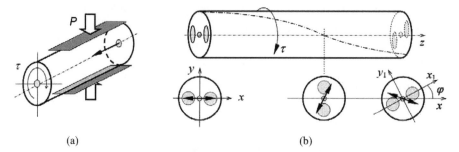

Figure 3.40   (a) A twisted fiber under lateral pressure and (b) a twisted birefringent fiber.

To overcome the contradiction, a new coupled mode equation for twisted birefringent fibers is deduced here, in which the degenerate polarization modes are taken as the eigenmodes in the unified lab coordinates, and the exact expression of dielectric constant in the strained core is used in the deduction from the basic Maxwell equation.

3.4.1.1 Dielectric Constant in the Twisted Birefringent Fiber
The coupled mode equation is deduced in the model of a twisted straight birefringent fiber, as shown in Figure 3.40(b), where a uniform twisting is considered with a constant twisting rate τ and a twisted angle of $\varphi = \tau z$. It is supposed reasonably that the normal strain and the shear strain act independently; the magnitude of birefringence is not affected by twisting, and keeps constant along the fiber. As explained in Section 2.4, the main mechanism of birefringence is the asymmetric normal strains in the fiber core, that is, the difference of normal strains $\Delta e = e_x - e_y$. As described in Section 3.1, the shear strains induced by torsion are deduced to be $e_{zx} = -\tau y$ and $e_{zy} = \tau x$; and the normal strain in z-direction e_z, and the shear strain component of e_{xy} are negligible.

The strains bring about an increment of the dielectric constant by the photoelastic effect, expressed as

$$
\varepsilon = \begin{pmatrix} \varepsilon_x & \varepsilon_{xy} & \varepsilon_{xz} \\ \varepsilon_{xy} & \varepsilon_y & \varepsilon_{yz} \\ \varepsilon_{xz} & \varepsilon_{yz} & \varepsilon_z \end{pmatrix} = \varepsilon_0 + \begin{pmatrix} p_{11}e_x + p_{12}e_y & 0 & p_{44}\tau y \\ 0 & p_{12}e_x + p_{11}e_y & -p_{44}\tau x \\ p_{44}\tau y & -p_{44}\tau x & p_{12}(e_x + e_y) \end{pmatrix}
$$

$$
= \left[\varepsilon_0 - \frac{(p_{11} + p_{12})e_\Sigma}{2}\varepsilon_0^2\right] + \varepsilon_0^2 p_{44} \begin{pmatrix} -\Delta e & 0 & \tau y \\ 0 & \Delta e & -\tau x \\ \tau y & -\tau x & e_\Sigma \end{pmatrix} = \bar{\varepsilon} + \tilde{\varepsilon}, \quad (3.160)
$$

where $e_\Sigma = e_x + e_y$, and relation of $p_{44} = (p_{11} - p_{12})/2$ is used. The dielectric constant is now a tensor, and is not homogeneous due to the spatial dependence of shear strains.

It is seen that the principal axis of birefringence rotates along with the twisting. The strains varying with the axial position z have to be converted to laboratory coordinates, as the resultant effect must be testified and utilized in the lab. The conversion between the local coordinate (x_1, y_1, z) and the lab coordinate (x, y, z) is expressed as

$$
\begin{aligned}
x &= x_1 \cos \varphi - y_1 \sin \varphi, \\
y &= x_1 \sin \varphi + y_1 \cos \varphi,
\end{aligned}
\tag{3.161}
$$

where $\varphi = \tau z$ is the principal axis rotation. By the definition of strain [6], $e_{ij} = (\partial u_i/\partial x_j + \partial u_j/\partial x_i)/2$, where u_i is the i-component of deformation vector, x_i stands for the three Cartesian axes, the strain conversion relations are deduced as (see Appendix 2)

$$
e_x = e_{x1} \cos^2 \varphi + e_{y1} \sin^2 \varphi - e_{xy1} \sin \varphi \cos \varphi, \tag{3.162a}
$$

$$
e_y = e_{x1} \sin^2 \varphi + e_{y1} \cos^2 \varphi + e_{xy1} \sin \varphi \cos \varphi, \tag{3.162b}
$$

$$
e_{xy} = e_{xy1} \cos 2\varphi + (e_{x1} - e_{y1}) \sin 2\varphi, \tag{3.162c}
$$

$$
e_{xz} = e_{xz1} \cos \varphi - e_{yz1} \sin \varphi, \tag{3.162d}
$$

$$
e_{yz} = e_{yz1} \cos \varphi + e_{xz1} \sin \varphi. \tag{3.162e}
$$

It is obtained from (3.161) that $e_{xz} = \tau y_1 \cos \varphi + \tau x_1 \sin \varphi = \tau y$ and $e_{yz} = e_{yz1} = -\tau x$, meaning that the shear strain does not change along the fiber, in agreement with the preassumption. The local shear strain is supposed to be negligible: $e_{xy1} \approx 0$ in local coordinates; the strain state in lab coordinates is expressed as

$$
e_x = e_{x1} \cos^2 \varphi + e_{y1} \sin^2 \varphi, \tag{3.163a}
$$

$$
e_y = e_{x1} \sin^2 \varphi + e_{y1} \cos^2 \varphi, \tag{3.163b}
$$

$$
e_{xy} = (e_{x1} - e_{y1}) \sin 2\varphi. \tag{3.163c}
$$

Thus the dielectric constant at position z in lab coordinates (x, y, z) is expressed as

$$
\varepsilon = \varepsilon_0 - \frac{(p_{11} + p_{12})e_\Sigma}{2} \varepsilon_0^2 + \varepsilon_0^2 p_{44}
\begin{pmatrix}
-\Delta e \cos 2\varphi & \Delta e \sin 2\varphi & \tau y \\
\Delta e \sin 2\varphi & \Delta e \cos 2\varphi & -\tau x \\
\tau y & -\tau x & e_\Sigma
\end{pmatrix}
$$

$$
= \bar{\varepsilon} + \tilde{\varepsilon}. \tag{3.164}
$$

The first two terms and the last term in (3.164) are denoted as $\bar{\varepsilon}$ and $\tilde{\varepsilon}$, respectively. This formula is a complete description of the twisted birefringent fiber.

3.4.1.2 The Coupled Mode Equation
As discussed in Section 3.1 for the torsion effect, the CME is derived to be

$$a_1' = j \langle E_1^* \hat{\Phi} E \rangle / 2\beta, \qquad (3.41a)$$

$$a_2' = j \langle E_2^* \hat{\Phi} E \rangle / 2\beta, \qquad (3.41b)$$

where $\langle E_i \hat{\Phi} E \rangle = \langle E_i \{ k^2 \tilde{\varepsilon} E + \nabla[\nabla \cdot (\tilde{\varepsilon} E)]/\bar{\varepsilon} \} \rangle$ is the coupling integral. It is necessary to notice that the zeroth-order eigenmodes are now the fundamental modes with the same propagation constant and perpendicular polarizations. The coupling integrals for the first perturbation term are deduced as

$$\langle E_2^* \tilde{\varepsilon} E_2 \rangle = -\langle E_1^* \tilde{\varepsilon} E_1 \rangle = \bar{\varepsilon}^2 p_{44} \Delta e \cos \theta, \qquad (3.165a)$$

$$\langle E_1^* \tilde{\varepsilon} E_2 \rangle = \langle E_2^* \tilde{\varepsilon} E_1 \rangle = \bar{\varepsilon}^2 p_{44} \Delta e \sin \theta. \qquad (3.165b)$$

The longitudinal component E_z is neglected in the integrals, as done in Section 3.1.

For the second perturbation term, the divergences are written as

$$\nabla \cdot (\tilde{\varepsilon} E_1) = \varepsilon_0^2 p_{44} \left[\Delta e(y \sin \theta - x \cos \theta) \frac{1}{r} \frac{\partial J}{\partial r} + j\beta \tau y J \right] e^{j\beta z}, \qquad (3.166a)$$

$$\nabla \cdot (\tilde{\varepsilon} E_2) = \varepsilon_0^2 p_{44} \left[\Delta e(y \cos \theta + x \sin \theta) \frac{1}{r} \frac{\partial J}{\partial r} - j\beta \tau x J \right] e^{j\beta z}. \qquad (3.166b)$$

It is noted that the last terms in (3.166a and 3.166b), which come from the derivative of term $\exp(i\beta z)$, contributes greatly in performing the operation of gradient; consequently the gradient has components in transverse directions, because the shear strains are linear functions of x or y. The first terms will vanish in the integrals due to the orthogonality of the Bessel functions. The integrals of the second perturbation term are then obtained as

$$\int E_2^* \cdot \nabla[\nabla \cdot (\tilde{\varepsilon} E_1)] dS = - \int E_1^* \cdot \nabla[\nabla \cdot (\tilde{\varepsilon} E_2)] dS \approx j\beta \varepsilon_0^2 p_{44} \tau, \tag{3.167}$$

$$\left| \int E_1^* \cdot \nabla[\nabla \cdot (\tilde{\varepsilon} E_1)] dS \right| = \left| \int E_2^* \cdot \nabla[\nabla \cdot (\tilde{\varepsilon} E_2)] dS \right| \approx 0. \tag{3.168}$$

The coupled mode equation is finally obtained to be

$$\begin{pmatrix} a_1' \\ a_2' \end{pmatrix} = \begin{pmatrix} -j\kappa_b \cos 2\tau z & \kappa_\tau + j\kappa_b \sin 2\tau z \\ -\kappa_\tau + j\kappa_b \sin 2\tau z & j\kappa_b \cos 2\tau z \end{pmatrix} \begin{pmatrix} a_1 \\ a_2 \end{pmatrix}, \tag{3.169}$$

where $\kappa_b = k^2 \varepsilon_0^2 p_{44} \Delta e / 2\beta \approx (n^3 p_{44} \Delta e / 2) k$ is denoted with approximation of $\bar{\varepsilon} \approx \varepsilon_0 \approx n^2 \approx n_{\text{eff}}^2$.

The CME (3.169) is compatible with the two limitation cases of pure torsion and pure normal strain. For a twisted fiber without any birefringence, the CME coincides with equation (3.44) obtained in Section 3.1.4:

$$\begin{pmatrix} a_1' \\ a_2' \end{pmatrix} = \frac{\varepsilon p_{44} \tau}{2} \begin{pmatrix} a_2 \\ -a_1 \end{pmatrix} = g\tau \begin{pmatrix} a_2 \\ -a_1 \end{pmatrix}. \tag{3.44a}$$

For a birefringent fiber without torsion ($\kappa_\tau = 0$), the equation is transformed to

$$\begin{pmatrix} a_1' \\ a_2' \end{pmatrix} = j \frac{\varepsilon^2 k}{2n_{\text{eff}}} p_{44} \Delta e \begin{pmatrix} -a_1 \\ a_2 \end{pmatrix}. \tag{3.170}$$

Its solution is expressed as

$$E_1 \propto a_1(0) \exp[j(\beta - \varepsilon^2 k^2 p_{44} \Delta e / 2\beta) z], \tag{3.171a}$$

$$E_2 \propto a_2(0) \exp[j(\beta + \varepsilon^2 k^2 p_{44} \Delta e / 2\beta) z]. \tag{3.171b}$$

The propagation constants with birefringence are written as

$$\beta_{1,2} = \beta \mp \frac{n^4 k^2 p_{44} \Delta e}{2\beta} \approx \beta \mp \kappa_b, \tag{3.171c}$$

giving a relation of $\kappa_b = \Delta\beta/2$, and the beat length of $L_B = \pi/\kappa_b = \lambda/(n^3 p_{44}\Delta e)$. Equation (3.169) is then rewritten as

$$\begin{pmatrix} a_1' \\ a_2' \end{pmatrix} = \begin{pmatrix} -j(\Delta\beta/2)\cos 2\tau z & g\tau + j(\Delta\beta/2)\sin 2\tau z \\ -g\tau + j(\Delta\beta/2)\sin 2\tau z & j(\Delta\beta/2)\cos 2\tau z \end{pmatrix} \begin{pmatrix} a_1 \\ a_2 \end{pmatrix}.$$

$$(3.172)$$

The Stokes vector and Poincaré sphere are often used to describe the polarization state. The equation of Stokes vector evolution is deduced from Equation (3.169):

$$\begin{pmatrix} S_1' \\ S_2' \\ S_3' \end{pmatrix} = 2 \begin{pmatrix} 0 & \kappa_\tau & -\kappa_b \sin 2\tau z \\ -\kappa_\tau & 0 & -\kappa_b \cos 2\tau z \\ \kappa_b \sin 2\tau z & \kappa_b \cos 2\tau z & 0 \end{pmatrix} \begin{pmatrix} S_1 \\ S_2 \\ S_3 \end{pmatrix}, \qquad (3.173)$$

where $S_1 = a_1 a_1^* - a_2 a_2^*$, $S_2 = a_1 a_2^* + a_2 a_1^*$, and $S_3 = j(a_1 a_2^* - a_2 a_1^*)$. It is proved that Equation (3.173) satisfies the condition of

$$\frac{d}{dz}(S_1^2 + S_2^2 + S_3^2) = 0, \qquad (3.174)$$

indicating that the DOP remains unchanged in the twisted birefringent fiber. Equation (3.173) is useful in numerical calculations by Runge–Kutta method or TMM.

Several features are noticed from the deduction of new CME:

1. The degenerate fundamental modes are taken as the eigenmodes and the lab coordinates are taken as a unified frame. The problem related to coordinate inconsistency is avoided.
2. The coefficients of new CME contain both contributions of linear birefringence and torsion-induced polarization mode coupling; and they coincide with the limitation cases of pure birefringence and pure torsion. There is no abrupt change from κ_τ to $\kappa_\tau - \tau$ for weak birefringence. The coefficients are functions of axial position with periodicity of $1/(2\tau)$, which reasonably corresponds to the principal axis rotated by angle $m\pi$.
3. The difference between the effects of shear strain and asymmetric normal strain is shown in the coefficients. Torsion causes coupling between the amplitudes of two degenerate modes; normal strains induce the phase difference of the polarization modes. The latter

is then a direct function of wavelength, whereas the former is basically independent of wavelength. They appear as the real part and imaginary part in the coefficients, respectively.

4. In a range between $z = z_1$ and z_2 with the interval $\Delta z = z_2 - z_1$ being small enough, the coefficients of equation (3.169) can be approximately regarded as constants; that is,

$$\begin{pmatrix} a' \\ b' \end{pmatrix} = \begin{pmatrix} -j\kappa_{b1} & \kappa_\tau + j\kappa_{b2} \\ -\kappa_\tau + j\kappa_{b2} & j\kappa_{b1} \end{pmatrix} \begin{pmatrix} a \\ b \end{pmatrix}, \qquad (3.175)$$

with $\kappa_{b1} = \kappa_b \cos 2\tau z_1$ and $\kappa_{b2} = \kappa_b \sin 2\tau z_1$. The equation can then be solved analytically, resulting in a propagation matrix of

$$\begin{pmatrix} a_1(z) \\ a_2(z) \end{pmatrix} = \begin{pmatrix} p_1 & q_1 \\ -q_1^* & p_1^* \end{pmatrix} \begin{pmatrix} a_1(z_1) \\ a_2(z_1) \end{pmatrix}, \qquad (3.176)$$

where

$$p_1 = \cos[\gamma(z - z_1)] - j\frac{\kappa_{b1}}{\gamma}\sin[\gamma(z - z_1)], \qquad (3.176a)$$

$$q_1 = \frac{\kappa_\tau + j\kappa_{b2}}{\gamma}\sin[\gamma(z - z_1)], \qquad (3.176b)$$

with parameter $\gamma = \sqrt{\kappa_\tau^2 + \kappa_b^2}$. The CME and propagation matrix (3.175) and (3.176) have similar forms to (3.156) and (3.157), but a major difference is noted: the matrix components here are periodic functions of local twisting angle τz_1, which reflects the coordinate rotation with propagation distance.

The difference between the new CME (3.173) and equations (3.156) and (3.159) can be demonstrated by numerical calculations. By the solved Stokes vector, the angle of polarization axis and the polarization extinction ratio are calculated as

$$\alpha = \frac{1}{2}\tan^{-1}\frac{S_2}{S_1}, \qquad (3.177)$$

$$R_{PE} = \sqrt{1 - S_3^2}. \qquad (3.178)$$

1. For conventional SMFs with weak birefringence, the calculated results indicate that equations (3.173) and (3.156) give similar

polarization rotation: $d\alpha/dz \simeq 0.074\tau$, whereas equation (3.159) gives much a large rotation rate near $|d\alpha/dz| \simeq \tau$, and with an opposite direction.

2. For typical polarization maintaining fibers, simulated results by Equation (3.173) show that the polarization will be rotating with a rate almost equal to the twist rate, $d\alpha/dz \simeq \tau$. Contrarily, almost no polarization rotation is obtained in simulation by equations (3.156) and (3.159). Obviously, the former result is reasonable since the principal axis of PMF is rotating along with the twisting.

3. For fibers with modest birefringence and comparable torsion, beat length of $L_B = 0.3$ m and twisting period of $L_T = 0.2$ m are taken in calculations as an example. Such a birefringence may occur in the fiber wound on a drum with diameter of about 30 mm. The polarization of the input wave is collimated at x-direction, and 1-m-long fiber is simulated. Figure 3.41 shows the output polarization rotation varied with the twist angles of the output end, where jumps from $-45°$ to $45°$ due to inverse tangents are kept in the figure to reduce its scale. It is shown that the characteristics simulated by the new CME are reasonable, in agreement with the experimental results.

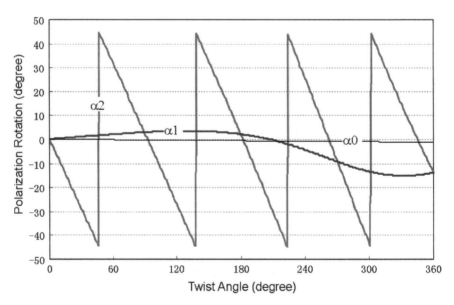

Figure 3.41 Output polarization rotation versus the twist angle of the output end: α 0 : calculated by (3.157), α 1 : by (3.159), α 2 : by (3.172).

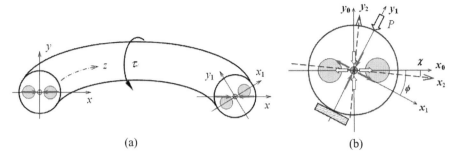

Figure 3.42   (a) A bent and twisted polarization maintaining fiber and (b) strain state composed of internal strain and lateral pressure induced strain.

3.4.1.3 *Effects of Lateral Pressure and Bending*

Fiber bending and lateral pressure applied on the fiber occur very often in practice; they are also the basic methods used to develop fiber sensors and fiber devices, such as PC. The new CME is also applicable in analyzing polarization evolutions of a twisted bent and/or pressed birefringent fiber, as shown in Figure 3.42(a). It is recognized that the fiber birefringence is caused mainly by the asymmetric normal strains in the core, no matter the normal strain is induced internally by SAP, or externally by bending and pressure. Their sum gives a composite effect for the resultant birefringence. It is not a scalar summation; their orientations may not coincide with each other as shown in Figure 3.42(b), where the empty arrows are strains induced by SAP, with the axes coincident with coordinate $x_0 - y_0$, and the solid line arrows stand for strains induced by lateral pressure, whose axes are oriented to $x_1 - y_1$. The bending-induced strains can also be depicted in the figure by the same way, as if the bending occurred in the plane of $y_1 - z$, resulting in compression strain in the same direction as the lateral pressure.

The internal strain and the externally induced strain have to be described in the same coordinate system to get their composite effect. By the conversion relations (3.162a–3.162e), the composite strains in $x_0 - y_0$ coordinates are expressed as

$$
\begin{aligned}
e_x &= e_{x0} + e_{x1} \cos^2 \phi + e_{y1} \sin^2 \phi, \\
e_y &= e_{y0} + e_{x1} \sin^2 \phi + e_{y1} \cos^2 \phi, \\
e_{xy} &= (e_{x1} - e_{y1}) \sin 2\phi.
\end{aligned}
\tag{3.179}
$$

The normal strain difference is then $\Delta e = \Delta e_0 + \Delta e_1 \cos 2\phi$, where Δe_0 is the SAP-induced strain and Δe_1 is the externally induced strain. By the coordinate conversion relation of a 2×2 tensor, the nominal shear strain e_{xy} in Equation (3.179) can be removed by rotating the co-ordinate to $x_2 - y_2$, as shown in Figure 3.42(b) by dash lines, provided that the rotation angle meets the condition of [14]

$$\tan 2\chi = \frac{2e_{xy}}{e_x - e_y} = \frac{2\Delta e_1 \sin 2\phi}{\Delta e_0 + \Delta e_1 \cos 2\phi}. \tag{3.180}$$

In coordinate $x_2 - y_2$, only normal strains appear without a shear strain term introduced by coordinate rotation. CME (3.172) is then valid to describe the polarization evolution with the normal strain difference in coordinate $x_2 - y_2$, expressed as

$$\Delta e_2 = e_{x2} - e_{y2} = \sqrt{(e_x - e_y)^2 + 4e_{xy}^2}$$

$$= \sqrt{(\Delta e_0 + \Delta e_1 \cos 2\phi)^2 + 4(\Delta e_1)^2 \sin^2 2\phi}. \tag{3.181}$$

The orientation and the amplitude of bending and pressure, and tor-sion rate as well, may vary randomly along the fiber in a practical fiber systems. The above formulas give a theoretical basis for the compos-ite strain state with different orientations and amplitudes in the unified coordinate, so long as the parameters are taken as functions of propaga-tion distance: $\Delta e_0(z)$, $\Delta e_1(z)$, $\phi(z)$, $\chi(z)$, and $\tau(z)$. On the basis of these formulas, the polarization evolution of a fiber with randomly varying strains can be calculated.

3.4.1.4 Model of Wave-Plate Concatenation and Poincaré Sphere Expression   The effects occurring in practical fiber system are more complicated; there may be factors causing mode coupling, other than the strains discussed above; and these effects are usually varying along the fiber randomly. A statistical method is presented [112] to describe the composite effect. And a model of wave-plate concatenation is pro-posed to cover various factors, in which the entire transmission matrix is expressed as a successive multiplication of birefringence phase re-tards and polarization rotations [73]:

$$\hat{T} = \prod_i \hat{B}(\delta_i)\hat{R}(\theta_i), \tag{3.182}$$

where the phase retard and polarization rotation are expressed as

$$\hat{B}(\delta_i) = \begin{pmatrix} \exp(-j\delta_i/2) & 0 \\ 0 & \exp(j\delta_i/2) \end{pmatrix}, \tag{3.183a}$$

$$\hat{R}(\theta_i) = \begin{pmatrix} \cos\theta_i & -\sin\theta_i \\ \sin\theta_i & \cos\theta_i \end{pmatrix}. \tag{3.183b}$$

Both of the matrices are unitary to meet the energy conservation condition with property $M^{-1} = M^{\dagger}$, where the superscript ($\dagger$) stands for the combined operation of complex conjugation and transposition. Fiber loss is omitted here for simplicity. It is proved that the resultant matrix $\hat{T}$ must be unitary too due to $(M_1 M_2)^{\dagger} = M_2^{\dagger} M_1^{\dagger}$. Therefore, the entirely combined effect can be regarded as a composite wave plate to link the output E_1 with input optical field E_0 as $E_1 = \hat{T} \cdot E_0$.

It is helpful to get intuitive pictures by using Poincaré sphere expression (see Appendix 3). Two vectors are introduced to describe the polarization evolution: b is the local birefringence vector, whose direction is parallel to the direction of the fast axis, and whose amplitude equals $\Delta\beta$; r stands for the polarization rotation vector with amplitude proportional to the rotation rate and direction parallel to S_3. Their combination makes the Stokes vector move on the sphere, described by differential equations [113]:

$$\frac{\partial S}{\partial z} = (b + r) \times S. \tag{3.184}$$

It is indicated that the polarization rotation makes the Stokes vector move in a circle parallel to the equator plane as shown in Figure 3.43(a), that is, the polarization axis rotates in x–y plane, while keeping the ellipticity unchanged. On the other hand, the birefringence causes the wave to periodically change in a circle on the sphere, as shown in Figure 3.43(b), unless the original Stokes vector is parallel (or antiparallel) to b. When both birefringence and rotation exist, the Stokes vector will move in a way shown in Figure 3.43(c) for constant b and r; and in an irregular way shown in Figure 3.43(d) for varying b and varying r.

The polarization maintaining fiber is a high birefringent fiber, as introduced in Section 2.4.2. It is indicated that the polarization direction can be maintained if the input polarization coincides with the principal axis of the PMF. However, the external conditions in practical fiber systems will make the fiber deviate from the regular state, leading to variations of the polarization and to serious impairment to sensors and measurement setups based on optical interference. Reference [113]

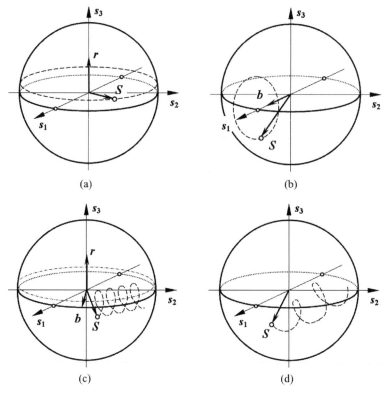

Figure 3.43 Stokes vector evolutions depictured on Poincaré sphere: (a) with a pure optical activity; (b) with pure birefringence; (c) with combination of two effects; and (d) with varying effects.

gives basic theoretical analyses and experimental measurements on the properties.

Apart from the polarization-dependent phase shift, the polarization-dependent loss (PDL) and/or the polarization-dependent gain (PDG) are also important effects, which are attributed mostly to particular device structures and mechanisms. Because the polarization characteristics of fibers and devices play key roles both in fiber communications and fiber sensors, it is of great importance to measure and to control the polarization state precisely. For this purpose, numerous polarization devices have been developed, which will be explained in Section 3.5.

3.4.2 Basic Characteristics of Polarization Mode Dispersion

The two polarization modes in conventional SMFs can be used to carry two signal channels if no coupling between them occurs this is

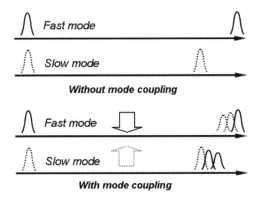

Figure 3.44 Conceptional picture of PMD.

utilized to develop the polarization multiplexing technology. But the cross polarization coupling exists inevitably, which has to be compensated for by polarization multiplexing. The coupling also causes signal pulse-width broadening, that is, the differential group delay (DGD), resulting in limitation of data speed and/or propagation distance. Such an effect is termed the polarization mode dispersion (PMD). Figure 3.44 gives an intuitive picture of PMD induced inter-symbol interference (ISI).

The PMD in SMFs comes from various irregularities, which are random spatially and vary temporally. As analyzed above the effects can be regarded as a concatenation of distributed phase retards and polarization rotations, and summed as a composite wave plate, if the loss can be ignored. This fact is the basis of the principal state of polarization (PSP), that is, the principal axes exist in any fiber system, at which no DGD is measured [73,113–115]. This is one of the reasons why PMD impairment can be compensated in some degree.

The spectral line-width of a propagating optical wave has not yet been considered in analyzing polarization evolution in Section 3.4.1, implying that the input is a continuous wave (cw) single-frequency light wave. For short pulse propagations, the spectral property of the propagating beam must be taken into consideration. By introducing a DGD vector τ_D, whose direction is parallel to the PSP, and whose amplitude equals DGD, the evolution of the Stokes vector obeys a differential equation of [114]

$$\frac{\partial S}{\partial \omega} = \tau_D \times S. \tag{3.185}$$

If the Stokes vector is parallel to τ_D, its derivative over frequency goes to zero, which is self-coincident with the concept of PSP. The

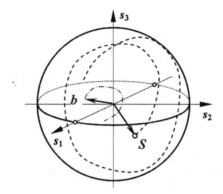

Figure 3.45   Stokes vector variation on Poincaré sphere caused by PMD.

DGD vector varies along the fiber length, described by [115]

$$\frac{\partial \boldsymbol{\tau}_D}{\partial z} = \frac{\partial \boldsymbol{b}}{\partial \omega} + \boldsymbol{b} \times \boldsymbol{\tau}_D, \tag{3.186}$$

where the birefringence vector $\boldsymbol{b}(z)$ contains its direction evolution. Equations (3.185) and (3.186) describe the Stokes vector moving on the Poincaré sphere, usually in a random way as shown in Figure 3.45.

Because PMD is a combined effect occurring randomly along the fiber, the PMD-induced DGD is a parameter with a statistical distribution. It is deduced theoretically and verified experimentally that the second momentum of DGD is expressed as [115,116]

$$\langle \tau_D^2 \rangle = \frac{1}{2(h\bar{v})^2} (2hL - 1 + e^{-2hL}), \tag{3.187}$$

where $\bar{v} = (\overline{d\Delta\beta/d\omega})^{-1}$ is the averaged group velocity difference of the two polarization modes; $h = P^{-1}dP/dz$ stands for the strength of cross-polarization coupling, where P is the mode power, dP/dz is the coupling rate to the other polarization mode. h^{-1} is a statistically averaged coupling length between the two modes, that is, the probability of cross-coupling in a distance of l is averaged to be hl. The probability density function of cross coupling is basically of Gaussian type.

It is seen that the DGD is such a function of propagation distance that it is linear for shorter propagation distance, whereas is proportional to the square root of the length for longer propagation:

$$\tau_{\text{rms}} = \sqrt{\langle \tau_D^2 \rangle} \approx \begin{cases} \sqrt{2}L/\bar{v} & (L \ll h^{-1}) \\ \sqrt{L/h}/\bar{v} & (L \gg h^{-1}) \end{cases}. \tag{3.188}$$

For optical fiber telecommunications the propagation distance is usually much longer than the mode coupling length; the PMD performance of a fiber is specified by a parameter with unit of $ps/\sqrt{km}$. The PMD issue is regarded as the last limitation of high-speed long-distance optical fiber telecommunications, and has attracted a lot of research interest. The fiber length in fiber sensor systems is usually much shorter than that in fiber communications, and group time delay is not a critical issue. However, interference detection is the most widely used technology in sensors, which is more sensitive than pulse energy detection in pulse code modulation (PCM) technology of conventional optical communication systems. Moreover, long-term stability and reliability are needed in sensors. Therefore, the polarization evolution effect is still an important factor; the theoretical method used in analyzing PMD is helpful in fiber sensor research and development.

3.4.3 Spun Fiber and Circular Birefringence Fiber

To reduce the influence of PMD of fibers, it is necessary to eliminate firstly the intrinsic noncircularity and residue stresses. An effective method is to twist the fiber in its drawing processing at high temperature with very high rotation speed. Any noncylindrically symmetric factors are smoothed and eliminated. Its effect is equivalent to decreasing the mode coupling length as much as possible, resulting in $\tau_{rms} = \sqrt{L/h}/\bar{v} \to 0$. Fibers fabricated by this technology are called *spun fibers* [117–119]. By this technology and other improvements, the PMD specification of manufactured fibers has been optimized to $0.1ps/\sqrt{km}$ or below.

It is necessary to maintain the circular polarization state in some fiber sensor applications, such as the current sensors by Faraday effect, where the physical quantity to be detected and measured is the polarization rotation angle induced by magnetic fields. Therefore, the extra polarization rotation induced by external disturbances has to be reduced as much as possible. It is proposed that circular polarization-maintaining fibers will help to maintain polarization stability against the usual external disturbances. A linearly polarized wave can be decomposed into two circularly polarized waves in the frame of a Jones vector:

$$E = \begin{pmatrix} A \\ Be^{j\varphi} \end{pmatrix} = \frac{A - jBe^{j\varphi}}{2} \begin{pmatrix} 1 \\ j \end{pmatrix} + \frac{A + jBe^{j\varphi}}{2} \begin{pmatrix} 1 \\ -j \end{pmatrix}, \qquad (3.189)$$

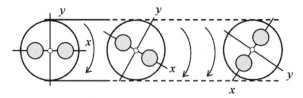

Figure 3.46   Structure of a circular birefringence fiber.

where vector $(1, j)^T / \sqrt{2}$ is the right-handed circularly polarized wave, and $(1, -j)^T / \sqrt{2}$ is the left one. NB: the phase factor φ is with ωt here, not with kz to keep coincident with the conventional observation in the time domain [19,20]. A pure rotation matrix will not change the circular polarization state besides a common phase factor, shown as

$$E = \begin{pmatrix} \cos\theta & \sin\theta \\ -\sin\theta & \cos\theta \end{pmatrix} \frac{E_\pm}{\sqrt{2}} \begin{pmatrix} 1 \\ \pm j \end{pmatrix} = \frac{E_\pm}{\sqrt{2}} \begin{pmatrix} 1 \\ \pm j \end{pmatrix} e^{\pm j\theta}. \qquad (3.190)$$

Polarization rotation can be induced by twisting fibers with some mechanical rotator. This is a postdraw method with the disadvantages of poor controllability and low stability, and need of inconvenient special mechanics. A special fiber, different from the conventional SMF and the conventional PMF, is then invented by twisting a high birefringence fiber during its drawing processing at high temperature (with low viscosity) by a high-speed rotator. A rotating birefringence induced by SAP is built in the structure, whereas the shear strain is released at high temperature. Such a fiber is called *circular birefringence fiber*, or *circular PMF*, or *polarization transforming fiber* by its main characteristics [107–109]. Figure 3.46 shows schematically the structure of a circular birefringence fiber.

The amplitude of birefringence here is basically regarded as constant along the fiber, but its direction is rotating with a controllable rate. It could be a constant rotation; could be minus for left rotation, or positive for right rotation; could vary from a very high rate to very low near zero, or vice versa; the rate variation could be a linear or nonlinear function of distance z, modeled by some forms. The built-in torsion can be formed also in a short section of postdraw *hi-bi* fiber by local heating and twisting processes [109].

As the twisting can be either right-handed or left-handed, the behavior of a section of circular birefringence is not bidirectional. This is an attractive difference between the uniform torsion-induced rotation and the conventional wave plate based on birefringence crystals. The former shares similar properties with the natural optical activity

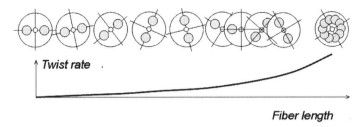

Figure 3.47   Circular birefringence fiber with an ascend twist rate.

materials, whereas the latter is bidirectional. The performance of a circular birefringence fiber can then be tailored by designing the twisting rate profile, to improve its temperature dependence and wavelength dependence. Figure 3.47 shows schematically a circular birefringence fiber fabricated with an ascending twist rate [109].

3.4.4 Faraday Rotation and Optical Activity

Because the movement of electrons is affected not only by an externally applied electric field, but also by a magnetic field, described by the Lorentzian force $F = -e(E + v \times B)$, which drives electron movement in a helical way, the electric polarization becomes a tensor, resulting in a difference of refractive indexes between the left-handed and right-handed circularly polarized waves. Therefore, the polarization of an optical wave will rotate if its propagation is parallel to the magnetic field. This effect was discovered by Michael Faraday in 1845. Faraday rotation is proportional to the magnetic field B, expressed as

$$\frac{\mathrm{d}\varphi}{\mathrm{d}z} = VB, \tag{3.191}$$

where φ is the azimuth angle, and V is called the Verdet constant. The Faraday effect is widely used in magnetic sensors, electric current sensors, and magneto-optic devices; and also as a powerful tool in the research of materials [120–122]. The Verdet constant of fused silica is quite small, compared with many magneto-optic materials. Nevertheless, fiber Faraday current sensors are attractive, because a fiber section long enough can be used. Many technical issues have to be considered for the sensor, which will be discussed in Chapter 6.

The polarization rotation is also observed in some crystals and liquids, such as the quartz crystal and glucose, termed optical activity. Such materials have chiral structures, or contain chiral molecules, showing different refractive indexes for the left-handed and right-handed circularly polarized waves [20]:

$$n_\pm = \sqrt{n^2 \pm G} \approx n \pm G/2n. \tag{3.192}$$

The polarization rotation is expressed as

$$\varphi = \frac{\pi}{\lambda}(n_- - n_+)z \approx -\frac{\pi G}{n\lambda}z. \tag{3.193}$$

Parameter G is a constant of the material, and dependent on concentration if it is a solution. It can be either positive or negative, corresponding to dextrorotation or levorotation, respectively. The optical activity is a natural property of material, independent of propagation direction, similar to the torsion-induced polarization rotation; but the rate of natural polarization rotation depends on the wavelength, whereas the rate of torsion-induced rotation depends on twist rate and the photoelastic coefficient, independent of the wavelength.

It is helpful to compare the polarization rotation in optical paths with different configurations and mechanisms: (1) with a wave plate, or a section of *hi-bi* fiber; (2) with a natural optical activity; (3) with torsion-induced rotation; (4) with Faraday rotation. Figure 3.48(a) shows the change of polarization state in a forward propagation and in a round-trip way propagation by phase retard of $\delta = \pi/2$. The output beam is generally an elliptically polarized wave. The reflected beam returns to a linearly polarized wave with angle deviation of 2φ. Figure 3.48(b) is for π retardation: the output is rotated $90°$; and the beam is reflected back to the input direction.

Figure 3.48(c) shows polarization rotations in forward and round-trip propagations passing through a material with natural optical activity. The polarization rotation caused by torsion-induced activity in SMFs has the same features, as analyzed in Section 3.1.4. The polarization azimuth does not change after a round-trip propagation because the activity changes its sign in the backward propagation.

Figure 3.48(d) shows the features of Faraday rotation, which depends not only on the amplitude of the magnetic field but also on its direction; in other words, if the rotation is right-handed for the forward wave, it is left-handed for the backward wave, because the projection of the magnetic field on the wave vector changes its sign. The rotation angle for the single trip is $\Delta\varphi = VBL$; it is doubled for the round trip propagation.

The input polarization depicted in Figure 3.48 is a linearly polarized wave for simplicity. It is not difficult to deduce the conclusions for

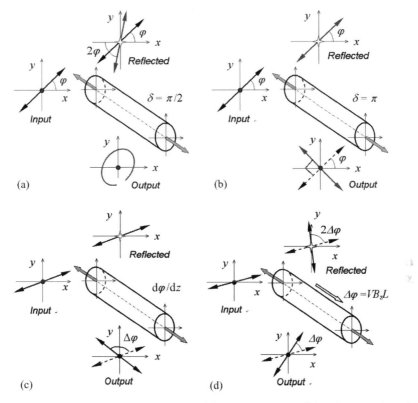

Figure 3.48 Polarization state change by (a) $\lambda/4$ wave plate; (b) $\lambda/2$ wave plate; (c) natural optical activity and torsion-induced activity; and (d) Faraday rotation (solid circles stand for the head of wave vector arrow, stars for its tail).

general cases with an elliptically polarized input wave. Some important features should be noticed:

1. The polarization evolution in wave plates and *hi-bi* fibers depends on the input azimuth to their principal axes, while no axis exists in the natural optical activity, torsion-induced activity, and Faraday rotation. That is to say, the vector of optical activity and Faraday rotation in the Stokes space is parallel to the axis of S_3, while the phase retardation vector is in S_1–S_2 plane.

2. The coefficients of torsion-induced rotation and Faraday rotation are not results of interference, and thus usually have weak wavelength dependence; while the polarization evolutions in wave plates, *hi-bi* fibers, and natural activity are attributed to the vectorial sum of two polarization modes, showing strong wavelength dependence.

3. The effects in wave plates, natural activities, and torsion-induced rotations are reciprocal, while Faraday effect is not reciprocal, because of its dependence on the direction of the magnetic field.

It is seen that if the input azimuth is adjusted at 45° to the λ/4 waveplate axis, the single trip output is circularly polarized, and the polarization of the reflected beam rotates 90°, that is, perpendicular to the input. Similarly, if Faraday rotation is adjusted to get $\Delta\varphi = \pi/4$, the polarization of the reflected wave is perpendicular to the input. These facts are used for optical isolation.

Another implied assumption should be mentioned that the reflection at the output end surface does not change the polarization state as in the usual cases. The possible exception is the phase conjugation, which is beyond the scope of this book.

3.5 FIBER POLARIZATION DEVICES

This section introduces several polarization devices used widely in fiber technologies, including the fiber polarizer, polarization controller, depolarizer and polarization scrambler, and optical isolators and optical circulators. Some are made of the fiber based on its polarization-dependent properties, called intrinsic fiber devices; others incorporate materials other than the fiber, such as birefringence crystals, but with fiber pigtails for connecting with the systems, called extrinsic fiber devices. Quite a number of devices are commercially available. This section gives just a brief description.

3.5.1 Fiber Polarizers

Quite a number of fiber polarizers are developed; some of them are introduced here as examples.

D-shaped fiber polarizer. The device is partly similar to the D-shaped fiber coupler as shown in Figure 3.8. The difference is that a metal film is coated on the polished surface, which is near the core, as shown in Figure 3.49 [123–125]. The basic mechanism of the polarizer is the PDL, induced by the metal coating. According to the well-known Fresnel formulae, the reflectivity at the interface between two dielectric mediums is polarization dependent. The light beam polarized inside the incident plane, called TM wave or p-component, has lower reflectivity than the TE wave (s-component), and is polarized perpendicular to the incident plane. At a certain incident angle, termed Brewster's angle, the

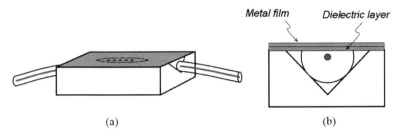

Figure 3.49   (a) Surface polariton D-shaped fiber polarizer and (b) its cross section.

reflectivity of the TM wave reaches zero. This property is used to make a polarizer with only a stack of glass plates. If the second medium is a metal layer, the refracted wave will suffer absorption; and the TM wave is absorbed more strongly than the TE wave. It is the basic mechanism of the polarizer.

The basic structure of the polarizer is regarded as a six-layer waveguide: the fiber core, the thinned cladding, a buffer layer, a metal layer, a top coating, and the air region, as shown in Figure 3.50(a). The structure is regarded as a doubled waveguide, composed of the fiber core and the metal layer, separated by the buffer (or the cladding) and the top coating (or the air). It is shown [126,127] that the buffer layer and the top coating are beneficial for enhancing the coupling between the fiber core and the metal layer, and thus for higher polarization extinction. The structure is simplified as a three-region waveguide for the theoretical analysis, with the metal layer in the middle, as shown in Figure 3.50(b).

The metal, such as gold and aluminum, has a complex refractive index, usually with a large imaginary. The properties of the interface between metal and dielectric media have been analyzed in detail in [19]. The index of metal is denoted as $\hat{n}_2 = n_2 + jn_{2i} = n_2(1 + j\kappa)$. According to Snell's law, $n_2(1 + i\kappa)\sin\theta_t = n_1\sin\theta_1$, meaning that the

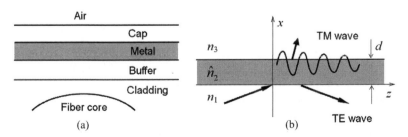

Figure 3.50   (a) Waveguide model of D-shaped polarizer and (b) polarizations.

refractive angle θ_t is now a complex variable. Reflections, expressed as $r_p = \tan(\theta_1 - \theta_t)/\tan(\theta_1 + \theta_t)$ and $r_s = -\sin(\theta_1 - \theta_t)/\sin(\theta_1 + \theta_t)$, for the p- and s-components respectively, are also complex, with different phase factors for the two components. For the three-region film, the composite reflection is expressed as

$$r_{123} = \frac{r_{12} + r_{23}\exp(i2\beta_x d)}{1 + r_{12}r_{23}\exp(i\beta_x d)}, \tag{3.194}$$

where $\beta_x = 2\pi n_2 \cos i_2/\lambda$. Consequently the composite reflections are complex and there is a difference between the two polarizations. In the limitation of infinitive conductivity of the metal, the s-component will be totally reflected, whereas the p-component penetrates into the metal layer and diminishes, as shown in Figure 3.50(b).

The polarization–suppression ratio depends not only on the different loss coefficients, but also on the effective length of the D-shaped waveguide [125]. Assuming the s-component suffers no loss except that due to mode conversion from s- to p-component, denoting as α_s, and denoting the loss coefficients of p-component as α_p, the suppression ratio S is obtained as

$$S = \frac{I_s}{I_p} = \left\{\frac{\alpha_s}{\alpha_p - \alpha_s} + \frac{\alpha_p - 2\alpha_s}{\alpha_p - \alpha_s}\exp[-(\alpha_p - \alpha_s)L]\right\}^{-1}. \tag{3.195}$$

It is obtained that a saturated suppression ratio reaches $S_{\text{sat}} = (\alpha_p - \alpha_s)/\alpha_s$ for $L \gg 1/\alpha_p$. Two approximated expressions are derived: $S \approx [\alpha_s/\alpha_p + \exp(-\alpha_p L)]^{-1}$ for $\alpha_p \gg \alpha_s$; and $S \approx [1 - (\alpha_p - 2\alpha_s)L]^{-1}$ for small length L.

Polarization split fiber coupler. The typical device is the PMF coupler as described in Figure 3.15. A distinct merit is that it can output two perpendicular polarization components simultaneously. However, it is necessary to understand that its characteristics are usually strong functions of working wavelengths, based on the interference essentials of a waveguide coupler.

Another scheme of polarizer is based on a null taper coupler [44], which was twisted through 45° after fusing and packaged with twisting kept. The polarization effect is due to the excitation of hybrid second modes in the coupler waist.

Polarizer with micro-optic components. Relying on the development of micro-optic components, polarizers are realized with small-size birefringence crystals, such as YVO_4, or dielectric films, packaged with other auxiliary components inside, and two fiber pigtails.

High birefringence (hi-bi) coiled fiber polarizer. On the basis of the bending losses a coiled birefringence fiber polarizer is developed [128]. The PDL is attributed to a coupling between the fundamental mode and so-called *whispering gallery* modes, which occurs at the curved cladding/coating interface. It is noticed that the PDL is basically a broadband effect, but still with certain wavelength dependence due to the mechanisms of the whispering gallery modes.

Tilted fiber Bragg grating (FBG) polarizer. The tilted fiber Bragg grating can be regarded as a shrinked-scale stack of glass plates with different reflections between *s*- and *p*-components near Brewster's Angle. Its principle and fabrication will be introduced in Section 4.3.

3.5.2 Fiber Polarization Controller

It is surely necessary in fiber systems and fiber devices to convert the polarization state of the input beam to a wanted state, which is just the task of a PC. Polarization conversion in a bulky optical system is usually realized by a $\lambda/2$ wave plate sandwiched between two $\lambda/4$ wave plates, and the three wave plates are rotatable to change their relative azimuths.

Such a configuration can be built by fiber components. Figure 3.51 shows a typical fiber polarization controller [129], in which the fiber is wound on three rotatable drums with different turns. The bending-induced birefringence occurs in the wounded fiber; and the phase retard is expressed as

$$\delta = Bk_0L = (p_{12} - p_{11})(1 + v)\frac{N\pi^2 n^3 d^2}{4\lambda R}, \qquad (3.196)$$

where N is the wound turns on the drum and R its radius. Similar to the polarization controller in bulky optics, the retards of three drums are set to $\pi/2$, π, and $\pi/2$, respectively, equivalent to the $\lambda/4$, $\lambda/2$, and $\lambda/4$ wave plates. The radius of the drum for $\pi/2$ retardence and $N = 2$

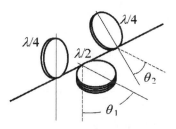

Figure 3.51 Fiber polarization controller based on bending-induced birefringence.

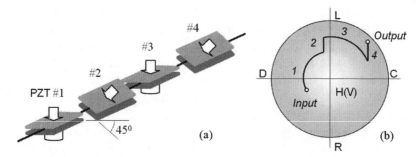

Figure 3.52 Fiber PC by (a) lateral pressures and (b) polarization state evolution.

is estimated to be about $R = 17$ mm. By stretching the fiber the retard can be increased and adjusted, as analyzed in Section 3.1.3. The azimuth between wave plates is realized by fiber twisting between the drums. The rotation in this PC structure is operated manually.

Figure 3.52(a) shows another PC scheme composed of four squeezers with $45°$ azimuth in between [130]. The squeezers are driven by PZT. By referring to formula (3.14), the lateral pressure-induced birefringence generates phase retardence of

$$\delta = BkL = \frac{8n^2}{\lambda Y d}(p_{12} - p_{11})(1 + v)PL, \qquad (3.197)$$

where L is the length of fiber section pressed laterally, and $F = PL$ is the pressure. Controllable retards are generated by the applied voltage. Figure 3.52(b) illustrates the procedure of adjusting in the Poincaré sphere, which is viewed from the H-V axis. Squeezer #1 and #3 makes the Stokes vector rotate round the H-V axis, whereas squeezer #2 and #4 makes it rotate round the CD axis, so that any wanted polarization state can be obtained from an arbitrary input state by four operations 1-2-3-4, so long as the maximum retard of the squeezers is enough. Faraday rotation caused by magnetic field applied codirectionally with the fiber is also utilized to adjust the azimuths in some other schemes [131].

3.5.3 Fiber Depolarizer and Polarization Scrambler

Most of the laser sources are polarized to a certain degree, causing various polarization-dependent effects, which may bring about undesired and harmful influences in communication and sensing applications, such as PDL, PDG, PMD, polarization-induced fading, polarization-induced noise, polarization hole burning in EDFA. Therefore, depolarizers are needed to convert the source light with a high degree of

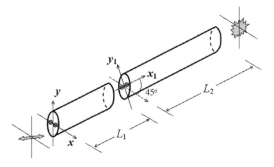

Figure 3.53 Schematic diagram of a Lyot filter-based depolarizer.

polarization (DOP) to that with low DOP as much as possible. As pointed by the general theory [132], DOP can not be reduced easily by conventional polarization components, such as wave plates, polarization rotators, especially for usual laser sources with narrow spectral line width; special structures have to be designed for depolarizers.

Lyot filter-based depolarizer. Similar to that in bulky component optical systems, a Lyot filter with sections of *hi-bi* fiber is a typical depolarizer, as shown in Figure 3.53, where the gap between the two sections is artificially enlarged just for clearness [133,134].

In case the azimuthal between the two sections is 45°, the transmission is deduced as

$$E = \begin{pmatrix} e^{j\delta_1/2}\cos\alpha\cos(\delta_2/2) - j\sin(\delta_2/2)e^{-j\delta_1/2}\sin\alpha \\ e^{-j\delta_1/2}\sin\alpha\cos(\delta_2/2) - j\sin(\delta_2/2)e^{j\delta_1/2}\cos\alpha \end{pmatrix}, \qquad (3.198)$$

where $\delta_1 = \Delta\beta L_1$ and $\delta_2 = \Delta\beta L_2$ are the phase retards of the two sections of hi-bi fiber; the input is a linear polarized wave with Jones vector of $(\cos\alpha \quad \sin\alpha)^T$. The components of output Stokes vector are then expressed as

$$s_1 = \cos 2\alpha \cos \delta_2 - \sin \delta_2 \sin 2\alpha \sin \delta_1,$$
$$s_2 = \sin 2\alpha \cos(\varphi - \delta_1), \qquad (3.199)$$
$$s_3 = \sin \delta_2 \cos 2\alpha + \cos \delta_2 \sin 2\alpha \sin \delta_1.$$

It is deduced from (3.199) that the DOP of the output is unity for an ideal single-frequency wave. For a practical optical wave with a finite spectral line width, the above expressions have to be averaged over its spectral profile $f(\omega - \omega_0)$, such as Gaussian, Lorentzian, and

Hyperbolic secant; and a coherence degree is defined as

$$\gamma(L) = \int_0^\infty f(\omega - \omega_0)\cos(\omega\tau_g)d\omega, \qquad (3.200)$$

with the group delay $\tau_g = L(d\Delta\beta/d\omega) = [\Delta n_{\text{eff}} + \omega(d\Delta n_{\text{eff}}/d\omega)]L/c$, where the birefringence and its dispersion are included. For Gaussian profile, it is deduced to be $\gamma(L) = \exp[-(\tau_g\Delta\omega)^2/4]$, where $\Delta\omega$ is the $1/e$ line width. The length ratio of the two sections, $L_2 = 2L_1$, is usually adopted in a practical depolarizer. By neglecting $\gamma(L_2)$ and $\gamma(L_1 + L_2)$, the DOP of output can be deduced approximately to be

$$\text{DOP} = \gamma(L_1)\left(\frac{3}{4} + \frac{\cos 2\delta_1}{2}\right)\sin 2\alpha. \qquad (3.201)$$

It decreases greatly with the coherence degree. Therefore, the fiber section length must be much longer than the coherence length of the input light signal for an effective depolarizer. Experimentally by using two sections of PMF with lengths of 0.5 m and 1 m and line width of a few nanometers, the DOP can be decreased down to $10^{-2} \sim 10^{-3}$ [133].

Cascaded fiber ring. As discussed in Section 3.3, when a light beam passes through a fiber ring its polarization state will be changed if some polarization-related effects exist in the ring intentionally or unintentionally. Similar to Lyot depolarizer, when the length of fiber ring is longer than the coherence length of the input light wave, its DOP will decrease. The case is equivalent to that passing through a Lyot filter recirculatory.

The depolarization of cascaded rings can be simulated based on the analysis for a single ring, as given in Section 3.3. Theoretical analyses and simulations show that the depolarization depends not only on the polarization effects inside the fiber, but also on the split ratio and polarization dependence of the coupler [85,133,134]. It is obvious that more rings can be used to reduce the DOP further [134–138]. Several configurations are proposed and developed. Figure 3.54(a) depicts a depolarizer with two rings cascaded, with an intentionally inserted polarization controller (PC), or a squeezer, and a twister for adjustable birefringence and polarization rotation. Figure 3.54(b) shows another structure incorporated with a PMF coupler and an orthogonal splice of PMF in the ring [138].

There are more depolarizers published in journals by utilizing other mechanisms, such as Mach–Zehnder interferometers incorporated with PMF, the magneto-optic effect, and special processing technology [139,140].

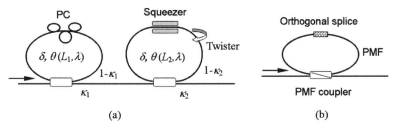

Figure 3.54 Fiber ring depolarizers: (a) cascaded rings and (b) PMF ring with orthogonal splice.

Polarization Scrambler. It is noticed that depolarizers are suitable for optical sources with low coherence. In case of a highly coherent source, the size of the above devices would be extended beyond the realistic range. Moreover some applications require sources with high coherence, such as in measurement and sensing applications by optical interferometry. Therefore, a special device, PS, is needed, which converts a linearly polarized coherent wave with DOP $\approx$ 1 into a wave with a time-averaged DOP $\ll$ 1, but with its coherence kept unchanged. The basic concept of PS is to introduce a temporally varying phase delay between the two perpendicular polarization modes, and make the Stokes vector travel on the Poincaré sphere quickly, resulting in reduction of the time averaged DOP. It is noticed that PS is an active device generally, whereas the depolarizer is passive.

Several schemes have been proposed and developed for the purpose. Figure 3.55 shows two examples. Figure 3.55(a) is composed of rotating wave plates driven mechanically. It can be regarded as a PC continuously varying in a wide range, covering the whole Poincaré sphere. Another structure is composed of a fiber coil fixed on a PZT cylinder driven electrically, which is also a varying phase retarder to make the state of polarization travel diversely. Generally, such structures have lower modulating speeds, and are suitable for slower signals.

Figure 3.55(b) shows another PS scheme made of electro-optic waveguide phase modulators [141,142], which has much higher modulating

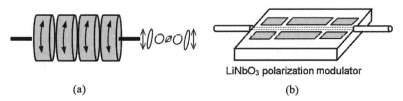

Figure 3.55 Structures of PS by using (a) rotating wave plates and (b) electro-optic polarization modulator.

speed, up to several gigahertz, suitable for high-speed signals. Faraday rotation is also utilized for the purpose [143]. The temporal response is one of the important performance indicators of PS, which should match with the characteristic time of related polarization-dependent effects. Other performance indicators include insertion loss, residue DOP, and its dependence on wavelength.

3.5.4 Fiber Optical Isolator and Circulator

It is no doubt that the OI is an important component in optical systems. For fiber systems, isolators with fiber pigtails as input and output ports are absolutely necessary to prevent undesired effects from impairing the transported optical signals, or sensed information. It is often set up at the output port of laser sources to cut off any returned optical wave that makes the laser work unstably. Basically, two effects are mostly used to develop isolators: Faraday rotation and birefringence, as discussed in Section 3.4.4.

The principle of Faraday isolation is shown in Figure 3.56(a), where the analyzer blocks the backward wave as its polarization is rotated $90°$. It is noticed that the transmission of this isolator depends on the input polarization; it is usually used for light waves with a known and constant input polarization, such as at the output of a laser. Practical applications require polarization-independent isolators, which allows inputs with any polarization direction to pass through without PDLs, and isolates the return wave without any polarization. Figure 3.56(b) shows such an isolator, which uses two birefringent wedges to split and combine the different polarization waves, and makes the backward wave transversely displaced and thus unable to be coupled into the input fiber pigtail by the same collimator. The different lines and symbols of the empty circle and short bar in the figure stand for the two

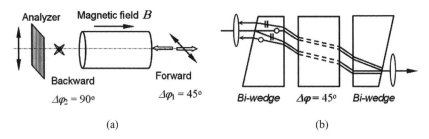

Figure 3.56 (a) Optical isolation by Faraday effect; A: analyzer, B: magnetic field. (b) Isolator by polarization beam split.

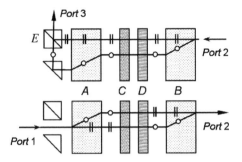

Figure 3.57 Optical circulator. *A* and *B*: birefringent walk-off block; *C*: Faraday rotator; *D*: phase retarder; and *E*: polarization beam combiner.

polarization directions. The middle block is a nonreciprocal element, mostly a Faraday rotator, to rotate the polarization direction 45° for a one way trip and 90° for a round trip. Details of design and technical issues in fabrication are not discussed in this book; interested readers can find related references in journals.

The OC is another useful device both in fiber systems and in bulk optics. It makes use of the backward wave in isolators as a third output. Its fundamental mechanism is the polarization beam splitting and polarization conversion, similar to the isolator. Figure 3.57 shows a schematic diagram of a three port circulator, where small circles and a short bar stand for the two perpendicular polarizations. On the basis of the same principle multiple port circulators are developed [144–146].

Another polarization device is the Faraday rotation mirror (FRM), which reflects the input optical beam, and rotates its polarization direction 90° simultaneously. Its Jones matrix is written as

$$M_{\text{FRM}} = \begin{pmatrix} 0 & 1 \\ 1 & 0 \end{pmatrix}. \tag{3.202}$$

FRM is useful in building fiber systems, especially in fiber interferometers, to eliminate the external disturbance with polarization dependence. Figure 3.58 gives an example to show the role of FRM in a fiber

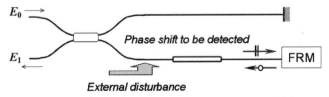

Figure 3.58 Fiber Michelson interferometer incorporated with an FRM.

Michelson interferometer, where the round-trip path cancels out the external polarization disturbance at the coupler, while leaving only the phase shift sensed.

The isolator, circulator, and FRM introduced here are called extrinsic fiber devices since the key components are not made of fiber. Nevertheless, all of them are packaged with fiber pigtails for easy connection with fiber systems. In the packaging microlenses are used to focus the light into the fiber core, and vice versa. The GRIN lens introduced in Section 2.3.4 is one of the most widely used components.

PROBLEMS

3.1 Describe the elastic properties and photoelastic effect of an isotropic medium. Derive the axial strain-induced index increment (3.5).

3.2 What is the difference between strains under a radial pressure and a diametrically unidirectional pressure? Qualitatively explain the birefringence induced by the latter.

3.3 Explain the bending-induced birefringence. Calculate the phase retard of a section of conventional SMF wound one turn on a cylinder with 3 cm diameter.

3.4 Derive the CME (3.41a and 3.41b) for the twisted fiber, explain how the torsion induces the coupling between the two polarization modes. Why does a twisted fiber have similar properties to an anisotropic medium?

3.5 What are the differences between torsion-induced polarization rotation in SMF and the polarization rotation in PMF?

3.6 Qualitatively explain bending-induced loss. Give an idea of designing a fiber sensor based on the bending loss.

3.7 Derive the transmission functions of the coupled waveguides (3.68a–3.68c).

3.8 Explain the three main applications of the fiber coupler respectively: (1) the beam splitter; (2) the beam combiner; and (3) the WDM. Where does the $\pi/2$ phase shift between the two outputs of the coupler comes from? Why is the 3 dB loss inevitable in the beam combiner?

3.9   What is the efficiency of axial coupling between two fiber sections with different mode properties? What is the adiabaticity criteria in a fiber taper?

3.10   Describe the characteristics of a 3×3 fiber coupler for the triangle and planar configurations. Describe and explain the advantages of a Mach–Zehnder interferometer incorporated with a 3×3 coupler, compared with that with a 2×2 coupler.

3.11   Derive the transmission of the fiber Sagnac loop (3.103). Why can it act as a mirror, and under what condition can it be a totally reflective mirror?

3.12   Why is the Sagnac interferometer able to remove the reciprocal disturbances? List possible nonreciprocal effects in practical loops, discuss their properties.

3.13   Discuss the polarization dependence of the Sagnac loop. Explain why a half wave plate can remove the polarization dependence.

3.14   Describe the characteristics of the fiber ring in two cases: (a) with ideal lossless components; (b) taking the loss into account. What is a critical coupling?

3.15   Compare the properties of a fiber ring and a GTI. Discuss their function of dispersion compensation.

3.16   Design an MZI for measuring the birefringence of a fiber section.

3.17   Describe the origins of the polarization dependence of the optical fiber. How do you describe the twisted birefringent fiber by means of an anisotropic dielectric constant?

3.18   Derive the Jones matrices of a wave plate and a fiber section with birefringence in laboratory coordinates. Explain the relations between the Jones vector and Stokes vector; how to use the Poincaré sphere to describe the different polarization states.

3.19   What roles do the cross-coupling between the two polarization modes play in polarization evolution and polarization mode dispersion?

3.20   Compare the characteristics of polarization states in propagation and reflection: (1) in a birefringent fiber; (2) in a twisted fiber; (3) in a medium with natural optical activity; (4) in a medium with Faraday rotation.

3.21   Explain the principle of a D-shaped fiber polarizer.

3.22 Explain the principles of PCs: by fiber coils, and by lateral pressure.

3.23 What is the difference between a depolarizer and PS? How are their functions realized?

3.24 Describe the functions of optical isolator, OC, and FRM, and explain their principles.

REFERENCES

1. Clark Jr. SP (Ed.): *Handbook of Physical Constants*. Boulder, CO: The Geological Society of America, Inc., 1966.
2. Iida S, Oono K, Kosaki H, Kumagaya H, Sawada S. *Tables of Physical Constants*. Second Edition. 1978. Asakura Publishing Co. Ltd., Tokyo. [In Japanese]
3. Namihira Y. Opto-elastic constant in single-mode optical fibers. *Journal of Lightwave Technology* 1985; 3: 1078–1083.
4. Bertholds A, Dändliker R. Determination of the individual strain-optic coefficients in single-mode optical fibers. *Journal of Lightwave Technology* 1988; 6: 17–20.
5. Jackson DA, Priest R, Dandridge A, Tveten AB. Elimination of drift in a single-mode optical fiber interferometer using a piezoelectrically stretched coiled fiber. *Applied Optics* 1980; 19: 2926–2929.
6. Landau LD, Lifshitz EM. *Theory of Elasticity*. Third English Edition. Butterworth-Heinemann, Oxford, Boston, Johannesburg, Melbourne, New Delhi, Singapore, 1997.
7. Timshenko S, Goodier JN. Theory of Elasticity. Engineering Societies Monographs, 1951.
8. Giallorenzi TG, Bucaro JA, Dandridge A, Sigel GH, Cole JH, Rashleigh SC, Priest RG. Optical fiber sensor technology. *IEEE Journal of Quantum Electronics* 1982; 18: 626–665.
9. Wagreich RB, Atia WA, Singh H, Sirkis JS. Effects of diametric load on fibre Bragg gratings fabricated in low birefringent fibre. *Electronics Letters* 1996; 32: 1223–1224.
10. Li Z, Wu C, Dong H, Shum P, Tian CY, Zhao S. Stress distribution and induced birefringence analysis for pressure vector sensing based on single mode fibers. *Optics Express* 2008; 16: 3955–3960.
11. Ulrich R, Rashleigh SC, Eickhoff W. Bending-induced birefringence in single-mode fibers. *Optics Letters* 1980; 5: 273–275.
12. Rashleigh SC, Ulrich R. High birefringence in tension-coiled single-mode fibers. *Optics Letters* 1980; 5: 354–356.

13. Smith AM. Birefringence induced by bends and twists in single-mode optical fiber. *Applied Optics* 1980; 19: 2606–2611.

14. Vessallo C. *Optical Waveguide Concepts*. Amsterdam, Oxford, New York, Tokyo: Elsevier, 1991.

15. Tsao C. *Optical Fiber Waveguide Analysis*. Oxford, New York, Tokyo: Oxford University Press, 1992.

16. Ulrich R, Simon A. Polarization optics of twisted single-mode fibers. *Applied Optics* 1979; 18: 2241–2251.

17. Zubia J, Arrue J, Mendioroz A. Theoretical analysis of the torsion-induced optical effect in a plastic optical fiber. *Optical Fiber Technology* 1997; 3: 162–167.

18. Tai H, Rogowski R. Optical anisotropy induced by torsion and bending in an optical fiber. *Optical Fiber Technology* 2002; 8: 162–169.

19. Born M, Wolf E. *Principles of Optics*. Seventh Edition. Cambridge: Cambridge University Press, 1999.

20. Saleh BEA, Teich MC. *Fundamentals of Photonics*. Hoboken: John Wiley & Sons, 2007.

21. Marcatili EAJ. Bends in optical dielectric guides. *Bell System Technical Journal* 1969; 48: 2103–2133.

22. Marcuse D. Loss analysis of single-mode fiber splices. *Bell System Technical Journal* 1977; 56: 703–718.

23. Marcuse D. *Light Transmission Optics*. New York, Cincinnati, Toronto, Melbourne: Van Nostrand Reinhold Company, 1982.

24. Snyder AW, Love JD. *Optical Waveguide Theory*. London, New York: Chapman and Hall, 1983.

25. Petermann K. Microbending loss in monomode fibers. *Electronics Letters* 1976; 12: 107–109.

26. Petermann K. Fundamental mode microbending loss in graded-index and W fibers. *Optical and Quantum Electronics* 1977; 9: 167–175.

27. Francois PL, Vassallo C. Comparison between pseudomode and radiation mode methods for deriving microbending losses. *Electronics Letters* 1986; 22: 261–262.

28. Petermann K, Kuhne R. Upper and lower limits for the microbending loss in arbitrary single-mode fibers. *Journal of Lightwave Technology* 1986; 4: 2–7.

29. Yang R, Wu C, Yip GL. Approximate formulas for the microbending loss in single-mode fibers. *Optics Letters* 1987; 12: 428–430.

30. Thyagarajan K, Shenoy MR, Ghatak AK. Accurate numerical-method for the calculation of bending loss in optical wave-guides using a matrix approach. *Optics Letters* 1987; 12: 296–298.

31. Li MJ, Tandon P, Bookbinder DC, Bickham SR, McDermott MA, Desorcie RB, Nolan DA, Johnson JJ, Lewis KA, Englebert JJ. Ultra-low

bending loss single-mode fiber for FTTH. *Journal of Lightwave Technology* 2009; 27: 376–382.

32. Parriaux O, Gidon S, Kuznetsov AA. Distributed coupling on polished single-mode optical fibers. *Applied Optics* 1981; 20: 2420–2423.

33. Newton SA, Bowers JE, Kotler G, Shaw HJ. Single-mode-fiber $1 \times N$ directional coupler. *Optics Letters* 1983; 8: 60–62.

34. Kawasaki BS, Hill KO. Low-loss access coupler for multimode optical fiber distribution networks. *Applied Optics* 1977; 16: 1794–1795.

35. Kim Y, Jeong Y, Oh K, Kobelke J, Schuster K, Kirchhof J. Multiport $N \times N$ multimode air-clad holey fiber coupler for high-power combiner and splitter. *Optics Letters* 2005; 30: 2697–2699.

36. Snyder AW. Coupled-mode theory for optical fibers. *Journal of the Optical Society America* 1972; 62: 1267–1277.

37. Yariv A. Coupled-mode theory for guided-wave optics. *IEEE Journal of Quantum Electronics* 1973; 9: 919–933.

38. Yariv A. *Optical Electronics in Modern Communications*. Fifth Edition. New York: Oxford University Press Inc., 1997.

39. Goure JP, Verrier I. *Optical Fiber Devices*. Institute of Physics Publishing, 2002.

40. Lacroix S, Gonthier F, Bures J. Modeling of symmetrical 2×2 fused-fiber couplers. *Applied Optics* 1994; 33: 8361–8369.

41. Birks TA, Russell PSJ, Culverhouse DO. The acousto-optic effect in single-mode fiber tapers and couplers. *Journal of Lightwave Technology* 1996; 14: 2519–2529.

42. Birks TA, Farwell SG, Russell PSJ, Pannell CN. Four-port fiber frequency shifter with a null taper coupler. *Optics Letters* 1994; 19: 1964–1966.

43. Culverhouse DO, Farwell SG, Birks TA, Russel PSJ. Four port fused taper acoustooptic devices using standard single mode telecommunications fiber. *Electronics Letters* 1995; 31: 1279–1280.

44. Birks TA, Culverhouse DO, Farwell SG, Russell PSJ. All-fiber polarizer based on a null taper coupler. *Optics Letters* 1995; 20: 1371–1373.

45. Snyder AW, Stevenson A. Anisotropic fiber couplers with nonaligned optical axes. *Journal of Lightwave Technology* 1988; 6: 450–462.

46. Snyder AW, Stevenson A. Polarization splitters and birefringent couplers. *Electronics Letters* 1985; 21: 75–76.

47. Wu TL, Chang HC. Rigorous analysis of form birefringence of fused fiber couplers. *Electronics Letters* 1994; 30: 998–999.

48. Yataki MS, Payne DN, Varnham MP. All-fiber polarizing beamsplitter. *Electronics Letters* 1985; 21: 248–251.

49. Bricheno T, Baker V. All-fiber polarization splitter combiner. *Electronics Letters* 1985; 21: 251–252.

50. Stolen RH, Ashkin A, Pleibel W, Dziedzic JM. Polarization-selective 3-dB fiber directional coupler. *Optics Letters* 1985; 10: 574–575.

51. Shafir E, Hardy A, Tur M. Mode-coupling analysis of anisotropic polarization-maintaining fibers. *Optics Letters* 1987; 12: 1041–1043.

52. Samir W, Garth SJ, Pask C. Theory of fused-tapered nonlinear-optical fiber couplers. *Applied Optics* 1993; 32: 4513–4516.

53. Chiang KS. Propagation of short optical pulses in directional couplers with Kerr nonlinearity. *Journal of Optical Society America B* 1997; 14: 1437–1443.

54. Wilson SJ, Bricheno T. Broad-band single-mode optical fiber couplers. *Optics Communications* 1990; 75: 106–110.

55. Sheem SK. Optical fiber interferometers with [3 × 3] directional-couplers—analysis. *Journal of Applied Physics* 1981; 52: 3865–3872.

56. Priest RG. Analysis of fiber interferometer utilizing 3 × 3 fiber coupler. *IEEE Journal of Quantum Electronics* 1982; 18: 1601–1603.

57. Sun L, Ye P. General-analysis of [3 × 3] optical-fiber directional-couplers. *Microwave and Optical Technology Letters* 1989; 2: 52–54.

58. Breguet J, Gisin N. Interferometer using a 3 × 3 coupler and Faraday mirrors. *Optics Letters* 1995; 20: 1447–1449.

59. Shih ST, Chen MH, Lin WW. Analysis of fibre optic Michelson interferometric sensor distortion caused by the imperfect properties of its 3 × 3 coupler. *IEE Proceedings—Optoelectronics* 1997; 144: 377–382.

60. Lacroix S, Bourbonnais R, Gonthier F, Bures J. Tapered monomode optical fibers- understanding large power transfer. *Applied Optics* 1986; 25: 4421–4425.

61. Marcuse D. Mode conversion in optical fibers with monotonically increasing core radius. *Journal of Lightwave Technology* 1987; 5: 125–133.

62. Gonthier F, Lapierre J, Veilleux C, Lacroix S, Bures J. Investigation of power oscillations along tapered monomode fibers. *Applied Optics* 1987; 26: 444–449.

63. Love JD, Henry WM, Stewart WJ, Black RJ, Lacroix S, Gonthier F. Tapered single-mode fibers and devices Part 1: adiabaticity criteria. *IEE Proceedings—J* 1991; 138: 343–354.

64. Black RJ, Lacroix S, Gonthier F, Love JD. Tapered single-mode fibers and devices Part 2: experimental and theoretical quantification. *IEE Proceedings—J* 1991; 138: 355–364.

65. Gupta BD, Singh CD. Evanescent-absorption coefficient for diffuse source illumination – uiform-fiber and tapered-fiber sensors. *Applied Optics* 1994; 33: 2737–2742.

66. Villatoro J, Monzon-Hernandez D, Mejia E. Fabrication and modeling of uniform-waist single-mode tapered optical fiber sensors. *Applied Optics* 2003; 42: 2278–2283.

67. Lefère HC. *The Fiber-Optic Gyroscope*. London, Boston: Artech House Inc., 1993.

68. Birks TA, Morkel P. Jones calculus analysis of single-mode fiber Sagnac reflector. *Applied Optics* 1988; 27: 3107–3113.

69. Mortimore DB. Fiber loop reflectors. *Journal of Lightwave Technology* 1988; 6: 1217–1224.

70. Liu F, Ye Q, Pang F, Geng J, Qu R, Fang Z. Polarization analysis and experimental implementation of PLZT electro-optical switch using fiber Sagnac interferometers. *Journal of Optical Society America B* 2006; 23: 709–713.

71. Fang X, Claus RO. Polarization-independent all-fiber wavelength-division multiplexer based on a Sagnac interferometer. *Optics Letters* 1995; 20: 2146–2148.

72. Kapron FP, Keck DB, Borrelli NF. Birefringence in dielectric optical waveguides. *IEEE Journal of Quantum Electronics* 1972; 8: 222–225.

73. VanWiggeren GD, Roy R. Transmission of linearly polarized light through a single-mode fiber with random fluctuations of birefringence. *Applied Optics* 1999; 38: 3888–3892.

74. Golub I, Simova E. π-shifted Sagnac interferometer for characterization of femtosecond first- and second-order polarization mode dispersion. *Optics Letters* 2002; 72: 3861–1861.

75. Gan J, Shen L, Ye Q, Pan Z, Cai H, Qu R. Orientation-free pressure sensor based on p-shifted single-mode-fiber Sagnac interferometer. *Applied Optics* 2010; 49: 5043–5048.

76. Doran NJ, Wood D. Nonlinear-optical loop mirror. *Optics Letters* 1988; 13: 56–58.

77. Sokoloff JP, Prucnal PR, Glesk I, Kane M. A terahertz optical asymmetric demultiplexer (Toad). *IEEE Photonics Technology Letters* 1993; 5: 787–790.

78. Stokes LF, Chodorow M, Shaw HJ. All-single-mode fiber resonator. *Optics Letters* 1982; 7: 288–290.

79. Gires F, Tournois P. An interferometer useful for pulse compression of a frequency modulated light pulse. *Comptes Rendus de l'Academie des Sciences Paris* 1964; 258: 6112. [In French]

80. Dingel BB, Izutsu M. Multifunction optical filter with a Michelson-Gires-Tournois interferometer for wavelength-division-multiplexed network system applications. *Optics Letters* 1998; 23: 1099–1101.

81. Yuan L, Zhou L, Jin W, Demokan MS. Multiplexing of fiber-optic white light interferometric sensors using a ring resonator. *Journal of Lightwave Technology* 2002; 20: 1471–1477.

82. Yariv A. Universal relations for coupling of optical power between microresonators and dielectric waveguides. *Electronics Letters* 2000; 36: 321–322.

83. Smith SP, Zarinetchi F, Ezekiel S. Narrow-linewidth stimulated Brillouin fiber laser and applications. *Optics Letters* 1991; 16: 393–395.

84. Choi JM, Lee RK, Yariv A. Ring fiber resonators based on fused-fiber grating add-drop filters: application to resonator coupling. *Optics Letters* 2002; 27: 1598–1600.

85. Martinelli M, Palais JC. Dual fiber-ring depolarizer. *Journal of Lightwave Technology.* 2001; 19: 899–905.

86. Xia B, Chen LR. A direct temporal domain approach for pulse-repetition rate multiplication with arbitrary envelope shaping. *IEEE Journal of Selected Topics in Quantum Electronics* 2005; 11: 165–172.

87. Chan K, Chan CK, Chen LK, Tong F. Mitigation of pattern-induced degradation in SOA-based all-optical OTDM demultiplexers by using RZ-DPSK modulation format. *IEEE Photonics Technology Letters* 2003; 15: 1264–1266.

88. Pang F, Han X, Cai H, Qu R, Fang Z. Characteristics of an add-drop filter composed of a Mach–Zehnder interferometer and double ring resonators. *Chinese Optics Letters* 2005; 3: 21–23.

89. Madsen CK. Integrated waveguide all pass filter tunable dispersion compensators. *Technical Digest of Optical Fiber Communication Conference*, 2002, Anaheim, CA, paper TuT1.

90. Dickinson G, Chapman DA, Gorham DA. Properties of the fiber reflection Mach–Zehnder Interferometer with identical couplers. *Optics Letters* 1992; 17: 1192–1194.

91. Weihs G, Reck M, Weinfurter H, Zeilinger A. All-fiber three-path Mach–Zehnder interferometer. *Optics Letters* 1996; 21: 302–304.

92. Huang C, Geng J, Cai H, Qu R, Fang Z. Wavelength interrogation based on a Mach–Zehnder interferometer with a 3×3 fiber coupler for fiber Bragg grating sensors. *Chinese Journal of Lasers* 2005; 32: 1397. [In Chinese].

93. Todd MD, Johnson GA, Althouse BL. A novel Bragg grating sensor interrogation system utilizing a scanning filter, a Mach–Zehnder interferometer and a 3×3 coupler. *Measurement Science & Technology* 2001; 12: 771–777.

94. Ren E, Lu H, Zhang B, Luo G. Optimization design of all-fiber 3×3 multiplexer based on an asymmetrical Mach–Zehnder interferometer. *Optics Communications* 2009; 282: 2818–2822.

95. Davies PA, Abdelhamid G. Four-port fiber-optic ring resonator. *Electronics Letters* 1988; 24: 662–663.

96. Ja YH. Analysis of four-port optical fiber ring and loop resonators using a 3×3 fiber coupler and degenerate two-wave mixing. *IEEE Journal of Quantum Electronics* 1992; 28: 2749–2757.

97. Ja YH. Phase sensitivities of optical-fiber ring and loop resonators using a nonplanar or planar 3×3 fiber coupler. *Journal of Modern Optics* 1995; 42: 117–130.

98. Machida S, Sakai J, Kimura T. Polarization conservation in single-mode fibers. *Electronics Letters* 1981; 17: 494–495.

99. Mcintyre P, Snyder AW. Light-propagation in twisted anisotropic media - application to photoreceptors. *Journal of Optical Society America* 1978; 68: 149–157.

100. Fujii Y, Sano K. Polarization coupling in twisted elliptical optical fiber. *Applied Optics* 1980; 19: 2602–2605.

101. Barlow AJ, Payne DN. Polarization maintenance in circularly birefringent fibers. *Electronics Letters* 1981; 17: 388–389.

102. Rashleigh SC, Burns WK, Moeller RP, Ulrich R. Polarization holding in birefringent single-mode fibers. *Optics Letters* 1982; 7: 40–42.

103. Okamoto K, Edahiro T, Shibata N. Polarization properties of single-polarization fibers. *Optics Letters* 1982; 7: 569–571.

104. Rashileigh SC. Origins and control of polarization effects in single-mode fibers. *Journal of Lightwave Technology* 1983; 1: 312–331.

105. Sakai JI, Kimura T. Polarization behavior in multiply perturbed single-mode fibers. *IEEE Journal of Quantum Electronics* 1982; 18: 59–65.

106. Wu C, Yip GL. Coupled-mode analysis of a bent birefringent fiber. *Optics Letters* 1987; 12: 522–524.

107. Huang HC. Fiber-optic analogs of bulk-optic wave plates. *Applied Optics* 1997; 36: 4241–4258.

108. Huang HC. Practical circular-polarization-maintaining optical fiber. *Applied Optics* 1997; 36: 6968–6975.

109. Rose AH, Feat N, Etzel SM. Wavelength and temperature performance of polarization-transforming fiber. *Applied Optics* 2003; 42: 6897–6904.

110. Winful HG, Hu A. Intensity discrimination with twisted birefringent optical fibers. *Optics Letters* 1986; 11: 668–670.

111. El-Khozondar HJ, Muller MS, El-Khozondar RJ, Koch AW. Polarization rotation in twisted polarization maintaining fibers using a fixed reference frame. *Journal of Lightwave Technology* 2009; 27: 5590–5596.

112. Tsubokawa M, Higashi T, Negishi Y. Mode couplings due to external forces distributed along a polarization-maintaining fiber—an evaluation. *Applied Optics* 1988; 27: 166–173.

113. Gordon JP, Kogelnik H. PMD fundamentals: polarization mode dispersion in optical fibers. *Proceedings of the National Academy of Sciences of the United States of America* 2000; 97: 4541–4550.

114. Poole CD, Wagner RE. Phenomenological approach to polarization dispersion in long single-mode fibers. *Electronics Letters* 1986; 22: 1029–1030.

115. Poole CD. Statistical treatment of polarization dispersion in single-mode fiber. *Optics Letters* 1988; 13: 687–689.

116. Gisin N, Vonderweid JP, Pellaux JP. Polarization mode dispersion of short and long single-mode fibers. *Journal of Lightwave Technology* 1991; 9: 821–827.

117. Barlow AJ, Payne DN, Hadley MR, Mansfield RJ. Production of single-mode fibers with negligible intrinsic birefringence and polarization mode dispersion. *Electronics Letters* 1981; 17: 725–726.

118. Shi Z. Experimental observation of spectral transmittance of highly bire-fringent fibre with high spin rate. *Optics Communications* 1999; 171: 61–64.

119. Barlow AJ, Ramskovhansen JJ, Payne DN. Birefringence and polariza-tion mode-dispersion in spun single-mode fibers. *Applied Optics* 1981; 20: 2962–2968.

120. Smith AM. Polarization and magneto-optic properties of single-mode op-tical fiber. *Applied Optics* 1978; 17: 52–56.

121. Papp A, Harms H. Magneto-optical current transformer. 1: principles. *Applied Optics* 1980; 19: 3729–3734.

122. Munin E, Roversi JA, Villaverde AB. Faraday effect and energy gap in optical materials. *Journal of Physics D: Applied Physics* 1992; 25: 1635–1639.

123. Yu T, Wu Y. Theoretical study of metal-clad optical waveguide polarizer. *IEEE Journal of Quantum Electronics* 1989; 25: 1209–1213.

124. Tseng SM, Chen CL. Side-polished fibers. *Applied Optics* 1992; 31: 3438–3447.

125. Dyott RB, Bello J, Handerek VA. Indium-coated D-shaped-fiber polar-izer. *Optics Letters* 1987; 12: 287–289.

126. Hsu KY, Ma SP, Chen KF, Tseng SM, Chen JI. Surface-polariton fiber polarizer: design and experiment. *Japanese Journal of Applied Physics Part 2* 1997; 36: L488–L490.

127. Zhu R, Wei Y, Scholl B, Schmitt HJ. In-line optical-fiber polarizer and modulator coated with Langmuir–Blodgett films. *IEEE Photonics Tech-nology Letters* 1995; 7: 884–886.

128. Donati S, Faustini L, Martini G. High-extinction coiled-fiber polarizers by careful control of interface reinjection. *IEEE Photonics Technology Letters* 1995; 7: 1174–1176.

129. Lefevre HC. Single-mode fiber fractional wave devices and polarization controllers. *Electronics Letters* 1980; 16: 778–780.

130. Walker NG, Walker GR. Polarization control for coherent communica-tions. *Journal of Lightwave Technology* 1990; 8: 438–458.

131. Ferguson BA, Chen CL. Polarization controller based on a fiber recircu-lating delay-line. *Applied Optics* 1992; 31: 7597–7604.

132. Lu SY, Chipman RA. Mueller matrices and the degree of polarization. *Optics Communications* 1998; 146: 11–14.

133. Bohm K, Petermann K, Weidel E. Performance of Lyot depolarizers with birefringent single-mode fibers. *Journal of Lightwave Technology* 1983; 1: 71–74.

134. Burns WK. Degree of polarization in the Lyot depolarizer. *Journal of Lightwave Technology* 1983; 1: 475–479.

135. Shen P, Palais JC, Lin C. Tunable single mode fiber depolarizer. *Electronics Letters* 1997; 33: 1077–1078.

136. Shen P, Palais JC, Lin CL. Fiber recirculating delay-line tunable depolarizer. *Applied Optics* 1998; 37: 443–448.

137. Shen P, Palais JC. Passive single-mode fiber depolarizer. *Applied Optics* 1999; 38: 1686–1691.

138. Lutz DR. A passive fiberoptic depolarizer. *IEEE Photonics Technology Letters* 1993; 5: 463–465.

139. Takada K, Okamoto K, Noda J. New fiberoptic depolarizer. *Journal of Lightwave Technology* 1986; 4: 213–219.

140. Yu SJ, Ohki Y, Fujimaki M, Awazu K, Tominaga J. Optical fiber depolarizer using birefringence induced by proton implantation. *Japanese Journal of Applied Physics* 2009; 48: 032404-1–032404-3.

141. Heismann F, Tokuda KL. Polarization-independent electrooptic depolarizer. *Optics Letters* 1995; 20: 1008–1010.

142. Heismann F. Compact electro-optic polarization scramblers for optically amplified lightwave systems. *Journal of Lightwave Technology* 1996; 14: 1801–1814.

143. Yamashita S, Hotate K. Polarization-independent depolarizers for highly coherent light using Faraday rotator mirrors. *Journal of Lightwave Technology* 1997; 15: 900–905.

144. Fujii Y. Compact high-isolation polarization-independent optical circulator. *Optics Letters* 1993; 18: 250–252.

145. Smigaj W, Romero-Vivas J, Gralak B, Magdenko L, Dagens B, Vanwolleghem M. Magneto-optical circulator designed for operation in a uniform external magnetic field. *Optics Letters* 2010; 35: 568–570.

146. Chen JH, Chen KH, Lin JY, Hsieh HY. Multiport optical circulator by using polarizing beam splitter cubes as spatial walk-off polarizers. *Applied Optics* 2010; 49: 1430–1433.

CHAPTER 4

FIBER GRATINGS AND RELATED DEVICES

The fiber grating is one of the important fiber devices and has been widely used in fiber communication systems and fiber sensor technologies. Four sections of Chapter 4 are devoted to fiber gratings: Section 4.1 for a brief introduction, Section 4.2 for its basic theory, Section 4.3 for descriptions of various fiber gratings, Section 4.4 for the sensitivities of grating and its applications in sensor technology.

4.1 INTRODUCTION TO FIBER GRATINGS

4.1.1 Basic Structure and Principle

Soon after the invention of fiber, people tried to fabricate gratings on the fiber to create desired spectral features to be used in important applications such as wavelength division multiplex technology. In earlier stages, fiber gratings were fabricated by photolithography on a lateral facet along the fiber that had been lapped and polished near its core, called D-shaped fiber grating [1]. The performance of D-shaped gratings is generally not good enough for practical applications; and its fabrication is not suitable for mass production.

Fundamentals of Optical Fiber Sensors, First Edition.
Zujie Fang, Ken K. Chin, Ronghui Qu, and Haiwen Cai.
© 2012 John Wiley & Sons, Inc. Published 2012 by John Wiley & Sons, Inc.

It was observed in 1978 [2] by K. O. Hill that an index grating inside the fiber core could be generated when a 488 nm argon laser beam was injected into a section of germanium (Ge)-doped fiber for some time, and the grating period coincided with that of the laser beam standing wave in the fiber section. The index change proved to be a permanent one. This observation is considered to be the discovery of fiber photo-sensitivity. It was found later in 1989 [3] that a UV illustration arranged laterally to the fiber could also induce index change, and a method of writing grating with different periods by holography was demonstrated, which provided the possibility of fabricating fiber grating as a prac-tical device with desired peak wavelengths. A few years later, phase masks were used in writing fiber grating by a homogeneous laser illus-tration [4,5], which made its fabrication suitable for mass production. Efforts were made to increase the photosensitivity, especially by using higher Ge doping levels. It was found [6] that photosensitivity could be greatly enhanced if a fiber with ordinary Ge doping had been soaked in a high-pressure hydrogen tank for some time. The method was then used widely, called hydrogen-loading sensitization. The design and fab-rication of various fiber gratings, and their applications have become mature technologies in the recent two decades [7].

The photoinduced grating is a section of fiber with its core index modulated periodically in the axial direction, as shown in Figure 4.1. It is a one-dimensional grating with two main parameters: index in-crement Δn and pitch period Λ. When the period equals half the light wavelength, the grating will couple the fundamental mode to the backward propagating fundamental mode, showing a function of wavelength-selective reflection. Such a grating is called fiber Bragg grating (FBG) because the mechanism is just the Bragg diffraction in crystal. Applying Bragg equation for X-ray diffraction to fiber grating, the basic equation for FBG is

$$\lambda = 2n_{co}\Lambda, \qquad (4.1)$$

where λ is the wavelength, and n_{co} is the effective index of the core mode. In the commonly concerned 1,550 nm band, Λ is around 500 nm.

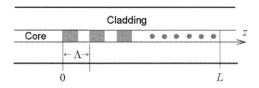

Figure 4.1 Index distribution in fiber grating.

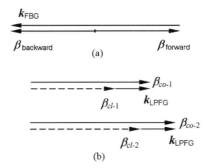

Figure 4.2 Wave vector relations for (a) FBG and (b) LPFG.

When the grating period is large in the range of sub-millimeter, the grating couples the fundamental core mode with the cladding modes, called the long-period fiber rating (LPFG) [8,9]. The resonant wavelength is deduced to meet conditions of

$$\beta_{\text{co}} - \beta_{\text{cl}} = 2\pi(n_{\text{co}} - n_{\text{cl}})/\lambda = 2\pi/\Lambda, \qquad (4.2\text{a})$$

$$\lambda = (n_{\text{co}} - n_{\text{cl}})\Lambda, \qquad (4.2\text{b})$$

where n_{cl} is the effective index of the corresponding cladding mode. The transmission spectrum shows several dips corresponding to the resonance between the two modes.

By means of diffraction optics, the input and diffracted light wave vectors, and the grating vector, defined as $k_{\text{FBG}} = 2\pi/\Lambda$, should satisfy the phase matching, as shown schematically in Figure 4.2. The input beam and reflected beam by the FBG have the same wavelength and opposite directions, as shown in Figure 4.2(a). The vector of LPFG is much shorter than that of FBG; and a few pairs of input and diffracted light waves may meet the phase relation, as shown in Figure 4.2(b) for two examples. (NB: the input beam and codirection coupled cladding mode beam in an LPFG have the same wavelength, but different wave vectors due to different effective indexes.) When the input mode is coupled to a cladding mode, it is often coupled further to a radiation mode, resulting in a loss peak in its transmission spectrum. Figure 4.3(a) gives a typical transmission spectrum of an FBG with $\Lambda = 0.535$ μm, showing a narrow downward peak with extinction ratio more than 30 dB and bandwidth around 0.2 nm; and Figure 4.3(b) gives a typical transmission spectrum of an LPFG with $\Lambda = 450$ μm, showing four peaks at different wavelengths corresponding to different cladding modes.

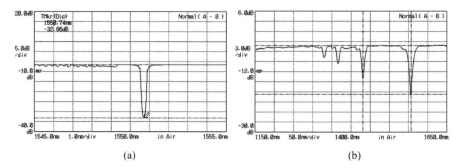

(a)   (b)

Figure 4.3 Typical transmission spectra of (a) FBG and (b) LPFG.

The grating period and the index modulation amplitude are functions of temperature and strain status; therefore, the measured transmission characteristics are sensitive to external conditions. This is the basic principle of fiber grating sensors.

4.1.2 Photosensitivity of Optical Fibers

Research on fiber photosensitivity have covered almost all kinds of fibers in the literature, especially fibers with different dopants, such as aluminum, boron, tin, rare-earth, and so on, as well as fibers with different structures, such as the multimode fiber, polarization maintaining fiber (PMF), photonic crystal fiber (PCF), and so on. Most fiber gratings are made in Ge-doped silica fibers because it is the most commonly used fiber in fiber communications and fiber sensors, with the lowest cost and optimized performance. Fiber grating made of standard fibers can be connected to conventional fiber networks compatibly. Many papers are devoted to investigating photosensitivity and its mechanisms, and several books give systematic and summarized elaboration [10,11].

4.1.2.1 Features of Photosensitivity Several features of the UV-induced photosensitivity in silica fibers are noticeable:

1. *Dependence on Ge doping.* The silica fiber photosensitivity is tightly linked with the doped germanium oxide, which is originally used to increase core index. It is observed experimentally that photosensitivity was hardly measured in fiber with pure silica core and lower index cladding; whereas higher sensitivity could be obtained in fibers with higher Ge doping. A special fiber with higher photosensitivity has been developed, which has a Ge-doping

concentration much higher than that in conventional single-mode fibers (SMFs).

2. *Condition of UV laser irradiations.* UV irradiation at special wavelengths is a necessary condition in fiber grating fabrication. The mechanism in the first experimental result [2] was later explained by a two-photon effect, which corresponds to 244 nm UV irradiation. The UV absorption spectra of Ge-doped silica fiber and its preform were studied experimentally in detail. It is found experimentally that irradiations at three UV bands give higher photosensitivity: 195, 242, and 256 nm. The measured refractive index change at 1,500 nm band is correlated with the UV absorptions, expressed by a fitted curve as [12]

$$\Delta n = (2.34\Delta\alpha_{242} + 4.96\Delta\alpha_{195} + 5.62\Delta\alpha_{256}) \times 10^{-7}, \qquad (4.3)$$

where the losses are in dB/mm. Fortunately, there are well-developed lasers working in those bands: ArF excimer laser (193 nm), frequency-doubled Ar^+ laser (244 nm), KrF excimer laser (248.5 nm), and frequency-doubled copper-vapor laser (256 nm).

3. *Dependence on UV intensity and irradiation dose.* It is of practical interest to understand in what way the index change depends on UV intensity and UV irradiation dose. The index change is found to be monotonic increasing with the UV intensity and the cumulated dose when the fiber is irradiated by low-intensity UV laser beams, of the order of 100 mJ/cm^2/pulse, which is usually used to obtain an index increment Δn of 10^{-4} to 10^{-3} for practical gratings. The photosensitivity in this intensity range is termed as Type-I photosensitivity [13].

It is observed experimentally in FBG imprinting [14] that the increasing rate of mean index decreases with UV exposure dose, and tends to saturate; further protracted exposure results in a negative rate. At the same time, index modulation changes similarly; the modulation disappears at a certain accumulated dose, and then becomes negative, as qualitatively shown in Figure 4.4. This phenomenon implies that two different mechanisms of positive and negative UV sensitivities may exist; the positive one is dominant in the range of lower doses, whereas the latter becomes dominant for higher dose. Such a negative UV sensitivity is termed Type-IIA sensitivity (or Type-III in some literature).

When the UV irradiation energy level is high, a single pulse may produce a large index change, even close to 10^{-2}, which is enough to generate a grating, called Type-II grating [15].

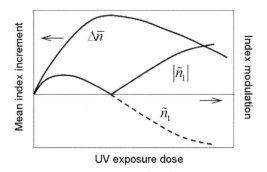

Figure 4.4   Mean index increment and index modulation versus UV exposure dose.

4. *Thermal stability and long-term stability.* Although the UV-induced index change is permanent, its stability in the long term and with temperature changes is still a concern for practical applications. Among the three types, Type-II is generally the most stable, Type-I is the least, and Type-IIA is in between. The properties are considered to be linked with different mechanisms. It is considered that in Type-II photosensitivity, the high-intensity laser beam induces some physical damage in the fiber core, whereas electronic defects play the main role in Type-I, as summarized in Section 2.2 of Reference [10]. However, Type-I sensitivity is most widely used for practical fiber gratings due to its easier fabrication and better controllability of specifications. It is found that the long-term stability can be improved by postprocessing measures such as annealing and flood exposures.

5. *Hydrogen loading sensitization.* During the investigation of photosensitivity mechanisms, there have been efforts to enhance the sensitivity, such as co-doping by boron, phosphor, tin, and so on together with germanium, and flame brushing [16]. Hydrogen loading is one of the most important discoveries and used widely as an effective and simple sensitization method. In the first reported experiment [6], gratings were written in ordinary 3 mol% Ge-doped silica fibers, which had been soaked in a high-pressure (typically 150 atm) hydrogen tank for several days to acquire 95% or over H_2 solubility in the fiber [17]. UV-induced index change up to $10^{-3} - 10^{-2}$ was achieved.

4.1.2.2 Mechanisms of Photosensitivity   It is important to understand what happens in the materials subject to UV irradiation, especially for manufacturing fiber gratings and for improving their performance. Several models have been proposed to explain the

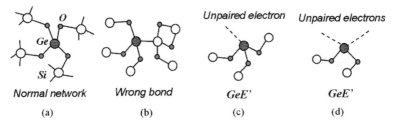

Figure 4.5 Normal network of Ge-doped silica (a) and some examples of defects (b, c, d).

experimental phenomena. However, the experimental results are so complicated that no one can explain all the phenomena, which are observed in more or less different experimental conditions and for fibers with different composition, and even with different fabrication processes. The basic mechanisms are determined by the molecular structures of the materials. For Ge-doped fused silica, SiO_2 molecules bond with each other to form a tetrahedral network, and doped GeO_2 molecules are included, as shown in Figure 4.5(a).

Oxygen vacancies often occur as fiber and its preform are processed at very high temperatures, at which GeO is more stable than GeO_2; and oxygen escapes outward, forming Ge oxygen-deficient centers (GODC). Figure 4.5(b) shows a possible wrong bond, which connects the silicon atom and Ge atom directly without oxygen. Several related defects are then generated, including *GeE'* centers, as shown in Figures 4.5(c) and (d) [18]. These defects are termed the color centers.

The electronic states of the color centers will change under UV exposure, such as transition of a wrong bond state to a *GeE'* color center, expressed as [19]

$$\underset{/}{\overset{\backslash}{\nearrow}}\text{Ge-Ge}\underset{\backslash}{\overset{/}{\swarrow}} \text{(or Ge-Si)} \xrightarrow{\text{hv}} \text{Ge E}' + \text{GeO}_3^+ (\text{or SiO}_3^+) + e^-.$$

Then the related absorptive color centers are bleached. According to the well-known universal *Kramers–Kronig* relations, the absorption change causes index change, expressed as [20,21]

$$\Delta n(\lambda') = \frac{1}{2\pi^2} P \int\limits_0^\infty \frac{\Delta\alpha(\lambda)}{1 - (\lambda/\lambda')^2} d\lambda, \tag{4.4}$$

where P stands for the principal part of the integral. Such a mechanism is usually termed a color center model.

It is concluded that the defects linking with doped germanium, especially the oxygen-deficient centers, are the main element. In H_2-loaded fibers, the hydrogen molecules, as a reducing agent, will react with GeO_2 under UV exposure to generate an OH^- radical, another type of oxygen-deficient center, by the following process [21,22]:

$$\underset{/}{\overset{\diagdown}{}} Ge - O - Si \underset{\diagdown}{\overset{\diagup}{}} + H_2/2 \longrightarrow \underset{/}{\overset{\diagdown}{}} Ge^{\diagup\, e^-} \quad OH - Si \underset{\diagdown}{\overset{\diagup}{}}.$$

The amount of such defects is usually much more than that of original GODC in ordinary fibers without H_2 loading, resulting in greatly enhanced photosensitivity. Although the OH^- radical is an important factor inducing propagation loss, it is negligible for very short fiber gratings.

There are also other mechanisms playing roles under different conditions of fiber grating fabrications. It has been observed that material densification occurs after larger doses of UV exposure, which is proposed as the main mechanism of Type-IIA photosensitivity [14]. Another possible mechanism is the photoinduced change in glass volume [23].

4.1.3   Fabrication and Classifications of Fiber Gratings

4.1.3.1   *Methods of Fiber Grating Fabrication*   Several methods have proved effective in fiber grating imprinting, such as by standing wave [2], point-by-point writing [24], by holography [25], and with phase mask [4,5]. The latter two methods have the merits of easier processing and better performance, and are suitable for mass production.

Phase mask used in fiber grating fabrication is simply a phase grating itself, which is transparent as a whole, but with a thickness difference of δ between groove top and bottom, as shown in Figure 4.6(a), where θ is

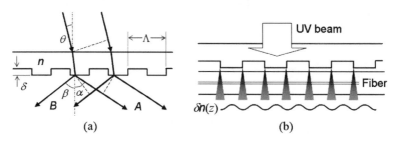

Figure 4.6   (a) Diffractions of phase mask and (b) periodic UV distribution.

the incident angle; A and B are diffractive beams at angles of α and β, corresponding to the order number of $\pm m$. By the theory of diffractive grating, the diffraction angle has to meet the equation of

$$\Lambda(\sin\theta - \sin\alpha) = m\lambda. \tag{4.5}$$

When the incident is normal to the phase mask, $\theta = 0$, symmetric diffractive beams are generated with a diffractive angle of $\sin\alpha = m\lambda/\Lambda$ with $m = \pm 1, 2\ldots$. The zeroth-order diffraction beam with $\alpha = 0$ is just the transmitted wave, which has to be reduced as much as possible, as it gives no contribution to building fiber grating except for increasing the index background. To suppress the zeroth-order diffraction, the thickness difference δ has to meet condition of $(n_{UV} - 1)\delta = \lambda_{UV}/2$ to get a phase difference between groove top and bottom to be π, so that the averaged total zeroth-order diffraction is eliminated. It is noticed that the index of mask material (usually fused silica) is a function of UV wavelength, so that the mask has to be designed according to the UV wavelength.

The diffractive beams of $m = \pm 1$, with the maximum and equal amplitudes, interfere in the area beneath the mask to provide periodic UV exposure on the fiber, as shown in Figure 4.6(b). The period of interfered pattern is $\Lambda_{fiber} = \Lambda/2$, just half of the period of mask grating, because every interface between groove top and bottom is the dark point due to π phase shift.

Holography is another important and useful method for fiber grating fabrication. In the setup, two beams from the same laser beam, divided by a beam splitter, are incident to the fiber with angle θ and interfered to generate a pattern with period of $\Lambda_{fiber} = \lambda/(2\sin\theta)$, as shown schematically in Figure 4.7(a).

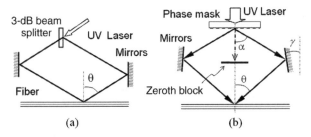

(a)   (b)

Figure 4.7 (a) Typical holographic setup for fiber grating imprinting. (b) Holographic setup incorporated with a phase mask.

The physical basis of both methods, by phase mask and by holography, is an optical interferometer. They have different features in fabrication. A phase mask is suitable for fiber gratings of fixed period, and for specific wavelength UV lasers. One of its merits is its lower requirement on laser coherence and is therefore suitable for excimer lasers. Another merit is higher reproducibility, which is important in mass production. Holographic method has the merit of flexible period adjustment, but it needs high coherent lasers such as frequency doubled Ar^+ laser. Figure 4.7(b) shows a combined set up, where the UV beam is divided into two by a phase mask, and the remaining zeroth diffractive wave can be blocked to reduce uniform index background, and the incident angle can be adjusted by mirror angle γ.

Practical applications require fiber gratings with long length and/or complicated structures; and then comprehensive facilities are invented, in which the interference period and radiation intensity are adjustable, and the fiber can move precisely and controllably, aided by a computer.

Long-period fiber grating (LPFG) has a large period in range of sub-millimeters; this makes the fabrication of LPFG simpler. The conventional methods include: UV exposure with an amplitude mask [26], point-by-point writing with CO_2 laser spot [27], electric arc discharge [28,29], and periodic corrugated bonding [30]. The former two methods are suitable for mass production and with better processing stability. The UV exposure with amplitude masks is just to duplicate the pattern of mask into the fiber core, as shown in Figure 4.8(a). The point-by-point writing with CO_2 laser spot is usually aided by a computer-controlled one-dimensional moving stage, as shown in Figure 4.8(b). The index change is induced by CO_2 laser heating upto over $1,000°C$; and some mechanisms are proposed: densification of the material, thermal stress, and release of the residue drawing stress or the drawing-induced defects (DID) [31,32].

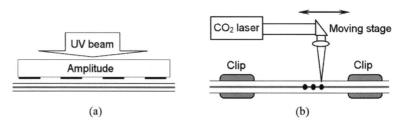

(a) (b)

Figure 4.8 (a) LPFG fabrication by UV exposure with amplitude mask. (b) Point-by-point scanning of CO_2 laser spot.

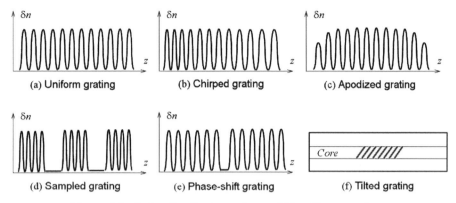

Figure 4.9 Index distributions of several fiber Bragg gratings.

4.1.3.2 Classifications of Fiber Gratings

Pushed by applications, various fiber gratings have been invented and produced to meet different demands. They can be classified according to several categories, by fiber materials: silica-based fibers and those other than silica, such as plastic optical fiber (POF), PCF; by working wavelength: 1,550, 980, 1,310, 1,064 nm, and visible range; and by period: FBG and LPFG. Among them, Bragg gratings in single-mode silica fibers are used mostly in optical fiber communications and fiber sensors.

Apart from the uniform FBG, several specially designed gratings are worth noting:

- *Chirped fiber Bragg grating.* It is a grating with period variation along the fiber, linearly (linearly chirped FBG, LCFG) or in other forms, shown in Figure 4.9(b). The chirping makes grating with a broadband spectrum, which is very useful for pulse compression in fiber communications, for pulse reshaping in laser technology, and for optical signal filtering in measurement.
- *Apodized fiber Bragg grating.* A grating with uniform index modulation usually shows an oscillating spectrum, not satisfactory in some applications, where a clear single reflection peak is needed. It is found that apodization of the index modulation amplitude with convex shapes in its length, shown in Figure 4.9(c), will help to improve the spectrum greatly.
- *Sampled fiber Bragg grating.* If the index modulation of a uniform grating is further modulated in superposition by a long-period square waveform, as shown in Figure 4.9(d), it is termed sampled FBG. Its spectrum becomes a multipeak one, and demonstrates interesting characteristics useful both in communication and sensing.

- *Phase-shift fiber Bragg grating.* If two uniform gratings are written in sequence with an interval corresponding to a phase shift between two gratings, shown in Figure 4.9(e), a very narrow transmission peak appears in its spectrum.
- *Polarization maintaining fiber gratings.* A grating imprinted in PMF demonstrates a polarization-dependent spectrum, which brings about new useful characteristics.
- *Tilted fiber Bragg grating.* In the above-mentioned gratings, the photoinduced index changes only in direction along the fiber axis, so that the Bragg vector is axial. It is found experimentally and theoretically that if the grating is written in a tilted direction, shown in Figure 4.9(f), some new features appear, such as coupling to radiation mode and polarization-dependent loss.

Characteristics of the various gratings and their mechanisms and applications will be analyzed in the following sections of this Chapter.

4.2 THEORY OF FIBER GRATING

It is necessary to understand the mechanism and characteristics theoretically. Section 4.2.1 expounds the theory of FBG with uniform period, based on the coupled mode theory (CMT). The second section is devoted to the theory of LPFG with uniform period. Section 4.2.3 presents basic theories of nonuniform fiber grating. Inverse engineering design involves designing the grating structure to meet the required characteristics, which is introduced in Section 4.2.4. The last section explains the apodization method to improve fiber-grating performance.

4.2.1 Theory of Uniform FBG

4.2.1.1 Coupled Mode Equation of FBG A great number of publications have dealt with the periodic optical structures, including books [10, 11,33–35] and papers [36–38]. The theory of fiber gratings is based on the CMT [37]. Here we give an explicit deduction. Its starting point is the Helmholtz equation:

$$\nabla^2 E + n^2(z)k_0^2 E = 0. \tag{4.6}$$

The refractive index is considered a periodic function of axial position z described as

$$n(z) = n_0 + \delta n(z) = n_0 + \sum_{m=0}^{\infty} \tilde{n}_m \cos[2\pi m z / \Lambda + \theta_m(z)]$$
$$\approx n_0 + \tilde{n}_0(z) + \tilde{n}_1 \cos(k_B z + \theta) \rightarrow \bar{n} + \tilde{n}_1 \cos k_B z. \quad (4.7)$$

In the approximation, higher order expansion terms of $m > 1$ are omitted because the higher order components are smaller and their effects are usually not in the concerned wave band. $\tilde{n}_0$ is the UV-induced background index change, which is a slowly varying function of position z. Bragg wave vector is defined as $k_B = 2\pi/\Lambda$. Phase factor θ is related with the coordinate starting point setting, which is a parameter to be considered in superstructure gratings. As the basic structure, a grating with uniform modulation period Λ and uniform modulation amplitude $\tilde{n}_1$ is considered here.

The index modulation depth is much smaller than the averaged index, $\tilde{n}_1 \ll \bar{n}$, so that its effect can be considered as a perturbation; whereas the eigenmode of SMF is taken as the zeroth-order solution. The CMT is then used to analyze the transmission characteristics of FBG. It is necessary to involve both the forward- and backward-propagation waves, and the trial solution of Equation (4.6) is expressed as

$$E(r) = F(r, \varphi)[A(z) \exp(j\beta z) + B(z) \exp(-j\beta z)], \quad (4.8)$$

where $F(r, \varphi)$ satisfies equation $\nabla_{xy}^2 F + (n_0^2 k_0^2 - \beta^2)F = 0$. By substituting it into (4.6), and normalizing the equation, that is, integrating the equation, multiplied by $F(r, \varphi)$, over the transverse cross section, we obtain

$$[A'' + j2\beta A' + \bar{n}\tilde{n}_1(e^{jk_B z} + e^{-jk_B z})k_0^2 A]e^{j\beta z},$$
$$+[B'' - j2\beta B' + \bar{n}\tilde{n}_1(e^{jk_B z} + e^{-jk_B z})k_0^2 B]e^{-j\beta z} = 0, \quad (4.9)$$

where $n^2 - n_0^2$ is approximated to $2\bar{n}\tilde{n}_1 \cos k_B z = \bar{n}\tilde{n}_1(e^{jk_B z} + e^{-jk_B z})$. Terms with $e^{j(k_B+\beta)z}$ and $e^{-j(k_B+\beta)z}$ are omitted because they are far beyond the concerned wave band. By denoting $\delta = \beta - k_B/2 = \beta - \beta_B$, the equation is simplified further to be

$$(A'' + j2\beta A' + \bar{n}\tilde{n}_1 k_0^2 B e^{-j2\delta z})e^{j\beta z},$$
$$+(B'' - j2\beta B' + \bar{n}\tilde{n}_1 k_0^2 A e^{j2\delta z})e^{-j\beta z} = 0. \tag{4.10}$$

The two second-order differentials are considered much smaller than other terms; thus a group of coupled mode equation (CME) is deduced as

$$A' = j\kappa B e^{-j2\delta z},$$
$$B' = -j\kappa A e^{j2\delta z}, \tag{4.11}$$

where the coupling coefficient $\kappa = \bar{n}\tilde{n}_1 k_0^2/2\beta = \pi\bar{n}\tilde{n}_1/n_{\text{eff}}\lambda \approx \tilde{n}_1 k_0/2$ is denoted. By combining the two equations, the equations are transformed to

$$A'' + j2\delta A' - \kappa^2 A = 0,$$
$$B'' - j2\delta B' - \kappa^2 A = 0. \tag{4.12}$$

The general solutions are obtained to be

$$A = [A_1 \exp(jsz) + A_2 \exp(-jsz)]\exp(-j\delta z),$$
$$B = [B_1 \exp(jsz) + B_2 \exp(-jsz)]\exp(j\delta z), \tag{4.13}$$

with $s = \sqrt{\delta^2 - \kappa^2}$. By using relations of $B_1 = A_1(s - \delta)/\kappa$ and $B_2 = -A_2(s + \delta)/\kappa$, derived from (4.11), the forward and backward optical fields are expressed as

$$E_+ = [A_1 \exp(jsz) + A_2 \exp(-jsz)]\exp(j\beta_B z),$$
$$E_- = [B_1 \exp(jsz) + B_2 \exp(-jsz)]\exp(-j\beta_B z). \tag{4.13a}$$

Considering a wave propagates in $+z$ direction and inputs to the grating with length of L, the constants should satisfy the boundary conditions at $z = 0$ and $z = L$:

$$A_1 + A_2 = 1,$$
$$B_1 + B_2 = r,$$
$$(A_1 e^{jsL} + A_2 e^{-jsL}) e^{j\beta_B L} = t, \tag{4.14}$$
$$(B_1 e^{jsL} + B_2 e^{-jsL}) = 0.$$

The axial variations of the field are then deduced to be

$$A = A(0) \frac{s \cos s(L-z) - j\delta \sin s(L-z)}{s \cos sL - j\delta \sin sL} e^{-j\delta z},$$

$$B = A(0) \frac{-j\kappa \sin s(L-z)}{s \cos sL - j\delta \sin sL} e^{j\delta z} \tag{4.15a}$$

for $\delta^2 \geq \kappa^2$; and

$$A = A(0) \frac{\sigma \cos \sigma(L-z) - j\delta \sin \sigma(L-z)}{\sigma \cos \sigma L - j\delta \sin \sigma L} e^{-j\delta z},$$

$$B = A(0) \frac{-j\kappa \sin \sigma(L-z)}{\sigma \cos \sigma L - j\delta \sin \sigma L} e^{j\delta z} \tag{4.15b}$$

for $\delta^2 \leq \kappa^2$ with $\sigma = \sqrt{\kappa^2 - \delta^2}$. The reflectance r and transmittance t of field amplitude are obtained:

$$r = \frac{E_-(0)}{E_+(0)} = \frac{-j\kappa \sin sL}{s \cos sL - j\delta \sin sL} \quad (\delta^2 \geq \kappa^2, s = \sqrt{\delta^2 - \kappa^2}),$$

$$= \frac{-j\kappa \sinh \sigma L}{\sigma \cosh \sigma L - j\delta \sinh \sigma L} \quad (\delta^2 \leq \kappa^2, \sigma = \sqrt{\kappa^2 - \delta^2}), \tag{4.16a}$$

$$t = \frac{E_+(L)}{E_+(0)} = \frac{s e^{j\beta_B L}}{s \cos sL - j\delta \sin sL} \quad (\delta^2 \geq \kappa^2, s = \sqrt{\delta^2 - \kappa^2}),$$

$$= \frac{\sigma e^{j\beta_B L}}{\sigma \cosh \sigma L - j\delta \sinh \sigma L} \quad (\delta^2 \leq \kappa^2, \sigma = \sqrt{\kappa^2 - \delta^2}). \tag{4.16b}$$

For the intensity reflectance and transmission, we have

$$R = |r|^2 = \frac{\kappa^2 \sin^2 sL}{\delta^2 - \kappa^2 \cos^2 sL} \quad \left(\delta^2 \geq \kappa^2, s = \sqrt{\delta^2 - \kappa^2}\right),$$

$$= \frac{\kappa^2 \sinh^2 \sigma L}{\kappa^2 \cosh^2 \sigma L - \delta^2} \quad \left(\delta^2 \leq \kappa^2, \sigma = \sqrt{\kappa^2 - \delta^2}\right),$$

(4.17a)

$$T = |t|^2 = \frac{s^2}{\delta^2 - \kappa^2 \cos^2 sL} \quad \left(\delta^2 \geq \kappa^2, s = \sqrt{\delta^2 - \kappa^2}\right),$$

$$= \frac{\sigma^2}{\kappa^2 \cosh^2 \sigma L - \delta^2} \quad \left(\delta^2 \leq \kappa^2, \sigma = \sqrt{\kappa^2 - \delta^2}\right).$$

(4.17b)

4.2.1.2 Basic Characteristics of Uniform FBG

It is seen from (4.17a) and (4.17b) that $R + T = 1$, meeting the requirement of energy conservation. The maximum reflection in the spectrum appears at $\delta = 0$, that is, $\beta = \beta_B = k_B/2$:

$$R_p = \tanh^2(\kappa L), \tag{4.18}$$

justifying the qualitative analysis in Section 4.1. And a series of side lobes exist at

$$\tan(sL) = sL, \tag{4.19}$$

with amplitudes of $R_m = \kappa^2 L^2/[1 + (\beta_m - \beta_B)^2 L^2]$, showing a Lorentzian-like envelope. It is also deduced that a series of zeros occur at $sL = m\pi$, where $m = \pm 1, \pm 2 \ldots$. The first two zeros at $\delta_{\pm 1} = \sqrt{\kappa^2 + \pi^2/L^2}$ gives the full width of the main peak, which decreases with an increase of the grating length L, but increases with the coupling coefficient κ. Figure 4.10 shows the calculated reflective spectra of four uniform Bragg gratings with $\kappa = (1, 2, 8, 5)$ cm^{-1}, corresponding to index increments of $\tilde{n}_1 \approx (0.5, 1, 4, 2.5) \times 10^{-4}$, and $L = (2, 1, 0.4, 0.5)$ cm, respectively.

The linewidth of the reflection peak is an important parameter. Because the reflectivity at the joining point of $s = \sigma = 0$, $R_j = \kappa^2 L^2/(1 + \kappa^2 L^2)$ is always larger than the half maximum $0.5 \tanh^2(\kappa L)$, the linewidth at half maximum, denoted by δ_h, is determined by equation of $\kappa^2 \sin^2(s_h L)/[\delta_h^2 - \kappa^2 \cos^2 s_h L] = \tanh^2(\kappa L)/2$, with $s_h = \sqrt{\delta_h^2 - \kappa^2}$, which can be reduced to $\sin(s_h L)/(s_h L) = \tanh(\kappa L)/[\kappa L\sqrt{2 - \tanh^2(\kappa L)}]$. For an FBG with a higher reflectivity of $R_p \simeq 1$, the equation is expressed approximately as $\mathrm{sinc}(s_h L) = 1/\kappa L$. For a

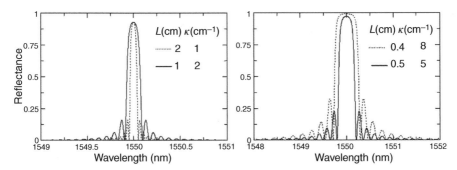

Figure 4.10 Calculated reflective spectra of uniform Bragg gratings.

very weak FBG with $\kappa L \ll 1$, it is $\text{sinc}(s_h L) = \sqrt{2}/2$. It is concluded that a grating with weak coupling and large length is better for high peak reflectivity and narrower linewidth.

The peak sharpness is a concerned parameter. A parabolic approximation at the peak of reflection spectrum is expressed to describe the sharpness of the peak:

$$R = R_p - C^2 L^2 \delta^2, \qquad (4.20)$$

where the top curvature is deduced to be $C = \sqrt{[\kappa L - \tanh(\kappa L)]\tanh(\kappa L)} /[(\kappa L)\cosh(\kappa L)]$. Figure 4.11 shows the variations of peak reflectance R_p and top curvature C with parameter κL, showing that the curvature varies with the peak reflectance, reaches its maximum at $R_p \approx 0.5$, and goes to zero at $R_p \to 0$ and $R_p \to 1$.

It is noticed from (4.16a) and (4.16b) that phase delays exist both in FBG reflection and transmission. The field amplitude reflectance can be written as $r = |r| \exp(i\phi_r)$ with the phase factor of

$$\begin{aligned} \phi_r &= \arctan\left[\frac{-s}{\delta \tan(sL)}\right] \qquad (s = \sqrt{\delta^2 - \kappa^2}), \\ &= \arctan\left[\frac{-\sigma}{\delta \tanh(\sigma L)}\right] \qquad (\sigma = \sqrt{\kappa^2 - \delta^2}). \end{aligned} \qquad (4.21)$$

It is shown that the field reflectance at the peak is a pure imaginary; that is, with phase delay of $\pi/2$. The group delay (GD) $\partial\phi_r/\partial\omega$ can be deduced from (4.21)

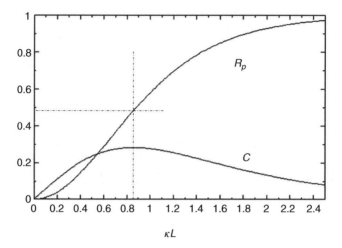

Figure 4.11 Variation of peak reflectivity R_p and curvature C with κL.

$$\tau_g = \frac{L}{v_g} \frac{\delta^2 - \kappa^2 \sin 2sL/(2sL)}{\delta^2 - \kappa^2 \cos^2 sL} \qquad (|\delta| \geq \kappa),$$

$$= \frac{L}{v_g} \frac{\delta^2 - \kappa^2 \sinh 2\sigma L/(2\sigma L)}{\delta^2 - \kappa^2 \cosh^2 \sigma L} \qquad (|\delta| \leq \kappa).$$

(4.22)

At the peak, $\tau_g = \tanh(\kappa L)/(\kappa v_g)$ corresponds to a time delay in a path with length of $\tanh(\kappa L)/\kappa$; it is a short part of the whole length of grating for large κL and high peak reflectance. Figure 4.12 shows the calculated GD, with $v_g \tau_g$ measured in millimeters. It is shown that two high delay peaks appear near the first two zeros in reflective spectrum; and high differential group delays (DGD) exist at two sides of the delay peaks. The dispersion characteristics must be taken into consideration in some applications of FBG.

4.2.1.3 Coupling between Core Mode and Backward Cladding Modes
The deduced formulas explain the characteristics of FBG near its main peak very well, but a series of smaller peaks at the shorter wavelength side of the main lobe are often observed in experiments [36], as shown in Figure 4.13. These peaks disappear and transform into a broader loss band when the grating is soaked in index matching liquid, indicating that these peaks are related to cladding modes.

The coupling between the core mode and cladding modes is omitted in the above deduction because of the consideration of mode

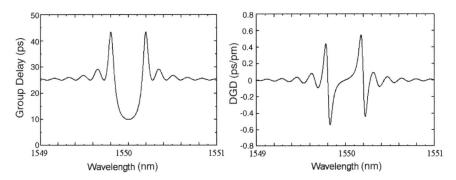

Figure 4.12   Group delay and DGD spectrum of uniform FBG.

orthogonality, expressed as $\int_{\infty} n_0(r, \varphi) F_{co} F_{cl}^{(m)} r\, dr\, d\varphi = 0$, which is the basic property of regular waveguides. However, it is not exact when the index perturbation is considered. The UV-induced index increment occurs mainly in the core, resulting in coupling between the core mode and the backward cladding modes, expressed as

$$\int_{\infty} (n - n_0) \boldsymbol{F}_{co} \boldsymbol{F}_{cl}^{(m)} r\, dr\, d\varphi = \tilde{n}_1 \int_{core} \boldsymbol{F}_{co} \boldsymbol{F}_{cl}^{(m)} r\, dr\, d\varphi, \qquad (4.23)$$

which may give nonzero results as the integral is limited in the core, not the whole transverse space.

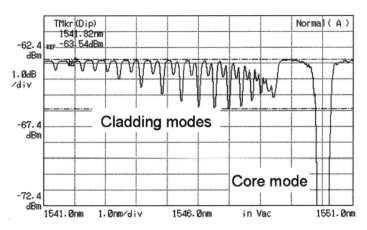

Figure 4.13   Measured spectrum of FBG showing cladding modes.

Figure 4.14   Coupling between core mode and backward cladding modes.

The coupling between the core mode and cladding modes obeys the wave vector relation of

$$k_B = 2\beta_B = \beta_{\text{co}} + \beta_{\text{cl}}^{(m)} = \frac{2\pi\, n_{\text{co}}}{\lambda} + \frac{2\pi\, n_{\text{cl}}^{(m)}}{\lambda}, \qquad (4.24)$$

as depicted in Figure 4.14 for two examples. The resonance wavelength is now written as $\lambda^{(m)} = [n_{\text{co}} + n_{\text{cl}}^{(m)}]\Lambda$.

The side lobes are usually not welcome because they cause cross-talk in dense wavelength division multiplex (DWDM) fiber communication applications. Reference [39,40] discusses the mechanism and presents a method to clear up the side lobe by doping germanium and fluorine in the cladding to make it photosensitive and proper index step to the core. Such a fiber is called cladding mode suppressed photosensitive SMF.

4.2.2   Theory of Long-Period Fiber Grating

As introduced in Section 4.1, the LPFG couples the fundamental core mode with the forward cladding modes, when the grating wave vector equals the wave vector difference between the core mode and the related cladding mode. Figure 4.15 shows a conceptional diagram of the coupling.

References [38,41] give a detailed theory of the LPFG. Here, a brief introduction is given. By the same concept as that for FBG, we start from Helmholtz equation (4.6) with a different index modulation, described as

Figure 4.15   Coupling between core mode and forward cladding modes.

$$n = \begin{cases} n_1 + \tilde{n}_1 \cos(k_L z) & (r < a), \\ n_2 & (r > a), \end{cases} \qquad (4.25)$$

where $k_L = 2\pi/\Lambda$ is the long-period grating wave vector. Based on the CMT, the core mode and related cladding modes are involved in the trial solution of Helmholtz equation with the form of

$$E(r) = F_{co}(r, \varphi) A(z) \exp(j\beta_{co} z) + \sum_m F_{cl}^{(m)}(r, \varphi) B_m(z) \exp(j\beta_m z), \qquad (4.26)$$

where the transverse function of the modes satisfy the zeroth-order equations of

$$\nabla_{xy}^2 F_{co} + (\bar{n}^2 k_0^2 - \beta_{co}^2) F_{co} = 0,$$
$$\nabla_{xy}^2 F_{cl}^{(m)} + (\bar{n}^2 k_0^2 - \beta_m^2) F_{cl}^{(m)} = 0. \qquad (4.27)$$

Substituting (4.25) and (4.26) into Helmholtz equation (4.6), it becomes

$$F_{co}(A'' + j2\beta_{co} A') e^{j\beta_{co} z} + (n^2 - \bar{n}^2) k_0^2 F_{co} A e^{j\beta_{co} z},$$
$$+ F_{cl}^{(m)}(B_m'' + j2\beta_m B_m') e^{j\beta_m z} + (n^2 - \bar{n}^2) k_0^2 F_{cl}^{(m)} B_m e^{j\beta_m z} = 0. \qquad (4.28)$$

By multiplying core mode and cladding mode field functions, and integrating over x–y space, respectively, it is further transformed to

$$A'' + j2\beta_{co} A' + \bar{n}\tilde{n}_1 k_0^2 I_m B_m e^{j(\beta_m + k_L - \beta_{co})z} = 0, \qquad (4.29a)$$

$$B_m'' + j2\beta_m B_m' + \bar{n}\tilde{n}_1 k_0^2 I_m^* A e^{j(\beta_{co} - k_L - \beta_m)z} = 0, \qquad (4.29b)$$

where the orthogonal property between the two modes is used, and terms with $\exp[j(\beta_m - k_L - \beta_{co})]$ in (4.29a) and that with $\exp[j(k_L + \beta_{co} - \beta_m)]$ in (4.29b) are omitted, by similar consideration in the deduction of (4.10). The overlap integral is

$$I_m = \int_{core} F_{co}^* \cdot F_{cl}^{(m)} r \, dr \, d\varphi. \qquad (4.30)$$

Modes F_{co} and $F_{cl}^{(m)}$ are normalized over the whole transverse space. The index modulation is limited in the core now, as described by (4.25), and the integral of I_m is carried in the core, giving a nonzero result. By omitting the second-order differentials, and denoting $2\delta_m = \beta_{co} - \beta_m - k_L$, the CME is deduced to be

$$
\begin{aligned}
A' &= j\kappa_{co}B_m e^{-j2\delta_m z}, \\
B'_m &= j\kappa_m A e^{j2\delta_m z},
\end{aligned}
\tag{4.31}
$$

with coupling coefficients of $\kappa_{co} = \bar{n}\tilde{n}_1 k_0^2 I_m / 2\beta_{co}$ and $\kappa_m = \bar{n}\tilde{n}_1 k_0^2 I_m^* / 2\beta_m$. The two coefficients have near equal values. Similarly to the FBG, the equations are transformed to

$$
\begin{aligned}
A'' - j2\delta_m A' - \kappa^2 A &= 0, \\
B''_m + j2\delta_m B' - \kappa^2 B_m &= 0,
\end{aligned}
\tag{4.32}
$$

with $\kappa^2 = \kappa_{co}\kappa_m$. Its general solution is

$$
\begin{aligned}
A &= (A_1 e^{js_m z} + A_2 e^{-js_m z})e^{-j\delta_m z}, \\
B_m &= (B_{1m} e^{js_m z} + B_{2m} e^{-js_m z})e^{j\delta_m z},
\end{aligned}
\tag{4.33}
$$

with $s_m = \sqrt{\delta_m^2 + \kappa^2}$. The boundary conditions are now $A_1 + A_2 = 1$ and $B_{1m} + B_{2m} = 0$. The mode amplitudes are then deduced to be

$$
\begin{aligned}
A(z) &= \left(\cos s_m z + j\frac{\delta_m}{s}\sin s_m z\right) e^{-j\delta_m z}, \\
B_m(z) &= j\frac{\kappa_m}{s_m} e^{j\delta_m z}\sin s_m z.
\end{aligned}
\tag{4.34}
$$

Intensities of the core mode and the related cladding mode are expressed as

$$
I_A(z) = I_0\left[\cos^2 s_m z + \frac{\delta_m^2}{\delta_m^2 + \kappa^2}\sin^2 s_m z\right],
\tag{4.35}
$$

$$
I_{Bm}(z) = I_0 \frac{\kappa_m^2}{\delta_m^2 + \kappa^2}\sin^2 s_m z.
\tag{4.36}
$$

It is shown that the optical energy is transferred between the core mode and the related cladding mode, forth and back. For most practical applications, the cladding mode will be coupled outward to the fiber jacket, or to the air by bending or so, resulting in a loss of the core mode. For a LPFG with length of L, the transmission spectrum of core mode is then expressed as

$$
I_A(L) = I_0\left[1 - \frac{\kappa^2}{\delta_m^2 + \kappa^2}\sin^2 s_m L\right],
\tag{4.37}
$$

with a resonant loss peak of $\cos^2 \kappa L$ at $\delta = 0$, that is, $\beta_{co} = \beta_m + k_L$. It is noticed that the peak amplitude is a periodic function of κL, different from the behavior of FBG. It increases with κL first, reaches the deepest at $\kappa L = \pi/2$, and then descends with κL increasing, where coupling from the cladding mode back to the core mode becomes dominant.

The 3 dB linewidth δ_H is determined by the equation of

$$\frac{\sin \sqrt{\delta_h^2 + \kappa^2}L}{\sqrt{\delta_h^2 + \kappa^2}L} = \frac{\sqrt{2}\cos \kappa L}{2\kappa L}, \tag{4.38}$$

implying that the linewidth decreases with grating length increase, and varies also with the peak amplitude.

Some features of the LPFG, different from FBG, should be noticed.

4.2.2.1 Mode Coupling Conditions It is shown that the coupling coefficient κ is proportional to the overlap integral between the core mode and the related cladding mode, as shown by (4.30). Obviously, the mode coupling must include condition of $I_m \neq 0$, in addition to the wave vector condition of $\beta_{co} = \beta_{cl} + k_L$. The cladding modes and the core mode have the same field distribution on the azimuth: $F_{co,cl}^{(v)}(r, \varphi) = R_{co,cl}^{(v)}(r) \exp(\pm jv\varphi)$. The overlap integral (4.30) for the fundamental core mode is then expressed as

$$I_m = \int_0^a R_{co}R_{cl}^{(m)}r\,dr \cdot \int_0^{2\pi} \exp[j(v-1)\varphi]d\varphi = I_m^{(1)}\delta_{v1}, \tag{4.39}$$

where δ_{v1} is the Kronecker delta function. Reference [38] analyzes the coupling between the core mode and the cladding modes in detail and calculates the coefficients up to $m \approx 170$.

As discussed in Section 2.2.5 by the two-region model, the field of cladding modes with $v = 1$ is expressed as a sum of $J_0(\beta_t r)$ and $J_2(\beta_t r)$ with different coefficients, which correspond to HE and EH modes, respectively. Obviously, the HE cladding modes have higher coupling coefficients than EH modes.

4.2.2.2 Waveguide Dispersion Factor The resonant wavelength formula $\lambda_L = (n_{co} - n_{cl,m})\Lambda$ gives the basic relation in analyzing dependences of the resonant wavelength on various parameters, such as on the index modulation depth, on the temperature and strain status, and on the index of medium surrounding the fiber. It is found that the dispersions of effective indexes affect the dependences greatly; in other words, functions of $n_{co} = n_{co}(\lambda)$ and $n_{cl} = n_{cl}(\lambda)$ must be taken into account [41,42].

Taking the dispersion of effective index into consideration, the derivative of the resonant wavelength λ_L over grating period Λ is derived as

$$
\frac{\partial \lambda_L}{\partial \Lambda} = [n_{co}(\lambda) - n_{cl,m}(\lambda)] + \Lambda \left(\frac{\partial n_{co}}{\partial \lambda} - \frac{\partial n_{cl,m}}{\partial \lambda} \right) \frac{\partial \lambda_L}{\partial \Lambda} , \qquad (4.40)
$$
$$
= \gamma_m (n_{co} - n_{cl,m})
$$

where a special parameter γ_m, called the waveguide dispersion factor, is introduced:

$$
\gamma_m = \left[1 - \Lambda \frac{\partial (n_{co} - n_{cl,m})}{\partial \lambda} \right]^{-1}. \qquad (4.41)
$$

It is noticed that the factor is not a constant, but a function of the mode number m, working wavelength λ, and the grating period Λ. Reference [43] proposed a method of measuring the waveguide dispersion factor.

The derivative of the resonant wavelength λ_L over the index modulation depth $\tilde{n}_1$, for a certain grating period Λ, is then obtained as

$$
\frac{\partial \lambda_L}{\partial \tilde{n}_1} = \Lambda \frac{\partial (n_{co} - n_{cl,m})}{\partial \tilde{n}_1} + \Lambda \frac{\partial (n_{co} - n_{cl,m})}{\partial \lambda} \frac{\partial \lambda_L}{\partial \tilde{n}_1} = \gamma_m \Lambda. \qquad (4.42)
$$

Such a relation exists in FBG also, but much weaker than that in LPFG. A similar factor for the FBG can be derived as

$$
\gamma_{FBG} = \left(1 - 2 \Lambda_{FBG} \frac{dn_{co}}{d\lambda} \right)^{-1}, \qquad (4.43)
$$

which is almost equal to 1 because of $\Lambda_{FBG} \ll \Lambda_{LPFG}$. It is important to notice in formula (4.41) that the waveguide dispersion factor γ_m of LPFG will be very large if the dispersion of effective index difference reaches the condition of

$$
\Lambda \frac{\partial (n_{co} - n_{cl,m})}{\partial \lambda} \simeq 1. \qquad (4.44)
$$

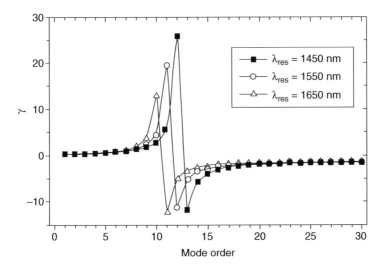

Figure 4.16 Calculated γ factor versus cladding mode order. (Reprinted with permission from reference [42].)

This property is beneficial to enhancing the sensitivity of sensors based on LPFG. Moreover, the factor changes its sign around the condition; beyond the point, γ_m becomes a large negative, which is impossible for γ_{FBG}. Reference [42] calculates the factor of LPFG versus the cladding mode number m for different resonant wavelengths, as shown in Figure 4.16, and also gives the variation of resonance wavelength with the grating period, as shown in Figure 4.17. It is seen that the peak

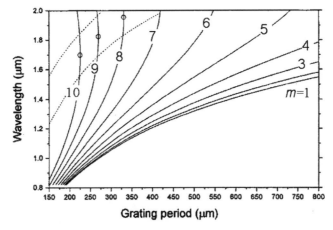

Figure 4.17 Calculated peak wavelength versus grating period. (Reprinted with permission from reference [42].)

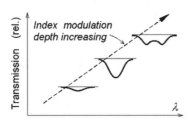

Figure 4.18 Evolution of transmission spectrum in LPFG fabrication.

wavelength is no longer proportional to the grating period, and its variation with the period depends on the mode number.

The flip-flop feature of γ_m versus m was verified experimentally [41,42]. It is shown that the behavior occurs usually in a region of higher mode number m, corresponding to larger index differences $(n_{co} - n_{cl,m})$. Therefore, to make use of the merit of higher sensitivity, a shorter grating period Λ should be adopted for a certain working wavelength.

The property explains the experimentally observed phenomenon in fabricating LPFG that the resonant peak is not only growing, but also shifting fast in the spectrum with the imprinting processing as shown schematically in Figure 4.18. Due to the property more attention has to be paid for precisely controlling the resonant wavelength position of LPFG.

4.2.3 Basic Theory of Nonuniform Fiber Gratings

As introduced in Section 4.1.3 quite many kinds of FBG with substructures are developed for various applications. Both of the grating period and the index modulation amplitude may have nonuniform distributions along the grating length. The CMEs are generally expressed as

$$
\begin{aligned}
A' &= j\kappa(z)B\mathrm{e}^{-j2\delta(z)z}, \\
B' &= -j\kappa(z)A\mathrm{e}^{j2\delta(z)z},
\end{aligned}
\tag{4.45}
$$

where $\delta(z) = \beta - \pi/\Lambda(z)$ and $\kappa(z) = \frac{1}{2}\tilde{n}_1(z)k_0$ are functions of the axial position of the fiber. By rewriting $k_B z = 2\pi z/\Lambda(z) = 2\pi z/\Lambda_0 + 2\theta(z)$, the wave vector deviation and the coupling coefficient are redefined as $\delta = \beta - \pi/\Lambda_0$ and $q(z) = \frac{1}{2}\tilde{n}_1(z)k_0 e^{j\theta(z)}$; the equation is then transformed to

$$A' = jq(z)Be^{-j2\delta z},$$
$$B' = -jq^*(z)Ae^{j2\delta z}. \quad (4.46)$$

To solve the equation, numerical methods have to be used since no analytic solutions exist generally. Several numerical methods of solving the equations of (4.45) are developed, showing effectiveness and different features in designing fiber gratings and understanding their characteristics. Four methods among them are introduced here: (1) transfer matrix method (TMM); (2) Runge–Kutta method; (3) Wentzel–Kramers–Brillouin (WKB) method; and (4) Rouard method.

4.2.3.1 Transfer Matrix Method

Among the various numerical methods the TMM is considered the most widely used method [44], due to its simplicity and universality. In the TMM, the whole length of nonuniform grating is divided into multiple sections, each section is considered as a uniform grating with the analytical solution given in Sections 4.2.1 and 4.2.2. Figure 4.19 shows the multisection model, where the ith section starts at z_{i-1}, ends at z_i with length of $\Delta z_i = z_i - z_{i-1}$. Two inputs and two outputs for each section are linked by matrix of

$$\begin{pmatrix} A_i \\ B_i \end{pmatrix} e^{j\beta z_k} = \begin{pmatrix} T_{11}^{(i)} & T_{12}^{(i)} \\ T_{21}^{(i)} & T_{22}^{(i)} \end{pmatrix} \begin{pmatrix} A_{i-1} \\ B_{i-1} \end{pmatrix} e^{j\beta z_{i-1}}. \quad (4.47)$$

By referring the deduction of uniform FBG formulas in Section 4.2.1, the matrix elements are obtained:

$$T_{11}^{(i)} = [\cos s\Delta z_i + j(\delta/s)\sin s\Delta z_i]e^{-j\delta\Delta z_i}$$
$$T_{12}^{(i)} = j(\kappa/s)\sin s\Delta z_i e^{-j\delta(z_i+z_{i-1})}$$
$$T_{21}^{(i)} = -j(\kappa/s)\sin s\Delta z_i e^{j\delta(z_i+z_{i-1})} \qquad (|\delta| \geq \kappa), \quad (4.48a)$$
$$T_{22}^{(i)} = [\cos s\Delta z_i - j(\delta/s)\sin s\Delta z_i]e^{j\delta\Delta z_i}$$

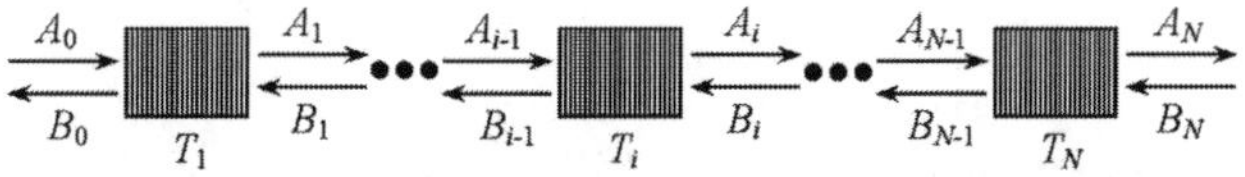

Figure 4.19 Multisection grating model for TMM.

$$T_{11}^{(i)} = [\cosh \sigma \Delta z_i + j(\delta/\sigma) \sinh \sigma \Delta z_i] e^{-j\delta \Delta z_i}$$

$$T_{12}^{(i)} = j(\kappa/\sigma) \sinh \sigma \Delta z_i e^{-j\delta(z_i + z_{i-1})}$$

$$T_{21}^{(i)} = -j(\kappa/\sigma) \sinh \sigma \Delta z_i e^{j\delta(z_i + z_{i-1})} \qquad (|\delta| \le \kappa). \quad (4.48b)$$

$$T_{22}^{(i)} = [\cosh \sigma \Delta z_i - j(\delta/\sigma) \sinh \sigma \Delta z_i] e^{j\delta \Delta z_i}$$

The transmission and reflection of entire grating is obtained by multiplying the concatenated matrices:

$$\begin{pmatrix} A_N \\ B_N \end{pmatrix} = \prod_{i=N}^{1} \begin{pmatrix} T_{11}^{(i)} & T_{12}^{(i)} \\ T_{21}^{(i)} & T_{22}^{(i)} \end{pmatrix} \begin{pmatrix} A_0 \\ B_0 \end{pmatrix} = \begin{pmatrix} T_{11} & T_{12} \\ T_{21} & T_{22} \end{pmatrix} \begin{pmatrix} A_0 \\ B_0 \end{pmatrix}. \quad (4.49)$$

By the boundary condition of

$$\begin{pmatrix} t \\ 0 \end{pmatrix} = \begin{pmatrix} T_{11} & T_{12} \\ T_{21} & T_{22} \end{pmatrix} \begin{pmatrix} 1 \\ r \end{pmatrix}, \quad (4.50)$$

the reflectance and transmission for the grating are obtained to be

$$r = -\frac{T_{21}}{T_{22}}, \quad (4.51a)$$

$$t = \frac{T_{11}T_{22} - T_{12}T_{21}}{T_{22}} = \frac{1}{T_{22}}. \quad (4.51b)$$

It is necessary to discuss the precision of TMM. The first issue of concern is how many sections the grating length should be divided into. Fewer sections mean a larger deviation from a uniform grating, whereas too short a section leaves too few spatial periods to regard it as a grating. It is obvious that the optimal number of N depends on the specific index distribution, that is, the degree of its nonuniformity. The condition of better approximation is roughly written as

$$\Lambda \ll \Delta z_i \ll |\kappa|/|\kappa|', \quad (4.52)$$

where the nonuniform coupling coefficient is considered, and the prime stands for the derivative to z. For CFBGs, the last term is replaced by Λ/Λ'.

The second issue is how to divide the whole grating; for example, by equal lengths or varying lengths? The answer depends also on the specific index distribution and the requirement of simulation. It is necessary to point out that formulas (4.48a) and (4.48b) are deduced from the results of Section 4.2.1, where the phase factor θ is omitted for simplicity. In the transfer matrix calculation if the entire length is divided in an arbitrary way, the phase factors will appear inevitably; then formulas (4.48a) and (4.48b) have to be revised to take the phase factor θ into account. Or otherwise serious errors would be generated in the calculated spectrum, like random noise. To avoid the complexity involved by the phase factor, the best way is to set the Bragg wave vector as $k_B(z_i)z_i = 2m_i\pi$, that is, $\theta = 0$, and $\bar{k}_B^{(i)}\Delta z_i = 2\pi(m_i - m_{i-1})$ for each section.

The TMM is also suitable for the LPFGs; but the matrix (4.48a) and (4.48b) have to be replaced by

$$
\begin{aligned}
T_{11}^{(i)} &= [\cos s\Delta z_i + j(\delta/s)\sin s\Delta z_i]e^{-j\delta\Delta z_i}, \\
T_{12}^{(i)} &= (j\kappa_{co}/s)\sin s\Delta z_i e^{-j\delta(z_i+z_{i-1})}, \\
T_{21}^{(i)} &= (j\kappa_{cl}/s)\sin s\Delta z_i e^{j\delta(z_i+z_{i-1})}, \\
T_{22}^{(i)} &= [\cos s\Delta z_i - j(\delta/s)\sin s\Delta z_i]e^{j\delta\Delta z_i}.
\end{aligned}
\tag{4.53}
$$

The transmitted core mode and cladding mode of entire grating is obtained by multiplying the concatenated matrices:

$$
\begin{pmatrix} A_N \\ B_N \end{pmatrix} = \prod_{i=N}^{1} \begin{pmatrix} T_{11}^{(i)} & T_{12}^{(i)} \\ T_{21}^{(i)} & T_{22}^{(i)} \end{pmatrix} \begin{pmatrix} A_0 \\ B_0 \end{pmatrix} = \begin{pmatrix} T_{11} & T_{12} \\ T_{21} & T_{22} \end{pmatrix} \begin{pmatrix} 1 \\ 0 \end{pmatrix} = \begin{pmatrix} T_{11} \\ T_{21} \end{pmatrix}.
\tag{4.54}
$$

4.2.3.2 Runge–Kutta Method The CME (4.45) is a combined group of the first-order differential equations. The typical and widely used method to solve such equations is the Runge–Kutta method. To express functions $A(z)$ and $B(z)$ in more symmetric forms, $u(z) = A(z)e^{j\delta z}$ and $v(z) = B(z)e^{-j\delta z}$ are introduced, and the equation group is transformed to

$$
\begin{aligned}
u' &= j\delta u + jqv, \\
v' &= -j\delta v - jq^*u.
\end{aligned}
\tag{4.55}
$$

A function of local reflection is defined as $\rho = v/u$ [45], which satisfies a differential equation in the form of a Racatti equation:

$$\rho' = -j(q^* + q\rho^2) - j2\delta\rho, \tag{4.56}$$

which can be decomposed into real part and imaginary parts, and numerically solved by using the Runge–Kutta method with known $\kappa(z)$. The reflection and transmission of the fiber grating are then obtained from the solved $\rho(z)$ with the boundary condition of $\rho(L) = 0$:

$$\begin{aligned} r &= \rho(0), \\ t &= u(L)/u(0). \end{aligned} \tag{4.57}$$

4.2.3.3 Graphical and WKB Analysis

It is shown that the grating spectrum depends strongly on the difference between κ and δ, which is a function of the axial position in nonuniform gratings. To understand its effect, combinations of the slowly varying amplitudes are introduced as $\Sigma(z) = u(z) + v(z)$ and $\Delta(z) = u(z) - v(z)$; they obey the equations of

$$\begin{aligned} \Sigma' &= j(\delta - \kappa)\Delta = j\mu\Delta, \\ \Delta' &= j(\delta + \kappa)\Sigma = j\varepsilon\Sigma, \end{aligned} \tag{4.58}$$

where $\varepsilon = \delta + \kappa$ and $\mu = \delta - \kappa$ are denoted as equivalent dielectric constant and magnetic permeability, as the equations are in form similar to the Maxwell equation. The equations are transformed to

$$\begin{aligned} \Sigma'' + n_e^2\Sigma &= (\mu'/\mu)\Sigma', \\ \Delta'' + n_e^2\Delta &= (\varepsilon'/\varepsilon)\Delta', \end{aligned} \tag{4.59}$$

with $n_e^2(z) = \varepsilon\mu = \delta^2 - \kappa^2$. The property of the local grating depends on the sing of $n_e^2(z)$. In the region of $\delta^2(z) \geq \kappa^2(z)$, $\Sigma(z)$ and $\Delta(z)$ stand for transmitting waves. In the region of $\delta^2(z) \leq \kappa^2(z)$, $\Sigma(z)$ and $\Delta(z)$ demonstrate a decayed evanescent field and stand for a reflected wave; the grating shows a local stop band.

Figure 4.20 gives an illustration of wave vector deviation δ and coupling coefficient κ varied along the fiber grating length, where a linear CFBG with apodized coupling coefficient is presented as an example. The idea of apodization will be explained later in Section 4.2.5. Three regions are depicted in the figure; the middle region corresponds to $\delta^2(z) \leq \kappa^2(z)$, showing an effective reflection; while the two end parts correspond to transmission. The ranges of three regions change with

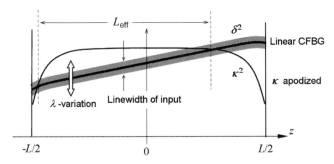

Figure 4.20   Illustration of δ and κ variation along fiber grating.

the wavelength of the input beam, as indicated in the figure by an arrow.

By introducing functions of $\tilde{\Sigma} = \Sigma/\sqrt{\mu}$ and $\tilde{\Delta} = \Delta/\sqrt{\varepsilon}$, Equation (4.59) can be further transformed to a typical form of

$$\tilde{\Sigma}'' + n_\mu^2 \tilde{\Sigma} = 0,$$
$$\tilde{\Delta}'' + n_\varepsilon^2 \tilde{\Delta} = 0, \tag{4.60}$$

with $\quad n_\mu^2 = \varepsilon\mu + \mu''/2\mu - 3\mu'^2/4\mu^2 \quad$ and $\quad n_\varepsilon^2 = \varepsilon\mu + \varepsilon''/2\varepsilon - 3\varepsilon'^2/4\varepsilon^2$, where the prime stands for the derivative to z. Equation (4.60) is similar to the Schrödinger equation for the wave function passing through a potential barrier, for which the WKB method is developed successfully. The solutions in WKB form are written as

$$\tilde{\Sigma}(z) = n_\mu^{-1/2}\left[\Sigma_1 \exp\left(j\int_0^z n_\mu dz\right) + \Sigma_2 \exp\left(-j\int_0^z n_\mu dz\right)\right],$$
$$\tilde{\Delta}(z) = n_\varepsilon^{-1/2}\left[\Delta_1 \exp\left(j\int_0^z n_\varepsilon dz\right) + \Delta_2 \exp\left(-j\int_0^z n_\varepsilon dz\right)\right]. \tag{4.61}$$

Attention has to be paid in dealing with the transition at $\delta^2(z) = \kappa^2(z)$; conditions for better approximation and asymptotic formulas are given in quantum mechanics. Detailed analysis can be found in reference [46]. The graphical and WKB method is suitable for investigating chirped fiber gratings especially.

4.2.3.4 Rouard Method   The fiber grating can be regarded as a one-dimensional multilayer structure. Reflection and transmission of an incident optical beam occur at every interface between the layers, described by Fresnel formulas; and multibeam interferences occur due to phase differences between the interfaces. Rouard presented a

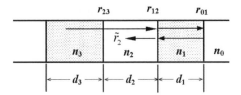

Figure 4.21   Illustration of thin film stack for recursive calculation.

recursive method to analyze multilayer films. The basic concept is to reduce two interfaces and the layer in between to one interface with an effective reflection and transmission; and then do the next layer and so forth, as shown in Figure 4.21 [47,48].

The reflection at the first interface is written as $\tilde{r}_1 = r_{01} = (n_1 - n_0)/(n_1 + n_0)$ according to Fresnel. By considering the first layer as one interface, the reduced reflection is written as [49] $\tilde{r}_2 = (\tilde{r}_1 + r_{12}e^{j2\delta_1})/(1 + \tilde{r}_1 r_{12}e^{j2\delta_1})$, with $r_{12} = (n_2 - n_1)/(n_2 + n_1)$ and $\delta_1 = 2\pi n_1 d_1/\lambda$. For the next layer, the composite reflection is reduced as

$$\tilde{r}_3 = \frac{\tilde{r}_2 + r_{23}e^{j2\delta_2}}{1 + \tilde{r}_2 r_{23}e^{j2\delta_2}}, \tag{4.62}$$

and so on, to cover the entire range of the grating.

In the fiber grating simulation by the Rouard method, each period of the grating can be divided into several layers, such as two layers for a step-index modulation. It is implied that the Rouard method is suitable for analyzing arbitrary nonuniform gratings, including nonuniform LPFG, although the calculation time may be longer than other methods.

4.2.4 Inverse Engineering Design

The characteristics of a fiber grating with known index modulation distribution are obtained by using the methods introduced above. Practical applications often put forward requirements of fiber gratings with specified reflection and transmission spectra, including their amplitudes and phases. It is needed to design the index modulation to synthesize and fabricate gratings meeting the required specifications. This is the task of inverse engineering design. Several algorithms are developed for the purpose, and have proven effective and satisfactory in practice. Among them genetic algorithm (GA) and layer peeling algorithm (LPA) are widely used.

4.2.4.1 Genetic Algorithm The GA method imitates the evolution of species in nature to optimize the designed gratings. The procedures of GA include [50,51] the following:

1. A population of "individuals" is created; each of them is a proposed grating, aided by physical conjecture but with certain randomness; and their characteristics are calculated by selected numerical methods, such as TMM and the Runge–Kutta method.
2. Assign a fitness function to characterize differences of each "individual" from the target, with some weighting factors.
3. Define operators of creating offspring, including crossover to generate better individuals, and mutation to expand the range of searching.
4. Estimate the fitness of the new generation; give higher chance of reproducing to the more fit individuals. Repeat the procedure. Set the least error to end the optimization.

Some technical issues are of concern to obtain higher precision and faster calculation for specific requirements, and the developed skills are given in reference [52]. An example of the fitness function for FBG design is [50]

$$d(R_{\text{calc}}, R_{\text{target}}) = \left\{ \sum_i \left(c_i |R_{\text{calc}}^{(i)} - R_{\text{target}}^{(i)}| \right)^p \right\}^{1/p}, \qquad (4.63)$$

where i is the number of discrete data of reflection spectra, c_i is the weighting factor, and p is a parameter to be optimized for specific requirements. Reference [53] shows an FBG with triangle reflectivity spectrum designed by GA.

4.2.4.2 Layer-Peeling Algorithm The input optical wave is regarded as a temporally incident beam in the layer-peeling algorithm [54–57]. The grating to be synthesized is divided into N layers with thickness of Δ. The required reflectance should be obtained at the surface of the first layer; by causality argument, it is the result of reflection at the second interface and propagation through the first layer in view

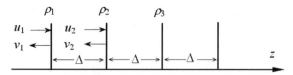

Figure 4.22   Conceptional illustration of layer peeling algorithm.

of temporal propagation. The matrix describing the effect of a layer is given in (4.48b) of TMM, written as:

$$\begin{pmatrix} \cosh(\sigma\Delta) + j\dfrac{\delta}{\sigma}\sinh(\sigma\Delta) & j\dfrac{q}{\sigma}\sinh(\sigma\Delta) \\ j\dfrac{q}{\sigma}\sinh(\sigma\Delta) & \cosh(\sigma\Delta) - j\dfrac{\delta}{\sigma}\sinh(\sigma\Delta) \end{pmatrix}.$$

The transfer matrix in LPA is modified to a product of $T_q \cdot T_\Delta$:

$$T_\Delta = \begin{pmatrix} e^{j\delta\Delta} & 0 \\ 0 & e^{-j\delta\Delta} \end{pmatrix},$$ (4.64a)

$$T_q = \begin{pmatrix} \cosh(|q|\,\Delta) & \dfrac{q}{|q|}\sinh(|q|\,\Delta) \\ \dfrac{q^*}{|q|}\sinh(|q|\,\Delta) & \cosh(|q|\,\Delta) \end{pmatrix} = \dfrac{1}{\sqrt{1-\rho\rho^*}}\begin{pmatrix} 1 & \rho^* \\ \rho & 1 \end{pmatrix},$$

(4.64b)

where the layer thickness is shrunk to a point in discrete PLA, while keeping $|q|\,\Delta = Const.$, the matrix elements are approximated as $\sigma \approx |q|$ for the peak reflection of $|\delta| \ll |q|$; a parameter $\rho = (q^*/|q|)\tanh(|q|\,\Delta)$ is introduced in (4.64b). The reflection at each layer is regarded as an impulse response, which are concatenated in sequence to form the reflection of the entire grating, as shown in Figure 4.22.

At the first point the input and reflected field are described as

$$\begin{pmatrix} u_1 \\ v_1 \end{pmatrix} = \begin{pmatrix} 1 \\ r(\delta) \end{pmatrix},$$ (4.65)

where $r(\delta)$ is the required reflectance of FBG to be synthesized. At the second layer the fields are expressed as:

$$\begin{pmatrix} u_2 \\ v_2 \end{pmatrix} = T_q T_\Delta \begin{pmatrix} 1 \\ r \end{pmatrix} \approx \begin{pmatrix} e^{j\delta\Delta} + r\rho_1^* e^{-j\delta\Delta} \\ re^{-j\delta\Delta} + \rho_1 e^{j\delta\Delta} \end{pmatrix},$$ (4.66)

resulting in

$$r_2 = \frac{v_2}{u_2} = e^{-j2\delta\Delta}\frac{r + \rho_1}{1 + r\rho_1^*}. \tag{4.67}$$

From the view of temporal propagation, the required reflection spectrum is decomposed as a Fourier series:

$$r(\delta) = \sum_{\tau=0}^{\infty} h(\tau)e^{j\delta\Delta\tau}, \tag{4.68}$$

where τ is a discrete time variable. The 0-th order coefficient is considered to be the impulse reflection ρ_1 as if only the first layer were present:

$$\rho_1 = h(0) = \frac{\Delta}{\pi}\int_{-\pi/2\Delta}^{\pi/2\Delta} r(\delta)d\delta, \tag{4.69}$$

which gives the coupling coefficient q of the first layer. After the first layer is peeled off, the same consideration is applied to the second layer. The whole procedure of PLA includes the following:

1. Start from a physical realizable reflection $r(\delta)$ as the target;
2. Compute its zeroth order coefficient of Fourier expansion ρ_1 by (4.69);
3. Compute reflection of the second layer r_2 by (4.67);
4. Repeat step (2) with

$$\rho_i = h_i(0) = \frac{\Delta}{\pi}\int_{-\pi/2\Delta}^{\pi/2\Delta} r_i(\delta)d\delta, \tag{4.70}$$

until the entire grating structure is determined.

LPA method is used for inversely designing FBG not only with specific reflection spectrum, but also with specific dispersion requirements [57].

4.2.4.3 Fourier-Transform Inverse-Scattering Technique The Fourier transform method has been used to design dielectric multilayer filters since early on and has also been developed for corrugated

waveguide filter designs [58]. Fourier transform gives a simple and explicit synthesis for weak gratings with low peak reflectance, for example, less than 10%. Racatti equation (4.56) is then approximated to be $\rho' \approx -j2\delta\rho - jq^*$ with solution of $\rho = -je^{-j2\delta z} \int_L^z q^* e^{j2\delta z} dz$. The reflectance at $z = 0$ is obtained as $r = \rho(0) = j \int_0^L q^* e^{j2\delta z} dz$. Inversely, if the reflectance $r(\lambda)$ is required, the coupling coefficient distribution along the grating length is synthesized by Fourier transform of

$$q(z) = \frac{1}{2\pi} \int_{-\infty}^{\infty} r(k)e^{-j2kz} dk. \tag{4.71}$$

Such a synthesis method for weak gratings is called the first-order Born approximation [49]. For gratings with stronger coupling, Fourier transform method should take a different form. With $q = |q|e^{j\theta}$ the Racatti equation is written as

$$\rho' = -j2\delta\rho - j|q|[\cos\theta(1 + \rho^2) - j\sin\theta(1 - \rho^2)]. \tag{4.72}$$

By using relations of $d(\tanh^{-1}\rho)/dz = \rho'/(1 - \rho^2)$ and $\rho/(1 - \rho^2) \approx \tanh^{-1}\rho$, the equation is rewritten and approximated as

$$\frac{d}{dz}(\tanh^{-1}\rho) = -j2\delta(\tanh^{-1}\rho) - |q|\sin\theta - j|q|\cos\theta\frac{1 + \rho^2}{1 - \rho^2}$$

$$\approx -j2\delta(\tanh^{-1}\rho) - jq.$$

Its solution is obtained to be $\tanh^{-1}\rho = -je^{-j2\delta z} \int_L^z q^* e^{j2\delta z} dz$, and inversely,

$$q(z) = \frac{1}{2\pi} \int_{-\infty}^{\infty} \tanh^{-1}[r(k)]e^{-j2kz} dk. \tag{4.73}$$

It is worth noticing that the peak reflectance of a uniform FBG is $r_p = -j\tanh(\kappa L)$, implying that the coupling coefficient can be regarded as the Fourier transform of the inverse hyperbolic tangent of peak reflectance. References [59–61] give detailed deduction and discussion of the method.

Apart from GA, LPA, and Fourier transform, other methods are also used to synthesize fiber gratings inversely. The CMEs can be converted to a pair of integral equations, with forms similar to

Gelfand–Levitan–Marchenko (GLM) coupled integral equations [44], which are investigated for scattering problems in quantum mechanics. The iterative solutions are given for the GLM algorithm [62]. Reference [63] gives a discussion by using the standard communication theory.

4.2.5 Apodization of Fiber Grating

It is shown in Figure 4.10 that the reflection of FBG has multiple side lobes. Such a feature is not satisfactory for usual applications where a filter with a single narrow band is needed.

It is noticed that the spacing between lobes is basically inversely proportional to the grating length L; and the side lobes present a Lorentzian-like amplitude distribution, which is similar to the sinc function describing diffraction from a rectangular aperture. From the basic principle of Fourier transform, it is reasonable to conjecture that if the rectangular amplitude of a fiber grating coupling coefficient is modified to near a parabolic function, the side lobes will be greatly reduced. It is conceived also that soft boundaries at the two ends of FBG, instead of a sharp interruption, will weaken the multiple reflections between them, leading to lower and fewer side lobes. This method of improving FBG's spectral characteristics is called apodization [64–67].

The widely used apodization functions are listed as follows:

Raise cosine: $\quad f(z) = 1 - a \cos^2(\pi z/L),$ $\qquad$ (4.74a)

Gaussian: $\quad f(z) = \exp \dfrac{-g(z - L/2)^2}{L^2},$ $\qquad$ (4.74b)

tanh: $\quad f(z) = \begin{cases} \tanh(2az/L)/\tanh a & (0 \le z \le L/2), \\ \tanh[2a(L - z)/L]/\tanh a & (L/2 \le z \le L), \end{cases}$ $\qquad$ (4.74c)

sinc: $\quad f(z) = \dfrac{\sin[\pi(z/2L - 1)]}{[\pi(z/2L - 1)]}.$ $\qquad$ (4.74d)

The effectiveness of different apodization functions can be simulated by TMM. The index distribution of Bragg grating is now expressed as

$$n(z) = \bar{n}(z) + \bar{n}_1 f(z) \cos k_B z. \qquad (4.75)$$

The action of apodization can be regarded as a grating with compositions of different spatial frequencies, according to Fourier transform of $f(z)$. For example, the raised cosine function apodized index modulation is expressed as

$$n(z) = \bar{n} + \tilde{n}_1[1 - a\cos^2(\pi z/L)]\cos k_B z$$

$$= \bar{n} + \left(1 - \frac{a}{2}\right)\tilde{n}_1\left[\cos k_B z + \frac{1}{2}\cos\left(k_B + \frac{2\pi}{L}\right)z\right. \quad (4.76)$$

$$\left. + \frac{1}{2}\cos\left(k_B - \frac{2\pi}{L}\right)z\right].$$

The oscillations outside the band are now composed of three spectra with different periods, resulting in smoothened oscillations.

Figure 4.23 shows the reflection spectra calculated by using TMM for three apodization functions of raised cosine (Figure 4.23b), with parameter $a = 0.5$; Gaussian function (Figure 4.23c), with $g = 3$; and sinc function (Figure 4.23d), compared with the original one without apodization (Figure 4.23a). The simulated grating length is 5 mm, and the maximum coupling coefficient at $z = L/2$ is 5 cm^{-1}. It is seen that the side lobes are removed greatly; in addition, the peak amplitude decreases

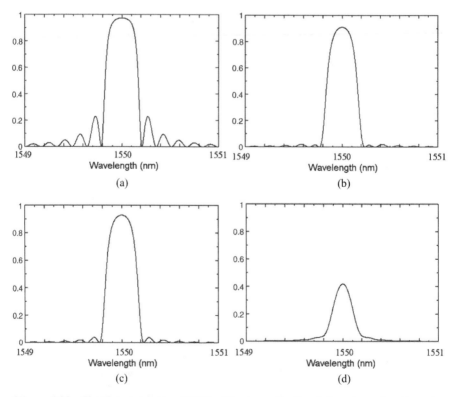

Figure 4.23   Simulated spectra of FBG with no apodization (a), and apodized by raise cosine function (b), Gaussian function (c), and sinc function (d).

in some degree with different apodization functions. It is also noticed that the linewidth and the peak sharpness are also changed by apodization, and different apodization functions give different peak reflections. For specific requirements of grating performance, the apodization parameters have to be designed and optimized carefully.

To carry out the apodization in fabrication of FBGs, several methods have been developed. The most reliable and convenient method is to use a specially designed phase mask with a UV transmission varying according to the desired function. This method is suitable for mass production. Two-step exposure with an amplitude mask scanning along the fiber, together with the phase mask, is often used to get desired index modulation. Flood exposure to reduce the index modulation depth gradually at the two end parts of the fabricated uniform FBG is a simple and effective method. Apodization is the effective technique to remove the ripples in CFBGs. Detailed analysis is given in Section 4.3.3.

The above-introduced apodization function $f(z)$ is an intentional design. In practice, nonuniformity of UV irradiation often occurs, resulting in an unintentional apodization. It is observed that the line shape of FBG spectrum is not symmetric, usually with higher side lobe at its shorter wavelength side than the other side. One of the mechanisms is attributed to nonuniform UV irradiations, resulting in a nonuniform index background $\bar{n}(z)$, usually in a convex form, and consequently a nonuniform Bragg wavelength along the grating length $\lambda_B(z) \propto 2\bar{n}(z)\Lambda$. An index background function is assumed to be $\bar{n} = n_0 + \Delta n \exp[-4(z - L/2)^2/L^2]$. Thus short wavelength light waves will be reflected at the two end parts of the grating, and an effect similar to Fabry–Perot (F-P) oscillation occurs, giving resonant peaks at the short wavelength side, whereas the main lobe comes from the middle part of the grating. Figure 4.24 shows a simulated spectrum of an FBG with length of 10 mm and $\Delta n = 5 \times 10^{-5}$.

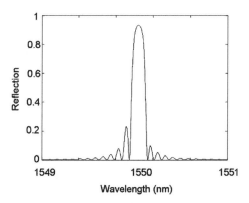

Figure 4.24 Simulated FBG spectrum with Gaussian index background.

4.3 SPECIAL FIBER GRATING DEVICES

A great number of fiber grating devices with various structures and attractive characteristics have been developed based on the UV-induced photosensitivity. Among them, several typical and useful gratings are described in this section. Section 4.3.1 is devoted to multisection FBGs. Section 4.3.2 describes the characteristics of CFBG. Section 4.3.3 explains the effect of tilted gratings. Section 4.3.4 states the issues related to PMF gratings. In the last section, two fiber grating devices with special structures are introduced.

4.3.1 Multisection FBGs

Structures composed of multisection uniform FBGs show attractive unique features; sometimes they are called superstructured fiber gratings. Among them, phase-shifted FBG, sampled FBG, FBG F-P cavity, and moiré fiber gratings are introduced in this subsection.

4.3.1.1 Phase-Shifted FBG The characteristics of waveguide gratings have been investigated sufficiently since being used in the distributed feedback (DFB) semiconductor laser. Phase shift technique is proved effective to overcome the stop-band problem in DFB lasers [35,68]. The physical principle of FBG is the same as that of DFB lasers, inspiring an idea to develop the phase-shifted FBG for a spectrum with transmission peak instead of reflection peak [69].

The structure of phase-shifted FBG is schematically shown in Figure 4.9(e). A phase-shifted grating can be considered as a three-section structure: a short phase-shift section sandwiched between two uniform gratings with different lengths of l_1 and l_2. Its transmission is described by multiplying matrixes as

$$
\begin{pmatrix} A \\ B \end{pmatrix} = \begin{pmatrix} T_{11}^{(2)} & T_{12}^{(2)} \\ T_{21}^{(2)} & T_{22}^{(2)} \end{pmatrix} \begin{pmatrix} e^{j\vartheta} & 0 \\ 0 & e^{-j\vartheta} \end{pmatrix} \begin{pmatrix} T_{11}^{(1)} & T_{12}^{(1)} \\ T_{21}^{(1)} & T_{22}^{(1)} \end{pmatrix} \begin{pmatrix} A_0 \\ B_0 \end{pmatrix}
$$

$$
= \begin{pmatrix} T_{11}^{(2)}T_{11}^{(1)}e^{j\vartheta} + T_{12}^{(2)}T_{21}^{(1)}e^{-j\vartheta} & T_{11}^{(2)}T_{12}^{(1)}e^{j\vartheta} + T_{12}^{(2)}T_{22}^{(1)}e^{-j\vartheta} \\ T_{21}^{(2)}T_{11}^{(1)}e^{j\vartheta} + T_{22}^{(2)}T_{21}^{(1)}e^{-j\vartheta} & T_{21}^{(2)}T_{12}^{(1)}e^{j\vartheta} + T_{22}^{(2)}T_{22}^{(1)}e^{-j\vartheta} \end{pmatrix} \begin{pmatrix} A_0 \\ B_0 \end{pmatrix}.
$$

$$(4.77)$$

The phase shift in practical gratings is fabricated in a section much shorter than the two uniform FBG sections, and can be regarded as occurring at a lumped point, expressed by a diagonal matrix. The optical phase shift at the point is approximated to a constant value 2ϑ with no dependence on the wavelength because the concerned band is very narrow near the peak. By the boundary condition of $\begin{pmatrix} t \\ 0 \end{pmatrix} = \begin{pmatrix} T_{11} & T_{12} \\ T_{21} & T_{22} \end{pmatrix} \begin{pmatrix} 1 \\ r \end{pmatrix}$, the reflectance and transmission of phase-shifted grating are obtained to be

$$r = -\frac{T_{21}^{(2)} T_{11}^{(1)} e^{j\vartheta} + T_{22}^{(2)} T_{21}^{(1)} e^{-j\vartheta}}{T_{21}^{(2)} T_{12}^{(1)} e^{j\vartheta} + T_{22}^{(2)} T_{22}^{(1)} e^{-j\vartheta}}, \tag{4.78a}$$

$$t = \frac{1}{T_{21}^{(2)} T_{12}^{(1)} e^{i\vartheta} + T_{22}^{(2)} T_{22}^{(1)} e^{-i\vartheta}}. \tag{4.78b}$$

Usually the two sections of FBG are with the same index modulation $\tilde{n}_1$ and the same period Λ, for the reasons of required performances and of fabrication easiness. By referring the formulas of matrix for uniform gratings (4.48a) and (4.48b), it is obtained that

$$r = -\kappa \frac{(s \cos sl_2 - j\delta \sin sl_2) \sin sl_1 + (s \cos sl_1 + j\delta \sin sl_1) \sin sl_2 e^{j2\vartheta}}{(s \cos sl_2 - j\delta \sin sl_2)(s \cos sl_1 - j\delta \sin sl_1) + \kappa^2 \sin sl_1 \sin sl_2 e^{j2\vartheta}},$$

$$\tag{4.79a}$$

$$t = \frac{s^2 e^{j[\vartheta + \delta(l_1 + l_2)]}}{(s \cos sl_2 - j\delta \sin sl_2)(s \cos sl_1 - j\delta \sin sl_1) + \kappa^2 \sin sl_1 \sin sl_2 e^{j2\vartheta}}.$$

$$\tag{4.79b}$$

Formulas (4.79a) and (4.79b) are for $\delta^2 \geq \kappa^2$; those for $\delta^2 \leq \kappa^2$ can be written accordingly.

A typical structure of phase-shifted FBG is with two sections of equal lengths, $l_1 = l_2 = l = L/2$, resulting in

$$r = -\kappa \frac{\sinh \sigma l [(1 + e^{j2\vartheta})\sigma \cosh \sigma l - j(1 - e^{j2\vartheta})\delta \sinh \sigma l]}{\sigma(\sigma \cosh 2\sigma l - j\delta \sinh 2\sigma l) - (1 - e^{j2\vartheta})\kappa^2 \sinh^2 \sigma l}, \tag{4.80a}$$

$$t = \frac{\sigma^2 e^{j(\vartheta + 2\delta l)}}{\sigma(\sigma \cosh 2\sigma l - j\delta \sinh 2\sigma l) - (1 - e^{j2\vartheta})\kappa^2 \sinh^2 \sigma l}, \tag{4.80b}$$

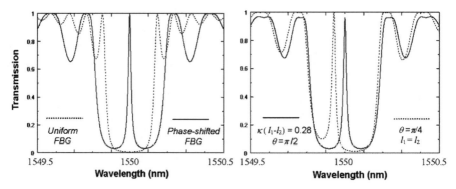

Figure 4.25   Transmission spectra of phase shifted FBG with different parameters.

which are for the region of $\delta^2 \leq \kappa^2$, as the behavior near the peak is the main concern for phase-shifted gratings. The reflectance and transmission of optical intensity are easily calculated by rr^* and tt^*. Figure 4.25 shows calculated transmission spectra with different phase shifts: $\vartheta = \pi/2, \pi/4$, and $\vartheta = 0$ for comparison. It is shown that a narrow transmission peak is obtained in $\pi/2$ phase-shifted FBG; whereas the peak moves off the central position when the phase deviates from $\pi/2$. The peak transmission reaches unity when $l_1 = l_2$, and decreases for $l_1 \neq l_2$.

To discuss its characteristics in detail, the transmission spectrum of intensity of $\pi/2$ phase-shifted FBG is deduced as

$$
\begin{aligned}
T &= \frac{(\kappa^2 - \delta^2)^2}{(\kappa^2 - \delta^2)^2 + 4\kappa^2\delta^2 \sinh^4 \sigma l} \\
&\approx \frac{1 - 2\delta^2/\kappa^2}{1 - 2\delta^2/\kappa^2 + 4\delta^2 \sinh^4 \kappa l/\kappa^2} \approx 1 - 4\frac{\sinh^4 \kappa l}{\kappa^2}\delta^2.
\end{aligned}
\tag{4.81}
$$

The last two approximations are for $\delta^2 \ll \kappa^2$; the top curvature is then given as

$$
T_\delta'' = -\frac{8 \sinh^4 \kappa l}{\kappa^2} \approx \frac{-1}{2\kappa^2} \exp(4\kappa l),
\tag{4.82}
$$

showing that the peak is quite sharp, its top curvature increases with the grating length. The full width of half maximum (FWHM) bandwidth

$\delta_{1/2}$ is deduced from (4.81) to be the root of equation

$$\frac{\sinh^4 \sqrt{\kappa^2 - \delta_{1/2}^2} l}{(\kappa^2 - \delta_{1/2}^2)^2} = \frac{1}{4\kappa^2 \delta_{1/2}^2} \tag{4.83}$$

with an approximate solution of

$$\delta_{1/2} \approx \kappa \exp(-\kappa l), \tag{4.84}$$

much smaller than that of uniform gratings, which is about κ for usual values of κL. It is seen from $\kappa L = \tanh^{-1} \sqrt{R_p}$ that the higher the reflection of the grating, the narrower the linewidth of phase-shift FBG.

For practical fabrications of phase-shifted FBG, deviations of the phase factor and the position of phase-shift insertion from ideal designs are often inevitable. The transmission of an equal-length segmented phase-shifted grating with a phase factor deviated from $\pi/2$ is deduced approximately as

$$T \approx \frac{1}{1 + 4\delta^2 \sinh^4 \kappa l / \kappa^2 + \Delta(\vartheta)}, \tag{4.85}$$

where term $\Delta(\vartheta)$ is a factor showing the effect of phase shift:

$$\Delta(\vartheta) \approx \sinh^2 2\kappa l \cos^2 \vartheta - 4 \sinh^3 \kappa l \cosh \kappa l \sin 2\vartheta \cdot \delta/\kappa. \tag{4.85a}$$

The spectral peak deviation is then deduced to be

$$\delta_\vartheta = \frac{\kappa \cosh \kappa l}{2 \sinh \kappa l} \sin 2\vartheta \approx \kappa \sin \vartheta \cos \vartheta, \tag{4.86}$$

where the approximation is for large κl. On the other hand, the section length difference $l_1 - l_2$ causes the peak transmission amplitude to decrease and is expressed as

$$T_M \approx \frac{1}{1 + \kappa^2 (l_1 - l_2)^2 / 2}. \tag{4.87}$$

To fabricate phase-shifted fiber gratings several techniques are adopted. For example:

1. With a phase-shifted phase mask as shown in Figure 4.26, where the doted line of index modulation is the original index distribution without phase shift, while the solid line is the phase-shifted index modulation.

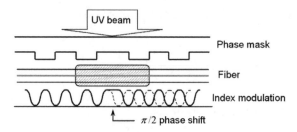

Figure 4.26 Phase shifted FBG fabricated with a phase mask.

2. Localized sheltering of UV radiation. During UV exposure over a phase mask for uniform FBG, a thin mask, such as a metal wire, is placed perpendicular to the fiber to shelter the UV laser beam at the point where the phase shift is to be induced.

3. Additional UV point exposure. After FBG fabrication, an additional UV beam radiates at the point of phase shift, by using a mask with a narrow gap, or by a focused UV beam.

In Figure 4.26 a dashed line square marks the localized sheltering, or the additional point exposure, which may cover several periods of grating, but makes an equivalent phase shift between the two sections:

$$\vartheta = k_0 \int_{\Delta l} \Delta n(z) \mathrm{d}z = \frac{2\pi}{\lambda} \Delta \bar{n} \Delta l. \qquad (4.88)$$

A $\pi/2$ phase shift means $\Delta \bar{n} \Delta l = \lambda/4$, which is why it is termed as a quarter wavelength phase shift. The spectrum is monitored during the UV exposure with localized sheltering and the additional point exposure to inspect the phase factor.

4. Dynamic introduction of the phase shift. One of the methods is by heating a short section in the middle of FBG to induce localized index change.

As a narrow bandwidth filter, the phase-shifted FBG is used in a variety of applications, such as for narrow linewidth laser sources, high-precision wavelength discriminator, high-sensitive sensors, and in DWDM technology.

4.3.1.2 Sampled FBGs A fiber grating with multiple blank sections is called a sampled FBG (SFBG). A typical SFBG has a periodic sampling function and a uniform grating period, as shown in Figure 4.27, where the sampling period is p, the grating section length is p_g, and the sampling ratio is defined as $\eta = p_g/p$. Many papers have been published on SFBG fabrications and characteristics [70–74].

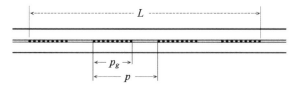

Figure 4.27   Basic structure of sampled FBG.

The unique feature of SFBG spectral characteristics is its multiple reflection peaks. In the SFBG, the index modulation is modified as

$$\tilde{n}(z) = \tilde{n}_1 f(z) \cos \frac{2\pi z}{\Lambda}, \tag{4.89}$$

with a periodic rectangular sampling function $f(z)$. Assuming the entire length L is so large that the number of sections is regarded as infinitive, Fourier decomposition gives

$$\tilde{n}(z) = \tilde{n}_1 \eta \left[\cos \frac{2\pi z}{\Lambda} + \sum_{n=1}^{\infty} \frac{\sin(\pi n \eta)}{\pi n \eta} \left(\cos \frac{2\pi z}{\Lambda_{n-}} + \cos \frac{2\pi z}{\Lambda_{n+}} \right) \right]. \tag{4.90}$$

It means that the SFBG is regarded as a composite grating combined by series of gratings with different periods of

$$\Lambda_{n\pm} = \Lambda/(1 \mp n\Lambda/p) \approx \Lambda(1 \pm n\Lambda/p). \tag{4.91}$$

They correspond to a series of peak wavelengths with spacing of

$$\Delta\lambda = 2n_{\text{eff}} \frac{\Lambda^2}{p} = \frac{\lambda_0^2}{2n_{\text{eff}} p}. \tag{4.92}$$

For example, if $\Delta\lambda = 0.8$ nm is needed in a 1,550 nm band, the sampling period is designed to be ~ 1 mm. It is noticed that the wavelength spacing does not depend on the sampling ratio η.

It is shown that the envelope of the multipeak is in form of a sinc function: $\sin(\pi n \eta)/(\pi n \eta)$. The number of reflective peaks in the central lobe of sinc function is inversely proportional to the sampling ratio: $N \le 2/\eta$, implying that lower η should be adopted if more peaks are needed; and the spectral width of the central envelop lobe is $N\Delta\lambda = \lambda_0^2/(n_{\text{eff}} p_g)$.

On the other hand, a longer entire grating length L should be used to obtain enough amplitude of reflection. The spectral width of every peak is also narrowed by a longer L, which is similar to the behavior of uniform FBG. The entire length is regarded as a single rectangular window, instead of the infinitive length assumption. Therefore, (4.90) must be multiplied by a sinc function of $\sin[(\beta - \beta_B)L]/[(\beta - \beta_B)L]$. Referring to the basic characteristics of a uniform FBG, the spectral width of mth peak is expressed as [75]

$$\Delta\lambda_m = \frac{\lambda^2}{\pi n_{\text{eff}}} \sqrt{\left(\frac{\tilde{n}_1}{\lambda} \sin \pi m \eta\right)^2 + \left(\frac{\pi}{L}\right)^2}, \qquad (4.93)$$

implying that the linewidth is inversely proportional to L. To obtain the detailed and exact transmission and reflective spectra of SFBG, numerical simulations are needed, mostly by TMM.

It is worth noticing in experiments and in simulations that lots of noise-like peaks exist in between the multiple peaks, which are attributed to the reflections back and forth at the boundaries of each section grating and blank part. The apodization is an effective technique

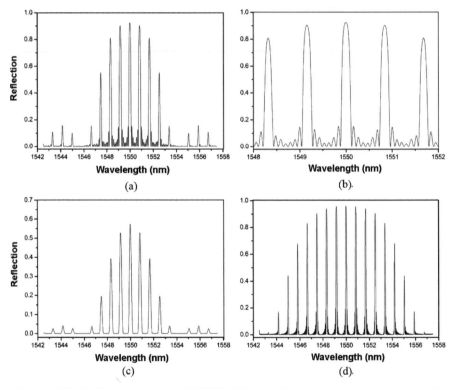

Figure 4.28 Reflection spectra of SFBG: (a) $L = 8$ mm, $p = 1$ mm, and $\eta = 1/5$; (b) magnified; (c) apodized; and (d) $L = 16$ mm, $p = 1$ mm, and $\eta = 1/10$.

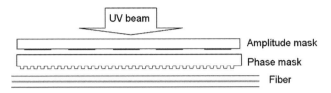

Figure 4.29   Schematic diagram of SFBG imprinting.

to suppress this kind of noise. Figure 4.28(a) shows simulated reflection spectra for $L = 8$ mm, $p = 1$ mm, and $\eta = 1/5$. Figure 4.28(b) shows its middle part; Figure 4.28(c) shows the spectrum of apodized SFBG with $f(z) = [1 - \cos(2\pi z/L)]/2$. Figure 4.28(d) is for $L = 16$ mm, $p = 1$ mm, and $\eta = 1/10$, showing that the linewidth has narrowed.

Fabrication of SFBG is not very difficult. A commonly used method is a combination of phase mask and amplitude mask, as shown in Figure 4.29. This method allows using separately designed amplitude masks for the same working wavelength. A phase mask with designed period is suitable for mass production of SFBG. Another possible technology is by scanning the UV laser beam intermittently over the phase mask along the fiber [70].

Figure 4.30 shows the experimentally measured reflection spectrum of an SFBG. The length of grating is 2 cm, the p and η are 508 μm and 1/11, respectively. The experimental result shows that the channel spacing is about 1.6 nm, in agreement with the above analysis.

The SFBG, as an effective comb filter, is useful because it is easier to match the peaks with wavelength channels for DWDM systems. More

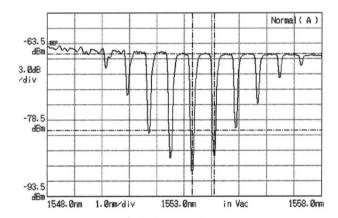

Figure 4.30   A measured transmission spectrum of SFBG.

sophisticated structures, such as SFBG with chirped sampling period, are developed for the purpose of multichannel dispersion compensation [75,76]. The SFBG is also expected to be used for WDM sensor system.

4.3.1.3 *FBG Fabry–Perot Cavity* Two concatenated FBGs with near the same reflection spectra compose an FBG F-P cavity with the composite reflectivity as

$$r_{FP} = \frac{-r_1 + r_2 e^{j2\Phi}}{1 - r_1 r_2 e^{j2\Phi}}, \tag{4.94}$$

where $\Phi = 2\pi n_{\text{eff}} d/\lambda = (n_{\text{eff}} d/c)\omega$ is the phase factor with the inner distance d, as shown in Figure 4.31, corresponding to a free spectral range $\Delta\lambda = \lambda^2/2n_{\text{eff}}d$; $r_1(\omega)$ and $r_2(\omega)$ are the reflectance of the two FBGs, with their respective amplitudes and phase factors. A particular case of $r_1 = r_2 = r(\omega)e^{j\phi(\omega)}$ is discussed here, giving the intensity reflectance:

$$R_{FP} = |r_{FP}|^2 = \frac{2R(1 - \cos 2\Phi)}{1 + R^2 - 2R\cos 2(\Phi + \phi)}, \tag{4.95}$$

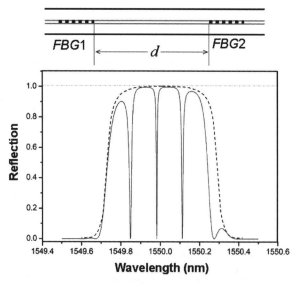

Figure 4.31 Structure of (upper) FBG-FP and (lower) simulated reflection spectrum.

where $R = |r(\omega)|^2$. Such an FBG–FP structure can be regarded as a simplified SFBG composed of two sections of FBG. Figure 4.31 shows its structure and simulated reflection spectrum.

Several features are worth noticing:

1. Fringes are induced by F-P oscillation in the envelope of the FBG reflection spectrum lobe. The number of peaks is proportional to the cavity length d, estimated by $N = \Delta\lambda_{FBG}/\Delta\lambda_{FP} = 2n_{eff}d\,\Delta\lambda_{FBG}/\lambda_B^2$.

2. The finesse of FBG–FP depends on the reflectance of FBG: $F(\lambda) = \pi\sqrt{R(\lambda)}/[1 - R(\lambda)]$, as a function of wavelength. For the peak of FBG spectrum, a sharp transmission peak is obtained with linewidth much narrower than that of FBG $\Delta\lambda_{FBG}$, for high $R(\lambda_B)$.

3. The main resonant peak of F-P cavity may deviate from the FBG peak. Their coincidence can be realized by adjusting the cavity length or by tuning FBG. It is favorable to have a higher and narrower main peak, and symmetric side peaks.

4. The FBG–FP has higher sensitivities to temperature and strains than the single FBG due to its narrower resonant peak. The FBG–FP has been used in sensor technologies [77,78] and laser technology.

4.3.1.4 Moiré Gratings
If the index is modulated by two sinusoidal waves with different spatial frequencies, the fiber grating will demonstrate respective diffractions simultaneously. Such a grating is called a moiré grating [79,80]. The index variation is described as

$$n(z) = n_0 + \tilde{n}_1 \cos\left(\frac{2\pi}{\Lambda_1}z + \theta_1\right) + \tilde{n}_2 \cos\left(\frac{2\pi}{\Lambda_2}z + \theta_2\right). \quad (4.96)$$

Assuming the two modulations have equal amplitudes, $\tilde{n}_1 = \tilde{n}_2$, and the modulation frequency difference is small, $\delta\Lambda = |\Lambda_1 - \Lambda_2| \ll \Lambda_1(\Lambda_2)$, the index is then expressed as

$$n(z) = n_0 + 2\tilde{n} \cos\left(\frac{2\pi}{\bar{\Lambda}}z + \bar{\theta}\right) \cos\left(\frac{2\pi}{\tilde{\Lambda}}z + \tilde{\theta}\right) \quad (4.97)$$

with

$$\bar{\Lambda} = \frac{2\Lambda_1\Lambda_2}{\Lambda_1 + \Lambda_2} \approx \frac{\Lambda_1 + \Lambda_2}{2}, \; \tilde{\Lambda} = \frac{2\Lambda_1\Lambda_2}{\Lambda_2 - \Lambda_1} \approx \frac{\bar{\Lambda}^2}{\Delta\Lambda},$$

$$\bar{\theta} = \frac{\theta_1 + \theta_2}{2}, \text{ and } \tilde{\theta} = \frac{\theta_1 - \theta_2}{2}.$$

From (4.97) the grating is regarded as having an averaged period $\bar{\Lambda}$, apodized by a cosine function with period of $\tilde{\Lambda}$. It is noticed that the phase shift $\tilde{\theta}$ of the two decomposed gratings strongly affects moiré fringes, showing different apodization effects. By properly adjusting the phase shift, the moiré grating can serve to design and fabricate fiber gratings with interesting performances. Figure 4.32 shows three

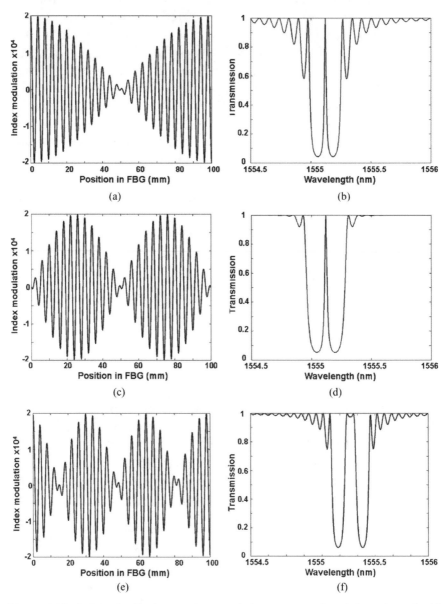

Figure 4.32 Index distributions and transmission spectra of moiré gratings. (a, b) $\Lambda_s = 2L, \varphi_1 - \varphi_2 = 0$; (c, d) $\Lambda_s = L, \varphi_1 - \varphi_2 = \pi$; and (e, f) $\Lambda_s = 2L/3, \varphi_1 - \varphi_2 = \pi/8$.

simulated moiré gratings with different phase shifts. It is worth noticing that $\pi/2$ phase shifting between the sections is ensured by the zero point of cosine apodization function, making it easier to set the transmission peak in coincidence with the reflection peak of FBG. Furthermore, a total transmission at the peak is obtained when the zero point of apodization function is fabricated at the midpoint of the FBG length similar to the phase-shift gratings.

Some technical methods are developed to fabricate moiré gratings. In holographic setup, the grating period is precisely controlled by the angles of two interference beams, whereas the phase factor $\tilde{\theta}$ can be adjusted by optics. In a setup with phase masks, a simple method was reported by using two-step exposure under the same mask, while the fiber is stretched differently, for example, unstretched at the first exposure, and then stretched at the second exposure, resulting in two FBGs with a little different period when the fiber is released. The wavelength difference and the spatial phase shift can be designed by the elastic property of the fiber, and controlled mechanically.

4.3.2 Chirped Fiber Bragg Grating

The CFBG with grating period varied along the grating length reflects light waves of different wavelengths at different positions, demonstrating properties of dispersion compensation, which is of particular importance in fiber optical communications and in ultrashort laser technologies [81–83]. Figure 4.33 shows pulse width evolution after reflection from CFBG, where the input optical pulse has a red-shifted front after propagation in a fiber with normal dispersion, as analyzed in Section 2.2.6. For the blue-shifted pulse, the opposite end of the CFBG should be taken as the input port.

The basic structure of CFBG is a linearly chirped FBG (LCFBG), whose period is a linear function of z with Bragg wavelength of

$$\lambda_B = 2\bar{n}(z)\Lambda(z) = 2\bar{n}(z)(\Lambda_0 + \Lambda_1 z). \tag{4.98}$$

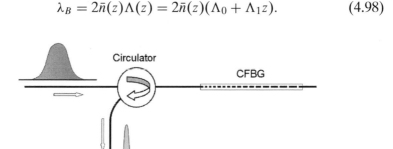

Figure 4.33 Dispersion compensation by a chirped fiber Bragg grating.

Its spectral band covers a range of $\Delta\lambda \approx 2\bar{n}\Lambda_1 L$ with grating length of L, corresponding to total traveling time of $\Delta\tau_{\text{tot}} = 2n_{\text{eff}}L/c$. The Bragg wavelength is now a function of position z, resulting in delay time of $\Delta\tau = 2n_{\text{eff}}z/c$ and wavelength shift of $\Delta\lambda = 2n_{\text{eff}}\Lambda_1 z$, the dispersion compensation coefficient is deduced as

$$\frac{\Delta\tau}{\Delta\lambda} = \frac{1}{\Lambda_1 c}. \tag{4.99}$$

The GD of an LCFBG is then expressed as

$$\tau_{\text{GD}}(\lambda) = \pm\frac{\lambda - \lambda_0}{\Delta\lambda}\frac{2n_{\text{eff}}L}{c}, \tag{4.100}$$

where sign $\pm$ is used to include two possible directions of the input optical beam. The Bragg wave vector varying along the fiber is now expressed as

$$\beta_B = \frac{\pi}{\Lambda_0 + \Lambda_1 z} \approx \frac{\pi}{\Lambda_0}\left(1 - \frac{\Lambda_1}{\Lambda_0}z\right). \tag{4.101}$$

The index modulation is written as $\delta n = 2\bar{n}_1\tilde{n}_1 \cos[k_B z + \theta(z)]$, where $k_B = 2\pi/\Lambda_0$, and $\theta(z) = -2\pi\Lambda_1 z^2/\Lambda_0^2$. For the varying coupling coefficient $\kappa(z) = \frac{1}{2}n_1 k_0 e^{j\theta(z)}$, numerical methods are necessary to solve CME, such as Runge–Kutta method and TMM. Figure 4.34(a) shows simulated the reflection spectrum and GD spectrum by using TMM with chirping rate of 1 nm/cm, and grating length of 5 cm.

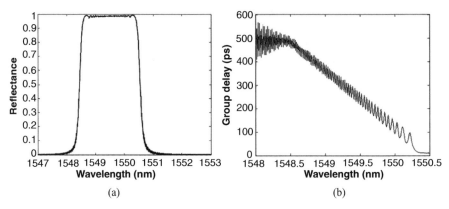

Figure 4.34 Calculated (a) reflection spectrum and (b) group delay of linear CFBG.

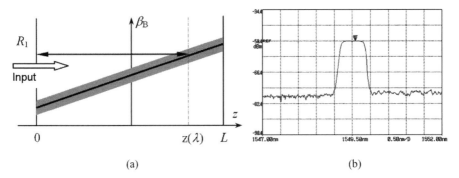

Figure 4.35 (a) Illustration of GD ripples; (b) measured spectrum of an apodized CFBG.

It is found that the GD spectrum is not an ideal straight line, but with a great number of ripples, as shown in Figure 4.34(b), which brings undesired evolutions of pulse profile and impairs the signal. It is indicated that the ripple is attributed to the reflection at the end point of CFBG by an additional reflection due to F-P interference of

$$R_{FP} = 1 - \frac{(1 - R_1)(1 - R_{\text{CFBG}})}{1 + R_1 R_{\text{CFBG}} - 2\sqrt{R_1 R_{\text{CFBG}}} \cos 2\Delta}, \quad (4.102)$$

where $\Delta = (2\pi/\lambda) \int_0^{z(\lambda)} \bar{n}(z) \mathrm{d}z \approx \Delta \bar{n} k z(\lambda)$; R_1 is the Fresnel reflection at the end of the grating caused by the difference of the averaged index. Figure 4.35(a) gives an illustration to explain the possible parasitic F-P interference, as analyzed in [45] in detail. The gray band stands for the reflection wave band. As analyzed in Section 4.2.4 the apodization can smoothen the grating boundary, and mitigate the residue reflection, thereby reducing the ripple [84–86]. Figure 4.35(b) shows a typical experimentally measured reflection spectrum.

It is helpful to investigate the properties of CFGB in a time domain, especially when it is used in applications of pulse reshaping. Taking GD and dispersion into account the reflection of chirped fiber grating is expressed as [87–89]

$$r(\omega) = |r(\omega)| \exp\left\{ j\left[\Phi(\omega_0) + \tau_{\text{GD}}(\omega - \omega_0) + \frac{\ddot{\Phi}}{2}(\omega - \omega_0)^2 \right] \right\}, \quad (4.103)$$

where $\ddot{\Phi} = (\partial \tau / \partial \omega)\big|_{\omega_0} = 4\pi n_{\text{eff}} L / (\omega_0^2 \Delta \lambda) = \lambda^2 D_{\text{CFBG}} / 2\pi c$, with $D_{\text{CFBG}} = \partial \tau / \partial \lambda = \pm 2L/(v_g \Delta \lambda)$. The signs here are for different chirping directions. As discussed in Section 2.2.6, the fiber dispersion causes pulse broadening, expressed as an additional phase factor $\frac{1}{2}\beta_2 z(\omega - \omega_0)^2$. It can be compensated by a linear CFGB with a parameter of

$$\ddot{\Phi} = \frac{4\pi n_{\text{eff}} L}{\Delta\lambda\omega_0^2} = \beta_2 z = \frac{2\pi c}{\omega_0^2} Dz, \qquad (4.104)$$

where D is the fiber dispersion parameter. A simple relation is deduced: $D_{\text{CFBG}} = Dz$. Formula (4.104) gives just a basic condition; quantitative dispersion compensation relation has to be analyzed in detail with concrete properties of input pulses, including their spectral linewidth and original spectral chirping. Moreover, a nonlinear CFBG can be designed and fabricated to compensate the higher order dispersions, or so-called dispersion slope compensation. Also the dispersions of multiple channels can be compensated by a structured CFBG [76,90].

The CFBG is now fabricated by well-developed techniques. Phase masks are available with customized designs, including chirping rate and apodization envelope.

4.3.3   Tilted Fiber Bragg Gratings

4.3.3.1   *Characteristics of Tilted Fiber Bragg Gratings*   In the preceding sections, the Bragg wave vector of fiber gratings is considered coincident with the axis of the fiber. However, it may deviate from the axial direction intentionally or unintentionally; such gratings are called tilted fiber gratings, or slanted fiber gratings. Unintentionally, introduced tilting may make impairments to the desired performance of FBG devices. Tilted FBGs (TFBG) can be intentionally fabricated by setting the fiber off the alignment of the phase mask, or that of holographic fringes. Because the fiber itself acts as a cylindrical lens, the tilt angle of the grating is different from that of the phase mask in imprinting. From Figure 4.36, their relations near the core are deduced as

$$\tan\theta \approx n\tan\theta_0 \qquad (4.105)$$

and the period of grating differs from that of the UV fringe outside the fiber as

$$\Lambda = \Lambda_0 \cos\theta_0 / \cos\theta. \qquad (4.106)$$

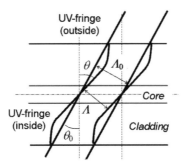

Figure 4.36 UV irradiation fringe patterns outside and inside the fiber.

Intentionally fabricated TFBGs have unique features and useful characteristics for special applications; they are called blazed fiber gratings sometimes. Figure 4.37 shows a typical TFBG structure with tilt angle θ and Bragg vector of $|k_B| = 2\pi/\Lambda$. The grating period in z direction is $\Lambda_z = \Lambda/\cos\theta$.

Figure 4.38 shows a typical transmission spectrum of TFBG [91]. It is seen that series peaks exist on the left side of the main peak, similar to those in Figure 4.13 for the coupling of the core mode to cladding modes in normal gratings. They are enhanced in the tilted gratings, even in the cladding mode suppressed fiber. This is especially so if a large side lobe is observed at 2–3 nm next to the main mode, which is usually called the ghost mode. The measurements indicate that the wavelength positions of these peaks depend on the tilt angle; and the cladding modes and the ghost mode are loss peaks, they do not appear in the measured reflection spectrum. These loss peaks cause impairment to the signal of related channels; therefore, they should be avoided as much as possible. In FBG fabrication, attention must be paid to adjust the imprinting apparatus precisely to ensure that the fringes are perpendicular to the fiber axis. On the other hand, such characteristics of TFBG can be utilized to develop special devices, for example, the gain flattening filter of EDFA [92,93], in-fiber polarizer [94], and in-fiber polarimeter [95].

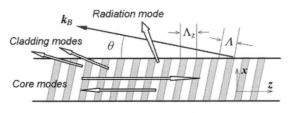

Figure 4.37 Basic structure of a tilted grating.

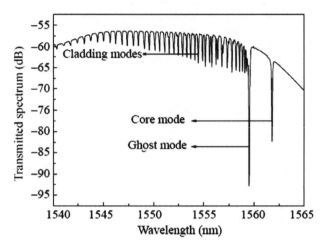

Figure 4.38 Typical transmission spectrum of a TFBG. (Reprinted with permission from reference [91].)

References [96–99] present comprehensive analyses and experimental investigations on the tilted gratings; [10] and [11] give summarized explanations. The CMEs should take three couplings into consideration here: (1) coupling between the forward and backward core modes; (2) coupling between the core mode and the cladding modes; and (3) coupling between the core mode and radiation mode. The index modulation is now written as

$$\Delta n = \tilde{n}_1 \cos[(2\pi/\Lambda)(z\cos\theta + x\sin\theta)]. \qquad (4.107)$$

The CME for the coupling (1) and (2) are then expressed as [38]

$$A'_{\text{co}} = j\kappa_{\text{co}-\text{co}}B_{\text{co}}e^{-j2\delta_{\text{co}}z} + j\sum_m \kappa_{\text{co}-m}B_m e^{-j2\delta_m z}, \qquad (4.108a)$$

$$B'_{\text{co}} = -j\kappa_{\text{co}-\text{co}}A_{\text{co}}e^{j2\delta_{\text{co}}z}, \qquad (4.108b)$$

$$B'_m = -j\kappa_{\text{co}-m}A_{\text{co}}e^{j2\delta_m z}, \qquad (4.108c)$$

where $\delta_{\text{co}} = \beta_{01} - \pi/\Lambda_z = \beta_{01} - \pi\cos\theta/\Lambda$, $\delta_m = (\beta_{01} + \beta_m)/2 - \pi\cos\theta/\Lambda$, corresponding to resonances of the core mode and cladding modes, respectively:

$$\lambda_{01} = 2n_{01}\Lambda/\cos\theta, \qquad (4.109)$$

$$\lambda_m^{\text{cl}} = (n_{01} + n_m^{\text{cl}})\Lambda/\cos\theta. \qquad (4.110)$$

The strengths of the couplings depend on the overlap integrals:

$$\kappa_{01-m}^{\text{co-cl}} \propto \tilde{n}_1 \int \cos\left[\frac{2\pi}{\Lambda}(z\cos\theta + x\sin\theta)\right] \Phi_{01}\Phi_m^{\text{cl}} r\,dr\,d\varphi,$$

$$(4.111)$$

$$= \frac{\tilde{n}_1}{2} e^{j(2\pi/\Lambda)z\cos\theta} \int (\cos\psi + j\sin\psi)\Phi_{01}\Phi_m^{\text{cl}} r\,dr\,d\varphi + c.c.,$$

where $\psi = 2\pi x \sin\theta/\Lambda = (2\pi r \sin\theta/\Lambda)\cos\varphi$ is denoted. Function $\cos\psi$ and $\sin\psi$ can be expanded into series of $\cos n\varphi$ as shown by (A1.38) in Appendix 1.

Because mode Φ_{0m}^{cl} has circular symmetry, and Φ_{1m}^{cl} is proportional to $\cos\varphi$, two series of integrals are generated: $\kappa_{01-0m}^{\text{co-cl}}$ and $\kappa_{01-1m}^{\text{co-cl}}$. Such results are different from the case of untilted gratings, where the mode orthogonality makes $\kappa_{01-1m}^{co-cl} = 0$.

The coupling between the core mode and the cladding modes is a two-way mutual process, similar to the case of LPFG. The effect is governed by their respective CMEs. The peak amplitude depends not only on the coupling coefficient but also on the grating length, with a factor, something like $\cos^2(\kappa_m L)$. References [97–99] give detailed discussions.

However, the above couplings are unlikely the mechanism of the ghost mode. Some explanations are proposed [98–102]. Coupling between the fundamental mode LP_{01} and the first-order core mode LP_{11} is considered in the literature. Because of the factor of $\cos\varphi$ and $\sin\varphi$ in the integral, the LP_{01} mode can couple with LP_{11} mode; and the following integral results in a significant value:

$$\kappa_{01-11}^{\text{co-co}} \propto \frac{\tilde{n}_1}{2} e^{j(2\pi/\Lambda)z\cos\theta} \int (\cos\psi + j\sin\psi)\Phi_{01}\Phi_{11}^{\text{co}} r\,dr\,d\varphi + c.c.$$

$$(4.112)$$

The LP_{11} mode is regarded as a leaky mode here. It is usually cut off in regular SMFs as its effective index is lower than the cladding index. When it is excited by the tilted grating, it will then dissipate into various cladding modes and radiation mode, displaying a large loss peak, even more than the main peak at times. The peak position is basically at

$$\lambda_{11}^{\text{co}} = (n_{01} + n_{11}^{\text{co}})\Lambda/\cos\theta \approx (n_{01} + n_2)\Lambda/\cos\theta. \qquad (4.113)$$

Its shift from the main peak is written as $\Delta\lambda_{11}^{\text{co}} \approx (n_{01} - n_2)\Lambda/\cos\theta$, which is coincident with the experimentally measured data.

Based on formulas (4.109), (4.110), and (4.113), the sensitivities of TFBG spectral characteristics to temperature, stains, and the index change of the medium outside fiber can be analyzed, and related sensors are studied and developed [103–105].

4.3.3.2 *Analysis of Radiation Mode Coupling*   As the tilt angle increases, the coupling between the core mode and radiation mode becomes dominant. It is then possible to utilize the coupling to develop functional devices. One of them is the tap of light wave [105], which can be used to monitor optical signals propagating in the fiber with negligible influence to the signals. It is also found experimentally that the coupling to radiation depends strongly on the polarization state of the propagating light wave in fiber for larger tilt angles, which is used to develop an in-fiber polarizer and polarimeter [94,95].

To develop such devices, we need to know not only the resonant wavelength position but also the spectrum and polarization dependence of the coupling coefficient; the fundamental diffraction theory is needed to understand light scattering behaviors in three-dimensional space. The relations of wave vectors are described in Figure 4.39 in x–y–z space, $\boldsymbol{k}_1 + \boldsymbol{k}_B = \boldsymbol{k}_2 + \boldsymbol{k}_n$, where $\boldsymbol{k}_1$ is the input wave vector; $\boldsymbol{k}_2$ is the radiation wave vector, $\boldsymbol{k}_n$ stands for wave vector offset in direction perpendicular to x–z plane. The vector's geometrical relations are deduced as: $k_{2z} = k_2 \cos \gamma = k_B \cos \theta - k_1$, $k_n^2 = k_B^2 \sin^2 \theta + k_{2t}^2 - 2 k_B k_{2t} \sin \theta \cos \phi$, $\tan \gamma = k_{2t} / k_{2z}$. Angle γ is obtained from $\cos \gamma \approx (k_B / k_1) \cos \theta - 1 = (\lambda / \Lambda) \cos \theta - 1$ with approximation of $k_2 \approx k_1 = n_{\mathrm{eff}} k_0$. For the optimum matching, $k_n = 0$, the resonant wavelength is deduced to be

$$\lambda_p = (n_{\mathrm{eff}} \cos \theta + \sqrt{n_1^2 - n_{\mathrm{eff}}^2 \sin^2 \theta}) \Lambda \approx 2 n_{\mathrm{eff}} \Lambda \cos \theta; \qquad (4.114)$$

and $\cos \gamma \approx 2 n_{\mathrm{eff}} \cos^2 \theta - 1$ is obtained.

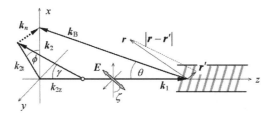

Figure 4.39   Wave vector relations in three-dimensional space.

The radiation causes a loss of the forward propagating mode in fiber. It is necessary to understand the dependence of coupling coefficients on the tilt angles, and its polarization-dependent characteristics. Three main theoretical methods are developed for the purpose: the CMT [97–99], the antenna theory based on Fraunhofer diffraction [106], and the volume current method (VCM) [107,108]. VCM is referred here to give a clear picture of the radiation mode. The basic idea of VCM is to take the index modulation of grating as a perturbation on the electromagnetic field of the regular waveguide. With $\delta\varepsilon(r) \ll \varepsilon(r)$, the electric and magnetic fields are written as $E = E_0 + E_1$ and $B = B_0 + B_1$, where the second term stands for the perturbations, which obey Maxwell equations as follows:

$$\frac{1}{\mu_0}\nabla \times B_1 = \varepsilon(r)\frac{\partial E_1}{\partial t} + J(r), \qquad (4.115a)$$

$$\nabla \times E_1 = -\frac{\partial B_1}{\partial t}, \qquad (4.115b)$$

with the equivalent polarization electric current density of $J(r) = \delta\varepsilon(r)(\partial E_0/\partial t) = -j\omega\delta\varepsilon(r)E_0$. Perturbations of electric and magnetic fields are then deduced by introducing the vector potential induced by the equivalent current, expressed as

$$A(r,t) = \frac{\mu_0}{4\pi}\int J(r',t)\frac{\exp(-ik|r - r'|)}{|r - r'|}dV'. \qquad (4.116)$$

The perturbation is now expressed as $\delta\varepsilon(r)k_0 = \kappa\{\exp[j(k_Bz\cos\theta + k_Br\sin\theta\cos\varphi)] + c.c.\}$, where κ is the coupling coefficient of grating. The core mode is $E_0(r') = E_0J_0(ur'/a)e^{j\beta z}\zeta$, with polarization vector of $\zeta = \hat{x}\cos\zeta + \hat{y}\sin\zeta$, as shown in Figure 4.39. r' and r in integral (4.116) are the positions of volume current and observation point, respectively, with $|r - r'| = \sqrt{(z - z')^2 + d^2}$, where $d = \sqrt{r^2 - 2rr'\cos(\varphi - \varphi') + r'^2}$ is the projection of $(r - r')$ on x–y plane. It is approximated that $d \approx r - r'\cos(\varphi - \varphi')$ in the phase term, and $d \approx r$ in other terms due to $|r'| \ll |r|$. Then the vector potential takes a form similar to the Fraunhofer diffraction formula. An analytic expression of vector potential (4.116) is deduced in reference [107] for an infinitive long grating; and the electric and magnetic fields of radiation mode are deduced consequently. The radiation Poynting vector is deduced [107]:

$$
\begin{aligned}
\boldsymbol{S} = \boldsymbol{E}_1 \times \boldsymbol{B}_1^* &= E_0^2 \frac{\pi \omega \kappa^2}{2\varepsilon r k_{2t}} f^2(k_n)[k_{2z}^2 + k_{2t}^2 \sin^2(\zeta - \phi)]\boldsymbol{k}_2 \\
&= E_0^2 \frac{\pi \kappa^2 k_0 k_2}{2c\varepsilon r \sin \gamma} f^2(k_n)[1 - \sin^2 \gamma \cos^2(\zeta - \phi)]\boldsymbol{k}_2,
\end{aligned}
\tag{4.117}
$$

where function $f(k_n)$ describes the contribution of offset vector k_n:

$$
f(k_n) = \int_0^a J_0(ur')J_0(k_n r')r'\mathrm{d}r' = \frac{k_n a J_0(u)J_1(k_n a) - u J_1(u)J_0(k_n a)}{k_n^2 a^2 - u^2}.
\tag{4.118}
$$

The loss of the propagating mode in fiber is then calculated by integrating $|\boldsymbol{S}|$ over the azimuth angle ϕ. The peak direction of the radiation beam and its intensity can be obtained. Refraction at the interface between fiber and the outer medium must be considered for the radiation beam outside the fiber.

Expression (4.117) gives a polarization dependence of the radiation beam. Figure 4.40 shows pictures of radiation lobes for the two polarized input beams by a TFBG with $\theta = \pi/4$ [107]. It is shown in Figure 4.40(b) that two radiation lobes at directions deviating from $90°$

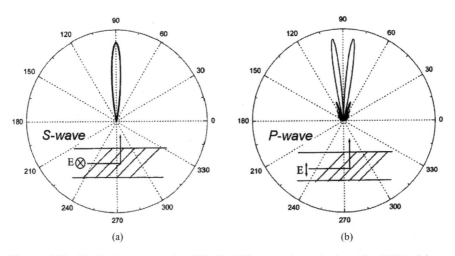

Figure 4.40 Radiation pattern for $45°$-tilted fiber grating calculated by VCM. (a) s-polarization, $\zeta = \pi/2$; (b) p-polarization, $\zeta = 0$. (Reprinted with permission from reference [107]. © 2001 IEEE.)

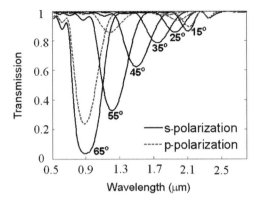

Figure 4.41   Transmission spectra of TFBGs with various tilting angles.

appear, different from Figure 4.40(a). It is because a beam polarized in x–z plane is contradictory to the property of transverse optical waves; and offset wave vectors in y direction are necessary. References [109–111] investigate radiation beam patterns, polarization characteristics and applications of TFBG.

An attractive function of the TFBG is to be used as a polarizer. It is known that the largest polarization is obtained at the incident angle equal to Brewster's angle, $\theta_B = \tan^{-1}(n_1/n_2)$ at the interface between media with index n_1 and n_2. In fiber gratings, the UV-induced index increment is so small that $\theta_B = \pi/4$ is expected. Figure 4.41 is a simulated result for tilted gratings with length of 20 mm, index modulation amplitude of 3×10^{-4} and $\Lambda = 0.761$ μm, showing the loss peaks of two polarized beams for different tilt angles, and the transmission dip of 45 degree tilted grating located at 1,500 nm band with high polarization suppression ratio.

4.3.4   Polarization Maintaining Fiber Gratings

As the PMF shows two different propagation constants for the two polarization modes, it is understandable that two sets of spectra with different resonant wavelengths are obtained from the grating imprinted in PMF. Figure 4.42 shows a typical measured spectrum of PMFBG, where the two peaks correspond to two polarization modes, respectively. Their amplitudes are limited to less than 3 dB when a broadband unpolarized source is used in the measurement. The spacing between the two peaks is determined by the birefringence B of PMF:

$$\Delta\lambda = B\bar{\lambda}/\bar{n}_{\text{eff}}. \tag{4.119}$$

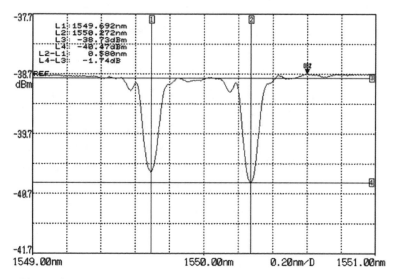

Figure 4.42   Typical transmission spectrum of FBG imprinted in PMF.

The polarization evolution characteristics of fiber grating imprinted in PMF are surely concerned for practical applications. Apart from the strain state in PMF caused by SAP, there are other factors inducing and affecting the fiber birefringence, such as nonuniform UV-exposure in fiber grating fabrication and externally applied forces [112–114]. References [115–118] give theoretical analyses on polarization-mode coupling in birefringent fiber gratings, and comparisons with experimental results.

The composite effect of bending, pressure, and twisting on PMF is analyzed in Section 3.4.4. To analyze these effects on FBG/PMF, four modes must be taken into consideration: the two polarization modes propagating in forward and backward directions [119–121]. Based on the theoretical frame presented in Section 3.4.4, the SAP-induced bire-fringence, the externally induced strains, and the UV-induced index modulation in axial direction are taken as perturbations on the degen-erate eigenmodes of the regular cylindrical waveguide. The strain states and index modulation are considered as acting independently of each other; except for the z direction strain e_z, which changes the grating period; a reasonable description is $\Delta\varepsilon = \tilde{\varepsilon}_{\text{elast}}(x, y) + \delta\varepsilon_{\text{UV}}(z)$.

It is recognized that the grating does not couple the two polarization modes with each other. However, if torsion exists, it will cause the four modes to couple with each other, described by a 4×4 CME:

$$
\begin{pmatrix} a_1' \\ a_2' \\ b_1' \\ b_2' \end{pmatrix} = \begin{pmatrix} -i\kappa_b\cos\theta & \kappa_\tau + i\kappa_b\sin\theta & i\kappa_g e^{-i2\delta z} & 0 \\ -\kappa_\tau + i\kappa_b\sin\theta & i\kappa_b\cos\theta & 0 & i\kappa_g e^{-i2\delta z} \\ -i\kappa_g e^{i2\delta z} & 0 & i\kappa_b\cos\theta & \kappa_\tau - i\kappa_b\sin\theta \\ 0 & -i\kappa_g e^{i2\delta z} & -\kappa_\tau - i\kappa_b\sin\theta & -i\kappa_b\cos\theta \end{pmatrix} \begin{pmatrix} a_1 \\ a_2 \\ b_1 \\ b_2 \end{pmatrix},
$$

$$(4.120)$$

where $b_{1,2}$ stand for the amplitudes of backward eigenmodes; $\kappa_g = \bar{n}_1 k_0/2$ is the coupling coefficient of grating; $\delta = \beta - \beta_B$, as defined in Section 4.2; parameters κ_b, κ_τ, and $\theta = 2\tau z$ are the same as those in Section 3.4.4. In case of no torsion, Equation (4.120) is simplified as

$$a_1' = -i\kappa_b a_1 + i\kappa_g e^{-i2\delta z} b_1, \qquad (4.121a)$$

$$b_1' = i\kappa_b b_1 - i\kappa_g e^{i2\delta z} a_1, \qquad (4.121b)$$

$$a_2' = i\kappa_b a_2 + i\kappa_g e^{-i2\delta z} b_2, \qquad (4.121c)$$

$$b_2' = -i\kappa_b b_2 - i\kappa_g e^{i2\delta z} a_2, \qquad (4.121d)$$

showing that the couplings occur between a_1 and b_1, and between a_2 and b_2, respectively, without coupling between mode 1 and mode 2. The two individual CME are solved with two peak wavelengths for the two polarizations:

$$\beta_{p1} = \beta_B - \kappa_b = \beta_B - \Delta\beta/2, \qquad (4.122a)$$

$$\beta_{p2} = \beta_B + \kappa_b = \beta_B + \Delta\beta/2. \qquad (4.122b)$$

Their spectral transmission and reflection spectra are given the same as that in Section 4.2. In case of no birefringence, the resonance occurs at $\beta = \beta_B$; the torsion has no influence on the Bragg diffraction.

Polarization mode coupling is more complicated when the fiber experiences forces varying along the fiber. Similar to the analysis in section 3.4.4, an approximate propagation matrix can be derived when the parameter $\theta = 2\tau z$ is regarded as a constant under condition of section length Δz small enough. Equation (4.120) can then be transformed to a constant coefficient equation. By using the trial solutions in forms of $a_1 = A_1 e^{i(\gamma - \delta)z}$, $a_2 = A_2 e^{i(\gamma - \delta)z}$, $b_1 = B_1 e^{i(\gamma + \delta)z}$, and $b_2 = B_2 e^{i(\gamma + \delta)z}$, the equation becomes a combined algebraic equation. Its determinant must be zero for a nonzero solution, that is,

$$
\begin{vmatrix}
\gamma - \delta + \kappa_c & -\kappa_s & -\kappa_g & 0 \\
-\kappa_s^* & \gamma - \delta - \kappa_c & 0 & -\kappa_g \\
\kappa_g & 0 & \gamma + \delta - \kappa_c & \kappa_s^* \\
0 & \kappa_g & \kappa_s & \gamma + \delta + \kappa_c
\end{vmatrix} = 0, \qquad (4.123)
$$

where $\kappa_c = \kappa_b \cos\theta$, and $\kappa_s = \kappa_b \sin\theta - i\kappa_\tau$. The equation can then be solved analytically to determine parameter γ.

From the trial solutions, the propagation matrix for the coupled four-mode equation, and consequently the reflection and transmission spectra, can be derived. Numerical methods, such as the TMM and the finite difference method, are usually used to investigate the characteristics of PMFBG under different conditions.

FBGs in PMF are often used in situations where both the polarization selection and wavelength selection are needed; for example, when used in frequency stabilized single polarization fiber lasers, and in diode lasers with a external fiber cavity [121–124].

4.3.5 In-Fiber Interferometers and Acoustic Optic Tunable Filter

4.3.5.1 LPFG Interferometers As analyzed in Section 4.2.2, the LPFG causes coupling between the input core mode and some cladding modes. The LPFG can be regarded as a double waveguide, and are used to build in-fiber interferometers [125–127].

Figure 4.43(a) shows a Mach–Zehnder interferometer (MZI) composed by a cascaded LPFG pair. A Michelson interferometer composed by an LPFG and a fiber-end reflector is shown in Figure 4.43(b). The interferometer is composed of two beams with an equal geometrical length, but different optical lengths due to the different indexes of core mode and cladding mode. For the MZI, denoting A and B as the amplitudes of core mode and cladding mode, the transmission of a normalized optical wave passing through the pair of LPFG is written as

$$
\begin{pmatrix} A \\ B \end{pmatrix} = \begin{pmatrix} t_2 & r_{co2} \\ r_{cl2} & t_2^* \end{pmatrix} \begin{pmatrix} e^{j\beta_{co}L} & 0 \\ 0 & \alpha e^{j\beta_{cl}L} \end{pmatrix} \begin{pmatrix} t_1 & r_{co1} \\ r_{cl1} & t_1^* \end{pmatrix} \begin{pmatrix} 1 \\ 0 \end{pmatrix}. \qquad (4.124)
$$

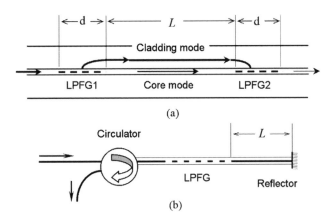

Figure 4.43 In-fiber MZI (a) and Michelson interferometer (b) composed by LPFG.

When the same period for the two gratings is assumed, the output is deduced to be $A = t_1 t_2 e^{j\beta_{co}L} + \alpha r_{cl1} r_{co2} e^{j\beta_{cl}L}$, and the intensity is

$$I = |t_1 t_2|^2 + \alpha^2 |r_{cl1} r_{co2}|^2 + 2\alpha |t_1 t_2 r_{cl1} r_{co2}| \cos \Phi, \qquad (4.125)$$

where α is the loss factor for the cladding mode; $t_{1,2}$, $r_{co1,2}$, and $r_{cl1,2}$ are expressed in (4.53); $\Phi = (\beta_{co} - \beta_{cl})L + \Delta\theta_{t-r}$ with $\Delta\theta_{t-r}$ standing for the phase factor of complex transmission and reflectance of two LPFG. The formula gives a description of interference fringes with modulation depth proportional to

$$2\alpha |t_1 t_2 r_{cl1} r_{co2}| \propto 2\alpha |t_1||t_2|\sqrt{1 - |t_1|^2}\sqrt{1 - |t_2|^2}, \qquad (4.126)$$

which takes the maximum when $|t_1|^2 = |t_2|^2 = 1/2$. Figure 4.44 shows an experimentally measured spectrum of paired LPFG MZI [43].

For the Michelson interferometer composed of LPFG, the transmission formula is similar to (4.125) in a simplified form with two identical LPFG and with beam length of $2L$. The reflector at the fiber end may introduce additional features. For example, a Faraday rotation mirror can mitigate the influence of externally induced birefringence.

The MZI and Michelson interferometer composed of LPFG give interference peaks much narrower than that of the original single LPFG; this feature is helpful in enhancing the sensitivity and precision in measurement and sensing applications. References [128,129] report several sensors based on LPFG pair.

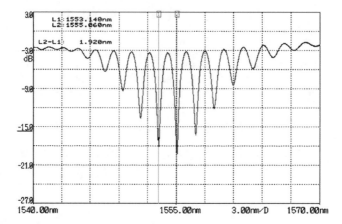

Figure 4.44   Transmission spectrum of MZI composed of two LPFG. (Reprinted with permission from reference [43].)

4.3.5.2 *In-Fiber Acoustic Optic Tunable Filter*   It is indicated [130] that the acoustic wave along the fiber can be regarded as a dynamic LPFG. Based on such an acoustic optic (AO) effect special devices are developed, such as frequency shifters, modal filters, tunable tapers and couplers, variable optical attenuators, tunable notch filter, and dispersion compensators [130–134]. Figure 4.45(a) shows a typical structure of an in-fiber acoustic optic tunable filter (AOTF), where the fiber with a thinned section is driven by a transverse/longitudinal vibrator and an acoustic wave is excited. At the other end an acoustic wave damper is used to make the acoustic wave a traveling wave. Otherwise, it will be a standing acoustic wave. The traveling wave has broader

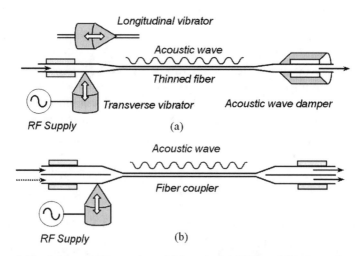

Figure 4.45   Schematic illustration of (a) typical AOTF and (b) dynamic coupler.

bandwidth, whereas the latter has separate longitudinal frequencies. Two transverse vibrators, set perpendicularly, can produce an acoustic wave composed of two dimensional transverse vibrations [135]. The acoustic wave can be excited in the waist of fiber coupler to construct a four-port fiber AO device, as shown in Figure 4.45(b).

The acoustic wave is regarded as an index modulation with period of the acoustic wavelength, due to the dynamically varying periodic distributed strain. As discussed in Section 3.1.6, the wavelengths of the longitudinal acoustic wave and transverse wave are deduced to be [130]

$$\Lambda_z = \sqrt{Y/\rho}/f, \qquad (4.127)$$

$$\Lambda_t = (Y/\rho)^{1/4}\sqrt{\pi d/2f}, \qquad (4.128)$$

where f is the vibration frequency, d is the diameter of the thinned fiber. The acoustic wavelength is controllable by adjusting the vibration frequency, thus a tunable LPFG is realized.

A difference between conventional LPFG and that induced by the transverse acoustic wave is noticed: the former has circular symmetry, whereas the latter can be driven in two perpendicular directions and interacts with two polarized optical waves. Reference [135] gives experimental results and theoretical explanations of the polarization dependent AOTF, showing potential for an all-fiber wavelength-tunable polarizer.

There are other interesting and useful fiber grating devices, such as FBG on photonic crystal fibers and on double cladding fibers. Interested readers can find references in journals.

4.4 FIBER GRATING SENSITIVITIES AND FIBER GRATING SENSORS

Since the earliest stage of their development, sensitivity of the fiber gratings to external conditions, especially to temperature and to stress, have been considered important and attractive topics of research and development. The stability of fiber grating devices is another topic of concern for applications. That said, these sensitivities make fiber gratings excellent sensor components. This section is devoted to the sensitivities of fiber gratings and related topics. The basic sensitivities to stress and to temperature are described in Section 4.4.1. Tunability of fiber gratings is then discussed in the next section. Section 4.4.3 describes packaging of fiber gratings for different purposes. In the last sub-section, the variety of fiber grating sensors and their applications are introduced with emphasis on sensor interrogation methods.

4.4.1 Sensitivities of Fiber Gratings

4.4.1.1 Sensitivities of FBG

The main parameter of fiber grating is its resonant wavelength. For the FBG, the Bragg wavelength is $\lambda_B = 2n_{eff}\Lambda$, its variations are determined by both the sensitivities of the effective index and the grating period, $\delta\lambda_B = 2n_{eff}\delta\Lambda + 2\Lambda\delta n_{eff}$. The axial strain of fiber occurs most often; the grating period changes accordingly; and the index also changes due to the photoelastic effect. By using (3.5) the increment of Bragg wavelength is proportional to the strain e_z:

$$\Delta\lambda_B = \lambda_B\left\{1 - \frac{n^2}{2}[(1-v)p_{12} - vp_{11}]\right\} = \lambda_B(1+\gamma)e_z. \qquad (4.129)$$

The effective elastic-optic coefficient is $\gamma = -0.22$ for silica fibers; it is estimated that $\Delta\lambda_B \approx 1.0 \times e_z\,\text{pm}/(\mu\varepsilon)$ and $1.2 \times e_z\,\text{pm}/(\mu\varepsilon)$ for $1,300$ nm and $1,550$ nm band, respectively, where $(\mu\varepsilon)$ stands for "*microstrain*," that is, e_z is measured in 10^{-6}.

For temperature sensitivity, both the thermo-optical effect of the index and the thermal expansion effect of the material have to be considered, giving a relation of

$$\Delta\lambda_B = \lambda_B\left(\alpha + \frac{1}{n}\frac{\partial n}{\partial T}\right)\Delta T, \qquad (4.130)$$

where $\alpha = \Lambda^{-1}(\partial\Lambda/\partial T)$ is the thermal expansion coefficient, which is about $5.5*10^{-7}\,°\text{C}^{-1}$ for silica. The temperature coefficient of resonant wavelength is obtained experimentally for silica FBG $\lambda_B^{-1}(\Delta\lambda_B/\Delta T) \approx 6.7 \times 10^{-6}/°\text{C}$, that is, $\Delta\lambda_B/\Delta T \approx 10$ pm/$°$C for 1,550 nm band. The data implies that the thermo-optical effect is dominant over the thermal expansion for silica fibers.

4.4.1.2 Sensitivities of LPFG

The resonance wavelength $\lambda_m = (n_{co} - n_{cl,m})\Lambda$ is determined by three parameters, that is, the core mode index, the cladding mode index, and the grating period. As discussed in Section 4.2.2 the dispersion of indexes plays important roles in determining the resonance wavelength; and a waveguide dispersion factor is defined:

$$\gamma_m = \left[1 - \Lambda\frac{\partial(n_{co} - n_{cl,m})}{\partial\lambda}\right]^{-1}. \qquad (4.41)$$

The derivative of LPFG resonant wavelength over temperature is deduced as

$$
\begin{aligned}
\frac{\partial \lambda_L}{\partial T} &= (n_{\text{co}} - n_{\text{cl}})\frac{d\Lambda}{dT} + \Lambda\frac{\partial(n_{\text{co}} - n_{\text{cl}})}{\partial T} \\
&+ \Lambda\frac{\partial(n_{\text{co}} - n_{\text{cl}})}{\partial \lambda}\frac{\partial \lambda_L}{\partial T} = \gamma_m[\alpha + \xi_m^{(T)}]\lambda_L,
\end{aligned} \tag{4.131}
$$

where $\xi_m^{(T)} = (n_{co} - n_{cl,m})^{-1}[\partial(n_{co} - n_{cl,m})/\partial T]$ is the equivalent thermo-optical coefficient. Similarly, the resonant wavelength varies with strain e in the fiber with coefficient of

$$
\begin{aligned}
\frac{\partial \lambda_L}{\partial |e|} &= (n_{\text{co}} - n_{\text{cl},m})\frac{\partial \Lambda}{\partial |e|} + \Lambda\frac{\partial(n_{\text{co}} - n_{\text{cl},m})}{\partial |e|} \\
&+ \Lambda\frac{\partial(n_{\text{co}} - n_{\text{cl},m})}{\partial \lambda}\frac{\partial \lambda_L}{\partial |e|} = \gamma_m[1 + \xi_m^{(e)}]\lambda_L,
\end{aligned} \tag{4.132}
$$

with the equivalent photoelastic coefficient of $\xi_m^{(e)} = (n_{co} - n_{cl,m})^{-1}$ $[\partial(n_{co} - n_{cl,m})/\partial |e|]$. Generally axial strains occur most frequently, $|e| = e_z$. Due to the waveguide dispersion factor, the temperature and strain coefficients of λ_L are usually much larger than those of FBG, and may vary greatly in complicated ways [42].

The characteristics of cladding modes depend also on the index of the surrounding medium n_3. This is a unique feature, different from the FBG, used as sensors for detecting index change of the surrounding medium. The sensitivity is also affected by the waveguide dispersion factor,

$$
\begin{aligned}
\frac{\partial \lambda_L}{\partial n_3} &= \Lambda\frac{\partial(n_{\text{co}} - n_{\text{cl},m})}{\partial n_3} + \Lambda\frac{\partial(n_{\text{co}} - n_{\text{cl},m})}{\partial \lambda}\frac{\partial \lambda_L}{\partial n_3} \\
&= \gamma_m \xi_m^{(n)}\lambda_L,
\end{aligned} \tag{4.133}
$$

with coefficient of $\xi_m^{(n)} = -(n_{co} - n_{cl,m})^{-1}(\partial n_{cl,m}/\partial n_3)$.

It is shown that the sensitivities of LPFG depend strongly on the waveguide dispersion factor; moreover, the factor itself is also a function of temperature and strain, making the sensitivities of LPFG more complicated. Different LPFG may show quite different behaviors due to the waveguide dispersion factor. Reference [42] measures the variations of two adjacent resonant wavelengths with temperature and with strain, showing curves with opposite slopes; and explains the behaviors by the effect of the waveguide dispersion factor.

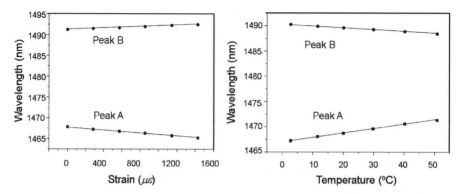

Figure 4.46   Measured resonance wavelength of an LPFG on PMF versus axial stain and temperature. (Reprinted with permission from reference [136]. © 2003 IEEE.)

Figure 4.46 shows experimental results of resonance wavelength versus axial strain and temperature of an LPFG imprinted in a PMF [136]. Two peaks, labeled *A* and *B*, are obtained due to the birefringence, with spectral spacing much larger than that of the two peaks measured in Bragg gratings imprinted in the PMF. Furthermore, their strain and temperature coefficients show not only different values, but also opposite signs.

4.4.2   Tunability of Fiber Gratings

It is obvious that the sensitivities of fiber gratings provide possibilities of developing sensors and tunable filters as well. Tunability is an attractive feature especially for an in-fiber device. Two basic methods are developed for the purpose.

4.4.2.1 Thermal Tuning   Thermal tuning methods are relatively simple. The fiber grating can be stuck onto a heater or a Peltier-cooler (thermal-electric, TE-cooler). A miniature heater is the metal layer as a resistor, coated directly on the fiber with the jacket stripped, demonstrating the highest heating efficiency. Combined with a TE-cooler, flexible tuning with a large range is realized. Furthermore, the coating thickness can be fabricated with variance along the fiber to generate a spatially variable temperature distribution, which is used for a tunable CFBG [137]. One of the merits of a metal coating heater is its faster response.

Figure 4.47 is the measured temporal variation of the peak wavelength of the metal-coated FBG under different heating power [138], showing that the heating and cooling rate of its peak wavelength

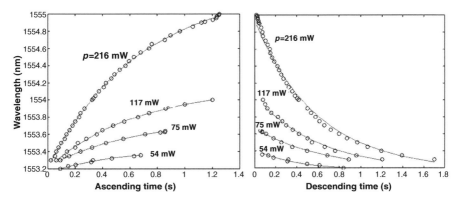

Figure 4.47 Temporal variations of peak wavelength of a metal coated FBG. (Reprinted with permission from reference [138]. © 2003 IEEE.)

reaches 4 ~ 5 nm/s; and the time constant is deduced from the fitted solid lines to be 0.6 second.

4.4.2.2 Stress Tuning Varieties of stress tuning mechanics are also proposed and demonstrated. The most widely used structure is to fix the fiber grating onto a cantilever with one end clamped, as shown in Figure 4.48(a). The deformation of a cantilever with transversely uniform sizes under a force F applied at its free end is described as (cf. A2.35)

$$\Delta x(z) = \frac{F}{2YI} z^2 (L - z/3), \tag{4.134}$$

where Δx is the x-direction displacement, $I = ba^3/12$ is the moment of inertia about y-axis of the cantilever with thickness a and width b, and Y is its Young's modules.

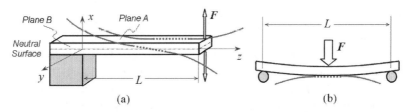

Figure 4.48 (a) Cantilever and (b) pure bending beam used for fiber grating tuning.

It is then obtained that the curvature radius of the bent cantilever is a function of z-position:

$$R_C \approx \frac{1}{\Delta x_z''} = \frac{YI}{F(L-z)}. \tag{4.135}$$

The strain distribution in x-z plane is deduced as

$$e_z = \frac{x}{R_C} = \frac{F(L-z)}{YI}x, \tag{4.136}$$

where x is measured from the neutral surface shown in the figure. The strain on surface A is obtained to be

$$e_z\big|_{x=a/2} = \frac{6F(L-z)}{Yba^2}. \tag{4.137}$$

The fiber grating to be tuned is stuck on plane A when the cantilever is free of external forces. The applied force F can thus bend the cantilever and tune the fiber grating. It is seen from (4.136) that the tuning rate is a function of position in z-direction, implying that a linear chirping is introduced, and a certain degree of linewidth broadening occurs. It is to be avoided for some applications. To overcome this undesired chirping, a pure bent beam is used, which is supported at two points, leaving its two ends freely moving, as shown in Figure 4.48(b). From elasticity analysis, its curvature radius is deduced to be $R_C = YI/FL$, which does not depend on the z-position. The strain on the surface at $x = a/2$ is obtained to be

$$e_z\big|_{x=a/2} = \frac{6FL}{Yba^2}. \tag{4.138}$$

The FBG stuck on the surface can then be tuned accordingly.

If the chirping rate of an LCFBG is to be adjusted while keeping the central wavelength unchanged, it is realized by sticking the grating onto plane B, the side surface of the cantilever. The different parts of CFBG are now stuck on to different positions in x direction, the grating will suffer an additional linearly varying strain, whereas the part at the neutral surface is kept unchanged. The chirping increment depends on the z-position of CFBG, and on the slant angle θ between the fiber and the cantilever neutral line, expressed as

$$\frac{\delta\lambda_B}{\lambda_B}(z_f) = \frac{3ax}{2L^3}(1-\gamma)(L-z_0) = \frac{3a}{2L^3}(1-\gamma)(L-z_0)z_f\sin\theta, \quad (4.139)$$

where γ is the effective photoelastic coefficient, z_f is the axial position of the fiber from the middle point, z_0 is the z-position of the beam, where the grating point with the wavelength to be fixed is aligned on the neutral surface.

There are quite a number of mechanics to tune the resonant wavelength, and to reshape the spectrum profile, such as by lateral pressure [139].

4.4.3 Packaging of Fiber Grating Devices

Various application situations require special packaging structures and technologies. The following considerations are important, but are not necessarily required simultaneously.

4.4.3.1 *Protection Packaging*
Protection is surely necessary for reliable fiber devices, especially since the fiber grating is usually fabricated onto a jacket-stripped fiber. In some applications the fiber sensor is buried into concrete structures, or a steel block; its protection is not a simple task. We will not go further into the technical details in this book.

4.4.3.2 *Decomposition of Temperature/Stress Cross-Sensitivity*
As the peak wavelength is sensitive to both temperature and strain, it is necessary to decompose the two factors, when the grating is used as a sensor. Several configurations are developed for the purpose [140,141, 143,144].

1. The simplest method is to use two gratings with different temperature and stress coefficients, based on different fiber materials or different packaging substrates. The two peak wavelength changes are coupled with each other, expressed as

$$\begin{pmatrix} \Delta\lambda_1 \\ \Delta\lambda_2 \end{pmatrix} = \begin{pmatrix} \zeta_{1\sigma} & \zeta_{1T} \\ \zeta_{2\sigma} & \zeta_{2T} \end{pmatrix} \begin{pmatrix} \Delta\sigma \\ \Delta T \end{pmatrix}. \quad (4.140)$$

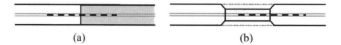

(a)   (b)

Figure 4.49 Discrimination of temperature and strain coefficients: (a) by a section fused of different fibers and (b) with a locally thinned section.

Thus the temperature and stress variations can be decomposed respectively by measurement of $\Delta\lambda_1$ and $\Delta\lambda_2$, so long as the coefficients are linearly independent.

2. Gratings with different periods can be imprinted into the same section, showing two peaks, such as in 1310 nm and 1550 nm bands. The thermo-optic coefficient is generally a function of wavelength, resulting in different temperature and strain coefficients. A combination of FBG and LPFG shows two sets of resonance wavelengths with different temperature and stress coefficients.

3. A fiber grating is imprinted into a fiber section fused of two fibers with different background indexes and different sensitivity coefficients, as shown in Figure 4.49(a). It can also be imprinted into a fiber section with a thinned part to make its stress coefficient different, as shown in Figure 4.49(b). Reference [142] proposes a scheme by use of a single FBG written in an erbium:ytterbium-doped fiber.

4.4.3.3 Temperature-Insensitive Packaging and Stress-Insensitive Packaging

Temperature-insensitive FBG are required in many applications, for example, used as a filter in DWDM fiber communications. Because the channel spacing is in sub-nanometer range, the temperature sensitivity has to be reduced to 1 pm/°C for a usual ten degrees temperature variation, about tenth of the original FBG sensitivity. For the purpose, a special ceramic with negative temperature coefficient (NTC) is used as the substrate of FBG packaging. To compensate for its temperature sensitivity the FBG is fixed onto the NTC substrate under a proper prestretching strain. The peak wavelength variation with temperature is now expressed as

$$\frac{\Delta\lambda_B}{\lambda_B} = \zeta_e e_z + \zeta_T \Delta T = \zeta_e(e_{z0} - \Delta\zeta_T \Delta T) + \zeta_T \Delta T, \qquad (4.141)$$

where ζ_e and ζ_T are the strain and temperature coefficients of the FBG; $\Delta\zeta_T$ is the thermal expansion coefficient difference between NTC

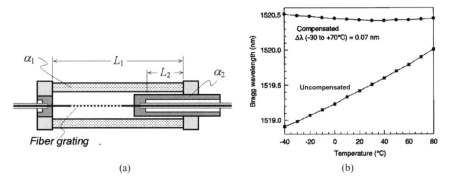

Figure 4.50 (a) Temperature-compensating package and (b) experimental results. (Reprinted with permission from reference [145].)

ceramic and fiber. It is then deduced that the temperature compensation condition is $\zeta_e \Delta \zeta_T = \zeta_T$. The prestrain has to be designed according to the coefficient and its linear range of NTC material.

Apart from the NTC ceramic, such an effect can be provided by a mechanism composed of two materials with different thermal expansion coefficients (TEC) and properly designed lengths. Figure 4.50(a) shows a cylindrical packaging by using two different materials of α_1 and α_2; for example, a silica tube and an aluminum tube ($\alpha_2 \approx 2.5 \times 10^{-6}/°C$). The composite TEC is $\tilde{\alpha} = (\alpha_1 L_1 - \alpha_2 L_2)/(L_1 - L_2)$, which may take a proper negative value by adjusting lengths L_1 and L_2. Figure 4.50(b) gives an example of experimental results [145], showing the temperature coefficient reduced to 0.7 pm/°C from 11 pm/°C of the uncompensated one.

Stress-insensitive packaging is also needed if the grating is used as a solitary temperature sensor. To avoid the thermal stress, the fiber grating sensor should be packaged in a small tube with strain isolated [146]. Another method is to fix the fiber grating, at temperatures high enough, in a case made of materials with a thermal expansion coefficient higher than silica. Thus the sensor will work at a thermal stress-free condition under temperatures lower than the package temperature.

4.4.3.4 *Sensitivity-Enhanced Packaging* Two functions are desired in sensor packaging: its sensitivity is enhanced, and the sensed parameter of the objectives is transferred to the characteristics of the grating sensor with specially designed structures [140]. Different sensed parameters, such as displacements, bending, torsion, vibrations, acoustic wave, pressure, acceleration, electric and/or magnetic field, water level, and chemical changes, are to be transferred to fiber strains.

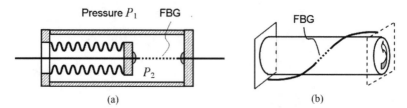

Figure 4.51   FBG sensors for (a) pressure and (b) torsion.

Quite a lot of structures have been proposed and demonstrated for the purpose [139,148]. Figure 4.51(a) shows a pressure sensor structure with an elastic bellow pushed by the pressure difference, and the FBG sensor suffers stretching accordingly; by using properly selected elastic coefficient of the bellow, the sensitivity of the FBG sensor can be enhanced and adjusted. Figure 4.51(b) is a torsion sensor with FBG stuck onto the side surface of the cylinder.

Figure 4.52(a) is another scheme of a pressure sensor, which utilizes a cylindrical drum with the sensor stuck on its surface. As discussed in Appendix 2 of this book, pressure change will cause deformation of the surface, resulting in a strain change in the fiber. On the other hand, the maximum curvature is located at the center; if an LPFG is stuck at the position, its loss peak will show high sensitivity. Figure 4.52(b) shows another packaging of a FBG pressure sensor.

Figure 4.53(a) shows a concept of sensors, where the FG is fixed on a special substrate with swelling virtue in absorbing certain chemicals [149]; or on a magnetostrictive material and other field-sensitive materials. Thus the fiber grating can detect the chemical concentration in the environment, such as petroleum leakage, humidity; or sense the magnetic field. The FG may be coated with the sensitive material to greatly reduce the volume of sensor. Figure 4.53(b) depicts a structure of corrosion sensor [150], where the diameter of spring wire may be

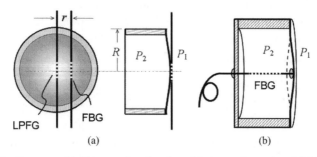

Figure 4.52   Sensors with fiber grating fixed on (a) the surface of and (b) inside drum.

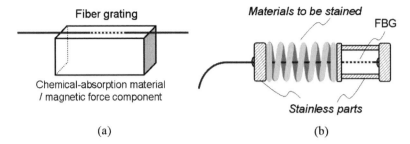

Figure 4.53 (a) Sensor with fiber grating stuck on a sensitive substrate and (b) corrosion sensor made of a metal spring.

etched in some corrosive environment and the FBG will detect the elastic coefficient change of the spring.

Figure 4.54(a) depicts an *FBG rosette* composed of three FBGs to provide stress status in a two-dimensional surface. Figure 4.54(b) shows an FBG sensor buried inside a composite prepreg tow piece (PTP) material, which is made of carbon fibers with high strength and pliability. These sensors are useful for structural health monitoring.

For higher temperature sensitivity, quite a number of materials with high thermal expansion coefficient are used as substrates to package fiber gratings, such as polymers and metals [147].

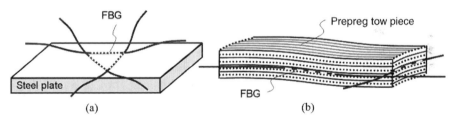

Figure 4.54 (a) Fiber Bragg grating rosette composed of three FBG and (b) FBG sensor buried inside PTP material.

4.4.4 Fiber Grating Sensor Systems and Their Applications

A great number of fiber grating sensor systems have been developed. This book will not introduce them in detail, but just give some discussions on several technical issues.

4.4.4.1 Interrogation of Peak Wavelength Variation
Fiber grating sensor technology has been widely studied and developed, and related systems have been commercialized. Because the sensed signal is carried

by the peak wavelength of fiber gratings, its demodulation is surely a key of the sensing technology. To interrogate the peak wavelength a variety of methods are presented.

1. By spectral analyzers:

    Although a spectral analyzer is the best device to read out the peak wavelength and its variation, it usually incurs higher costs and is not as convenient in field applications. Therefore alternative methods are needed to act as special spectral analyzers. Fortunately, several compact spectral analyzers have been invented in the development of fiber communication technologies, especially the arrayed waveguide grating (AWG), which is regarded as a *spectrometer on a chip*, and thin-film filters (TFF), which is based on the narrow band interference filter made of multilayer dielectric film, and CCD (Charge Coupled Device) array spectrometer [151].

2. By fiber F-P filters:

    The fiber F-P filter is a commercially available, cost-effective, and narrow-band filter, usually with controllable scanning functions. Its working wavelength can be designed and adjusted by its cavity length. It can be used as a spectral analyzer with high resolution; and its tunability allows for coverage over a broad spectral range. The detailed characteristics of fiber F-P filters will be introduced in Chapter 7 of this book.

3. By MZIs:

    The MZI incorporated with a 3×3 fiber coupler is discussed in Section 3.3.4. The outputs of three photodetectors, given by formulas (3.143a–c), are rewritten as

$$I_1 = \frac{1}{3}\left[1 - \sin\left(\frac{2\pi}{\lambda^2}\Delta L \delta\lambda + \frac{\pi}{3}\right)\right]I_0, \qquad (4.142a)$$

$$I_2 = \frac{1}{3}\left[1 - \sin\left(\frac{2\pi}{\lambda^2}\Delta L \delta\lambda - \frac{\pi}{3}\right)\right]I_0, \qquad (4.142b)$$

$$I_3 = \frac{1}{3}\left[1 + \sin\left(\frac{2\pi}{\lambda^2}\Delta L \delta\lambda\right)\right]I_0, \qquad (4.142c)$$

where $\Delta L = n_{\text{eff}}(L_1 - L_2)$ is the optical path difference (OPD). Then the FBG peak wavelength variation $\delta\lambda$ is deduced from the three outputs; and the data processing can do the phase unwrapping to expand the measurement range. By using FBG with

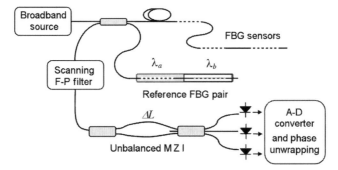

Figure 4.55   FBG sensor system with an unbalanced MZI as signal interrogator.

linewidth narrow enough, high sensitivities have been demonstrated. Figure 4.55 shows a multipoint quasi-distributed sensor system with several FBG sensor heads connected in series, with an unbalanced MZI incorporated with a 3×3 fiber coupler, as the wavelength interrogation device [152,153]. To interrogate the multiple signals a scanning F-P filter is inserted in front of the unbalanced MZI to eliminate the uncertainty due to the multipass property of MZI.

The spectrum drift of fiber interferometers is a common problem. The fiber lengths of MZI beams are affected easily by environmental conditions, such as temperature variation, internal strain and its relaxation (creep), vibration and sound wave, and other effects. To overcome the drift problem, a reference FBG pair shown in the figure is used as a phase drift compensator [154], which is composed of two FBGs, stuck on two substrates, for example made of aluminum and silica, resulting in temperature coefficients properly different from each other, but packaged together in the same temperature environment. Their temperature coefficients are calibrated in laboratory coordinates to be $\gamma_a = \partial \lambda_a / \partial T$ and $\gamma_b = \partial \lambda_b / \partial T$. After the compensator is connected in the system, their wavelength change can be read by the interferometer, expressed as

$$\delta \lambda_a = \delta \lambda_{\mathrm{MZI}} + \gamma_a \delta T,$$
$$\delta \lambda_b = \delta \lambda_{\mathrm{MZI}} + \gamma_b \delta T. \qquad (4.143)$$

Therefore the interferometer drift $\delta \lambda_{\mathrm{MZI}} = (\gamma_a \delta \lambda_b - \gamma_b \delta \lambda_a)/ (\gamma_a - \gamma_b)$ is calculated in real time by measured $\delta \lambda_a$ and $\delta \lambda_b$, and used to correct the signals of FBG sensors.

Interferometers other than MZI are also used as interrogation methods, as described in references [155,156].

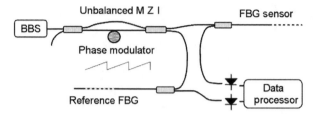

Figure 4.56 FBG sensor system with an MZI modulated source.

4. By modulated sources:

The unbalanced MZI, combined with a broad band source (BBS), can serve as a tunable light source. Figure 4.56 shows its schematic structure, where the phase modulator, driven by a triangle wave, scans the interference fringes of the unbalanced MZI; and a reference FBG provides a comparison for sensor signals.

5. By Matched Fiber Grating Filters:

The fiber grating itself, as a spectral filter, is actually a candidate of the wavelength discriminator, when the edge of its spectral profile is selected and adjusted to coincide with the input wavelength. This scheme is called edge filtering demodulation, as shown in Figure 4.57(a), in which a linearly varying slope edge of its spectrum is preferable. Therefore, an FBG with a reflection spectrum in triangle profile is designed to serve as an edge filter [157], because the LPFG usually has resonant peaks much broader than FBG, it is suitable to serve as the edge filter [158], shown in Figure 4.57(b).

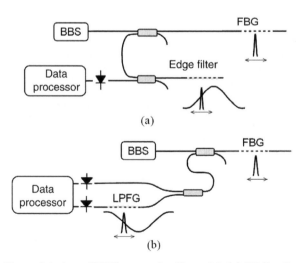

Figure 4.57 Demodulation of FBG sensor by filter with (a) FBG reflective edge and (b) LPFG loss edge.

There is also interest in active interrogation methods, which is to put the FBG sensor into a fiber laser cavity. The lasing wavelength will be determined by the reflection peak of the FBG. The much higher laser intensity makes it helpful in picking up the sensed signals [159,160].

4.4.4.2 Features and Merits of Fiber Grating Sensor Systems

The fiber grating sensor has its unique features and merits when compared with other sensors:

1. Has memory of sensed parameters. The property of gratings provides the sensors with a built-in self-referencing capability. Because the sensed information is encoded directly into the wavelength, the datum does not depend on the light source power; the signal does not relate with its historical evolution, as that of usual interferometers. This advantage is quite attractive over the interferometer-based sensors.

2. Capability of wavelength division multiplex [161,162]. The wavelength encoding of sensed signals also facilitates wavelength division multiplexing by assigning each sensor to a specified "slice" of the available source spectrum.

3. Capability of building quasi-distributed sensor systems. FBGs with different Bragg wavelengths can be connected with fibers and extended to long distances. In addition, the FBG quasi-distributed sensor system is capable of dynamic sensing for a vibration and sound sensor, with high spatial resolution down to millimeter range [163].

4. Small size and possibility of incorporation with various materials and structures to realize multiparameter sensors.

References [164–166] present and summarize the fiber grating sensor systems and related technologies. It is shown that the fiber grating sensors have been widely used in a variety of applications, including civil engineering, environmental monitoring and protection, energy source technology, and national security.

PROBLEMS

4.1 Referring to the original Bragg Equation in X-ray diffraction of crystal, explain the condition of Bragg wavelength.

4.2   Read out the peak wavelengths from Figure 4.3; estimate the indexes of core mode and related cladding modes from Equations (4.1.1) and (4.1.2), if $\Lambda_{\mathrm{FBG}} = 1.071$ μm and $\Lambda_{\mathrm{LPFG}} = 440$ μm.

4.3   What is the Kramers–Kronig relation? Explain photosensitivity using the relation.

4.4   Design holographic optics for writing a 1,550 nm FBG. How much should the incident angle on the fiber be set, if a 244 nm frequency doubling Ar^+ laser is available.

   Design a phase mask for writing FBG at 1,550 nm band by 248.5 nm XeF excimer laser. How much should the period of groove be? How deep should the groove for null zeroth-order diffraction be? The index of quartz at 248.5 nm is 1.508.

4.5   Write the CME of FBG, deduce the expressions of the transmission and reflection wave of an FBG. What is the relation between the peak wavelength of reflection and the grating period?

4.6   Write the formulas of peak reflection and linewidth, discuss their relation with the index modulation depth and with the grating length.

4.7   Write the CME of LPFG; deduce the expression of the transmission wave of LPFG. What is the difference of coupling mechanisms between FBG and LPFG?

4.8   What factors and parameters does the wavelength of the LPFG loss peak depend on? What kind of cladding modes can couple with the core mode by the grating? What parameters does the amplitude of the LPFG loss peak depend on?

4.9   What is the TMM? Write programs of TMM for FBG and for LPFG respectively; calculate and compare the spectra of FBG and LPFG with typical parameters.

4.10 What do the side lobes of the reflection spectrum of FBG come from? Why does the apodization method improve the spectrum? Simulate the apodization effectiveness by TMM program for a raised cosine function and a Gaussian function.

4.11 What is the inverse engineering design of FBG; describe the basic concept of GA and layer-peeling algorithm methods.

4.12 Describe the characteristics of phase-shifted FBG. What requirements should be satisfied for a narrow linewidth and highest transmission?

4.13 Describe the characteristics of sampled FBG. What parameters determine the period and the duty cycle of the comb filter? If the period of the reflection peaks of 0.8 nm is required in 1,550 nm band, what size of the sample period should be used?

4.14 What is the moiré FBG? How is the moiré grating fabricated?

4.15 Describe the characteristics of an FBG F-P interferometer.

4.16 How do you use a CFBG to compensate the dispersion of an optical pulse propagating in fiber? If the dispersion of the conventional fiber is 17 ps/nm/km, how much should the chirping rate of a CFBG be for compensating pulse broadening after 40 km propagation?

4.17 Why do ripples occur in the reflection spectrum of CFBG? How can the ripples be removed?

4.18 What differences from a normal FBG will be observed in the reflection spectrum of a TFBG? What useful functions and devices could be obtained from the TFBG?

4.19 Describe the characteristics of Bragg grating imprinted in a PMF.

4.20 Describe the sensitivities of an FBG to strains and temperature change, and their relations with the FBG parameters. If a temperature-insensitive FBG is needed, what packaging structures can be used?

4.21 Describe the sensitivities of an LPFG to strains and temperature change. What is the waveguide dispersion factor of LPFG? How does the waveguide dispersion factor affect the sensitivities of LPFG?

4.22 Give structure designs of FBG strain sensors used for a pressure measurement.

4.23 Describe the characteristics of a rectangular cantilever. How do you use the cantilever to calibrate the strain sensitivity of FBG?

4.24 How do you demodulate the cross-sensitivities of FBG to strain and to temperature?

4.25 Describe the method of interrogation of a FBG sensor by an MZI? What are the advantages of an MZI incorporated with a 3×3 fiber coupler?

4.26 Can you propose any new schemes for FBG sensor interrogation?

REFERENCES

1. Russell PSJ, Ulrich R. Grating-fiber coupler as a high-resolution spectrometer. *Optics Letters* 1985; 10: 291–293.

2. Hill KO, Fujii Y, Johnson DC, Kawasaki BS. Photosensitivity in optical fiber waveguides: application to reflection filter fabrication. *Applied Physics Letters* 1978; 32: 647–649.

3. Meltz G, Morey WW, Glenn WH. Formation of Bragg gratings in optical fibers by a transverse holographic method. *Optics Letters* 1989; 14: 823–825.

4. Hill KO, Malo B, Bilodeau F, Johnson DC, Albert J. Bragg grating fabricated in monomode photosensitive optical fiber by UV exposure through a phase mask. *Applied Physics Letters* 1993; 62: 1035–1037.

5. Malo B, Johnson DC, Bilodeau F, Albert J, Hill KO. Single-excimer-pulse writing of fiber gratings by use of a zero-order nulled phase mask: grating spectral response and visualization of index perturbations. *Optics Letters* 1993; 18: 1277–1279.

6. Lemaire PJ, Atkins RM, Mizrahi V, Reed WA. High pressure H_2 loading as a technique for achieving ultrahigh UV photosensitivity and thermal sensitivity in GeO_2 doped optical fiber. *Electronics Letters* 1993; 29: 1191–1193.

7. Hill KO, Meltz G. Fiber Bragg grating technology fundamentals and overview. *Journal of Lightwave Technology* 1997; 15: 1263–1276.

8. Vengsarkar AM, Lemaire PJ, Judkins JB, Bhatia V, Erdogan T, Sipe JE. Long-period fiber gratings as band-rejection filters. *Journal of Lightwave Technology* 1996; 14: 58–65.

9. Bhatia V, Vengsarkar AM. Optical fiber long-period grating sensors. *Optics Letters* 1996; 21: 692–694.

10. Kashyap R. *Fiber Bragg Gratings*. San Diego, London, Boston, New York, Sydney, Tokyo, Toronto: Academic Press, 1999.

11. Othonos A, Kalli K. *Fiber Bragg Gratings Fundamentals and Applications in Telecommunications and Sensing*. London, Boston: Artech House, 1999.

12. Dong L, Archambault JL, Reekie L, Russell PS, Payne DN. Photo-induced absorption change in germanosilicate preforms: evidence for the color-center model of photosensitivity. *Applied Optics* 1995; 34: 3436–3440.

13. Poumellec B, Kherbouche F. The photo refractive Bragg gratings in the fibers for telecommunications. *Journal de Physics III France* 1996; 6: 1595. Quoted from Reference [10].

14. Riant I, Haller F. Study of the photosensitivity at 193 nm and comparison with photosensitivity at 240 nm influence of fiber tension: type IIa aging. *Journal of Lightwave Technology* 1997; 15: 1464–1469.

15. Archambault JL, Reekie L, Russell PS. 100% reflectivity Bragg reflectors produced in optical fiber by single excimer laser pulses. *Electronics Letters* 1993; 29: 453–455.

16. Bilodeau F, Malo B, Albert J, Johnson DC, Hill KO, Hibino Y, Abe M, Kawachi M. Photosensitizing of optical fiber and silica-on-silicon/silica waveguides. *Optics Letters* 1993; 18; 953–955.

17. Lemaire PJ. Reliability of optical fibers exposed to hydrogen: prediction of long-term loss increases. *Optical Engineering* 1991; 30: 780–789.

18. Russell PS, Hand DP, Chow YT, Poyntz-Wright LJ. Optically-induced creation, transformation and organization of defects and color-centers in optical fibers. *Proceedings of SPIE*. 1991; 1516: 47–54.

19. Nishii J, Fukumi K, Yamanaka H, Kawamura K, Hosono H, Kawazoe H. Photochemical reactions in GeO_2-SiO_2 glasses induced by ultraviolet irradiations: comparison between Hg lamp and excimer lasers. *Physics Review B* 1995; 52: 1661–1665.

20. Saleh BEA, Teich MC. *Fundamentals of Photonics*. Hoboken : John Wiley & Sons, 2007.

21. Araújo FM, Joanni E, Marques MB, Okhotnikova OG. Dynamics of infrared absorption caused by hydroxyl groups and its effect on refractive index evolution in ultraviolet exposed hydrogen loaded GeO_2-doped fibers. *Applied Physics Letters* 1998; 72: 3109–3111.

22. Grubsky V, Starodubov DS, Feinberg J. Effect of molecular water on thermal stability of gratings in hydrogen-loaded optical fibers. *Technical Digest of Optical Fiber Communication Conference*, 1999, San Diego, CA, paper ThD2.

23. Douay M, Xie WX, Taunay T, Bernage P, Niay P, Cordier P, Poumellec B, Dong L, Bayon JF, Poignant H, Delevaque E. Densification involved in the UV-based photosensitivity of silica glasses and optical fibers. *Journal of Lightwave Technology* 1997; 15: 1329–1342.

24. Hill KO, Malo B, Vineberg KA, Bilodeau F, Johson DC, Skinner I. Efficient mode conversion in telecommunication fiber using externally written grating. *Electronics Letters* 1990; 26: 1270–1272.

25. Askins CG, Tsai TE, Williams GM, Putnam MA, Bashkansky M, Friebele EJ. Fiber Bragg reflectors prepared by a single excimer pulse. *Optics Letters* 1992; 17: 833–835.

26. Bhatia V, Vengsarkar AM. Optical fiber long-period grating sensors. *Optics Letters* 1996; 21: 692–694.

27. Davis DD, Gaylord TK, Glytsis EN, Kosinski SG, Mettler SC, Vengsarkar AM. Long-period fiber grating fabrication with focused CO2 laser pulses. *Electronics Letters* 1998; 34: 302–303.

28. Hwang IK, Yun SH, Kim BY. Long-period fiber gratings based on periodic microbends. *Optics Letters* 1999; 24: 1263–1265.

29. Malki A, Humbert G, Ouerdane Y, Boukhenter A, Boudrioua A. Investigation of the writing mechanism of electric-arc-induced long-period fiber gratings. *Applied Optics* 2003; 42: 3776–3779.

30. Jiang Y, Li Q, Lin CH, Lyons E, Tomov I, Lee HP. A novel strain-induced thermally tuned long-period fiber grating fabricated on a periodic corrugated silicon fixture. *IEEE Photonics Technology* 2002; 14: 941–943.

31. Davis DD, Gaylord TK, Glytsis EN, Mettler SC. Very-high-temperature stable CO_2-laser induced long-period fiber gratings. *Electronics Letters* 1999; 35: 740–742.

32. Liu Y, Chiang KS. CO_2 laser writing of long-period fiber gratings in optical fibers under tension. *Optics Letters* 2008; 33: 1933–1935.

33. Agrawal GP. *Nonlinear Fiber Optics*. Third Edition. Singapore: Elsevier Sciences, 2001.

34. Marcuse D. *Light Transmission Optics*. New York, Cincinnati, Toronto, Melbourne: Van Nostrand Reinhold Company, 1982.

35. Yariv A. *Optical Electronics in Modern Communications*. Fifth Edition. New York: Oxford University Press, 1997.

36. Mizrahi V, Sipe JE. Optical properties of photosensitive fiber phase gratings. *Journal of Lightwave Technology* 1993; 11: 1513–1517.

37. Erdogan T. Fiber grating spectra. *Journal of Lightwave Technology* 1997; 15: 1277–1294.

38. Erdogan T. Cladding-mode resonances in short- and long-period- fiber grating filters. *Journal of Optical Society America A* 1997; 14: 1760–1773.

39. Dong L, Reekie L, Cruz JL, Caplen JE, de Sandro JP, Payne DN. Optical fiber with depressed claddings for suppression of coupling into cladding modes in fiber Bragg gratings. *IEEE Photonics Technology Letters* 1997; 9: 64–66.

40. Dong L, Qi G, Marro M, Bhatia V, Hepburn LL, Swan M, Collier A, Weidman DL. Suppression of cladding mode coupling loss in fiber Bragg gratings. *Journal of Lightwave Technology* 2000; 18: 1583–1590.

41. MacDougall TW, Pilevar S, Haggans CW, Jackson MA. Generalized expression for the growth of long period gratings. *IEEE Photonics Technology Letters* 1998; 10: 1449–1451.

42. Shu X, Zhang L, Bennion I. Sensitivity characteristics of long-period fiber gratings. *Journal of Lightwave Technology* 2002; 20: 255–266.

43. Gao K, Fang Z. A new method of measuring the waveguide dispersion factor and the thermo-optic coefficient of long-period fiber gratings. *Optics Communications* 2005; 244: 227–231.

44. Yamada M, Sakuda K. Analysis of almost-periodic distributed feedback slab waveguides via a fundamental matrix approach. *Applied Optics* 1987; 26: 3474–3478.

45. Song GH, Shin SY. Design of corrugated waveguide filters by the Gel'fand–Levitan–Marchenko inverse-scattering method. *Journal of Optical Society America A* 1985; 2: 1905–1915.

46. Poladian L. Graphical and WKB analysis of nonuniform Bragg gratings. *Physical Review E* 1993; 48: 4758–4767.

47. Weller-Brophy LA, Hall DG. Analysis of waveguide gratings: application of Rouard's method. *Journal of Optical Society America A* 1985; 2: 863–871.

48. Weller-Brophy LA, Hall DG. Analysis of waveguide gratings: a comparison of the results of Rouard's method and coupled-mode theory. *Journal of Optical Society America A* 1987; 4: 60–65.

49. Born M, Wolf E. Principles of Optics. Seventh Edition. Cambridge University Press, 1999.

50. Skaar J, Risvik KM. A genetic algorithm for the inverse problem in synthesis of fiber gratings. *Journal of Lightwave Technology* 1998; 16: 1928–1932.

51. Gill A, Peters K, Studer M. Genetic algorithm for the reconstruction of Bragg grating sensor strain profiles. *Measurement Science and Technology* 2004; 15: 1877–1884.

52. Casagrande F, Crespi P, Grassi AM, Lulli A, Kenny RP, Whelan MP. From the reflected spectrum to the properties of a fiber Bragg grating: a genetic algorithm approach with application to distributed strain sensing. *Applied Optics* 2002; 41: 5238–5244.

53. Huang R, Zhou Y, Cai H, Qu R, Fang Z. A fiber Bragg grating with triangular spectrum as wavelength readout in sensor systems. *Optics Communications* 2004; 229: 197–201.

54. Feced R, Zervas MN, Muriel MA. An efficient inverse scattering algorithm for the design of nonuniform fiber Bragg gratings. *IEEE Journal of Quantum Electronics* 1999; 35: 1105–1115.

55. Skaar J, Wang L, Erdogan T. On the synthesis of fiber Bragg gratings by layer peeling. *IEEE Journal of Quantum Electronics* 2001; 37: 165–173.

56. Li H, Kumagai T, Ogusu K, Sheng Y. Advanced design of a multichannel fiber Bragg grating based on a layer-peeling method. *Journal of the Optical Society of America B* 2004; 21: 1929–1938.

57. Ouyang Y, Sheng Y, Bernier M, Paul-Hus G. Iterative layer-peeling algorithm for designing fiber Bragg gratings with fabrication constraints. *Journal of Lightwave Technology* 2005; 23: 3924–3930.

58. Winick KA, Roman JE. Design of corrugated waveguide filters by Fourier transform techniques. *IEEE Journal of Quantum Electronics* 1990; 26: 1918–1829.

59. Jaggard DL, Kim Y. Accurate one-dimensional inverse scattering using a nonlinear renormalization technique. *Journal of the Optical Society of America A* 1985; 2: 1922–1930.

60. Azaña J, Muriel MA. Technique for multiplying the repetition rates of periodic trains of pulses by means of a temporal self-imaging effect in chirped fiber gratings. *Optics Letters* 1999; 24: 1672–1674.

61. Azaña J, Chen LR. Synthesis of temporal optical waveforms by fiber Bragg gratings: a new approach based on space-to-frequency-to-time mapping. *Journal of Optical Society America B* 2002; 19: 2758–2769.

62. Peral E, Capmany J, Marti J. Iterative solution to the Gel'fand–Levitan–Marchenko coupled equations and application to synthesis of fiber gratings. *IEEE Journal of Quantum Electronics* 1996; 32: 2078–2084.

63. Brinkmeyer E. Simple algorithm for reconstructing fiber grating from reflectometric data. *Optics Letters* 1995; 20: 810–812.

64. Albert J, Hill KO, Malo B, Theriault S, Bilodeau F, Johnson DC, Erickson LE. Apodisation of the spectral response of fiber Bragg gratings using a phase mask with variable diffraction efficiency. *Electronics Letters* 1995; 31: 222–223.

65. Pastor D, Capmany J, Ortega D, Tatay V, Marti J. Design of apodized linearly chirped fiber gratings for dispersion compensation. *Journal of Lightwave Technology* 1996; 14: 2581–2588.

66. Ennser K, Zervas MN, Laming RI. Optimization of apodized linearly chirped fiber gratings for optical communications. *IEEE Journal of Quantum Electronics* 1998; 34: 770–778.

67. Fröhlich H, Kashyap R. Two methods of apodisation of fiber-Bragg-gratings. *Optics Communications* 1998; 157: 273–281.

68. Kogelnik H, Shank CV. Coupled-wave theory of distributed feedback lasers. *Journal of Applied Physics* 1972; 43: 2327–2335.

69. Agrawal GP, Radic S. Phase-shifted fiber Bragg gratings and their application for wavelength demultiplexing. *IEEE Photonics Technology Letters* 1994; 6: 995–997.

70. Eggleton BJ, Krug PA, Poladian L, Ouellette F. Long periodic superstructure Bragg gratings in optical fibers. *Electronics Letters* 1994; 30: 1620–1622.

71. Othonos A, Lee X, Measures RM. Superimposed multiple Bragg gratings. *Electronics Letters* 1994; 30: 1972–1974.

72. Rabin MW, Swann WC, Gilbert SL. Interleaved, sampled fiber Bragg gratings for use in hybrid wavelength references. *Applied Optics* 2002; 41: 7193–7196.

73. Yamashita S, Yokoojia M. Channel spacing-tunable sampled fiber Bragg grating by linear chirp and its application to multiwavelength fiber laser. *Optics Communications* 2006; 263: 42–46.

74. Qu R, Ding H, Zhao H, Fang Z. Photoimprinting of sampled fiber Bragg gratings. *Technical Digest Third Optoelectronics and Communication Conference*, 2002, Chiba, Japan, p. 156.

75. Jayaraman V, Chuang Z, Coldren LA. Theory, design, and performance of extended tuning range semiconductor lasers with sampled gratings. *IEEE Journal of Quantum Electronics* 1993; 29: 1824–1834.

76. Chen XF, Luo Y, Fan CC, Wu T, Xie SZ. Analytical expression of sampled Bragg gratings with chirp in the sampling period and its application in dispersion management design in a WDM system. *IEEE Photonics Technology Letters* 2000; 12: 1013–1015.

77. Slavík R, Doucet S, LaRochelle S. High-performance all-fiber Fabry–Pérot filters with superimposed chirped Bragg gratings. *Journal of Lightwave Technology* 2003; 21: 1059–1065.

78. Guan ZG, Chen D, He S. Coherence multiplexing of distributed sensors based on pairs of fiber Bragg gratings of low reflectivity. *Journal of Lightwave Technology* 2007, 25: 2143–2148.

79. Albert J, Hill KO, Johson DC, Bilodeau F, Rooks MJ. Moiré phase masks for automatic pure apodisation of fiber Bragg gratings. *Electronics Letters* 1996; 32: 2260–2261.

80. Kashyap R, Swanton A, Armes DJ. Simple technique for apodising chirped and unchirped fiber Bragg gratings. *Electronics Letters* 1996; 32: 1226–1228.

81. Byron KC, Sugden K, Bricheno T, Bennion I. Fabrication of chirped Bragg gratings in photosensitive fiber. *Electronics Letters* 1993; 29: 1659–1660.

82. Williams JAR, Bennion I, Sugden K, Doran NJ. Fiber dispersion compensation using a chirped in-fiber Bragg grating. *Electronics Letters* 1994; 30: 985–987.

83. Kashyap R, McKee PF, Campbell RJ. A novel method of producing photo-induced chirped Bragg gratings in optical fibers. *Electronics Letters* 1994; 30: 996–998.

84. Pastor D, Capmany J, Ortega D, Tatay V, Marti J. Design of apodized linearly chirped fiber gratings for dispersion compensation. *Journal of Lightwave Technology* 1996; 14: 2581–2588.

85. Ennser K, Zervas MN, Laming RI. Optimization of apodized linearly chirped fiber gratings for optical communications. *IEEE Journal of Quantum Electronics* 1998; 34: 770–778.

86. Pisco M, Iadicicco A, Campopiano S, Cutolo A, Cusano A. Structured chirped fiber Bragg gratings. *Journal of Lightwave Technology* 2008; 26: 1613–1625.

87. Poladian L. Group-delay reconstruction for fiber Bragg gratings in reflection and transmission. *Optical Letters* 1997; 22: 1571–1573.

88. Poladian L. Resonance mode expansions and exact solutions for nonuniform gratings. *Physical Review E* 1996; 54: 2963–2975.

89. Azaña J, Chen LR. Synthesis of temporal optical waveforms by fiber Bragg gratings: a new approach based on space-to-frequency-to-time

mapping. *Journal of the Optical Society of America B* 2002; 19: 2758–2769.

90. Loh WH, Zhou FQ, Pan JJ. Sampled fiber grating based-dispersion slope compensator. *IEEE Photonics Technology Letters* 1999; 11: 1280–1282.

91. Liu B, Miao YP, Zhou HB, Zhao QD. Pure bending characteristic of tilted fiber Bragg grating. *Journal of Electronic Science and Technology of China* 2008; 6: 470–473.

92. Kashyap R, Wyatt R, Campbell RJ. Wideband gain flattened erbium fiber amplifier using a photosensitive fiber blazed grating. *Electronics Letters* 1993; 29: 153–156.

93. Mihailov SJ, Walker RB, Stocki TJ, Johnson DC. Fabrication of tilted fiber-grating polarisation-dependent loss equalizer. *Electronics Letters* 2001; 37: 284–286.

94. Zhou K, Simpson G, Chen X, Zhang L, Bennion I. High extinction ratio in-fiber polarizers based on 45 tilted fiber Bragg gratings. *Optics Letters* 2005; 30: 1285–1287.

95. Westbrook PS, Strasser TA, Erdogan T. In-line polarimeter using blazed fiber gratins. *IEEE Photonics Technology Letters* 2000; 12: 1352–1354.

96. Erdogan T, Sipe JE. Radiation-mode coupling loss in tilted fiber phase gratings. *Optics Letters* 1995; 20: 1838–1840.

97. Erdogan T, Sipe JE. Tilted fiber phase grating. *Journal of Optical Society America A* 1996; 13: 296–313.

98. Dong L, Ortega B, Reekie L. Coupling characteristics of cladding modes in tilted optical fiber Bragg gratings. *Applied Optics* 1998; 37: 5099–5105.

99. Lee KS, Erdogan T. Fiber mode coupling in transmissive and reflective tilted fiber gratings. *Applied Optics* 2000; 39: 1394–1404.

100. Morey WW, Meltz G, Love JD, Hewlett SJ. Mode-coupling characteristics of UV-written Bragg gratings in depressed-cladding fiber. *Electronics Letters* 1994; 30: 730–732.

101. Kang SC, Kim SY, Lee SB, Kwon SW, Choi SS, Lee B. Temperature-independent strain sensor system using a tilted fiber Bragg grating demodulator. *IEEE Photonics Technology Letters* 1998; 10: 1461–1463.

102. Hewlett SJ, Love JD, Meltz G, Bailey TJ, Morey WW. Coupling characteristics of photo-induced Bragg gratings in depressed- and matched-cladding fiber. *Optical and Quantum Electronics* 1996; 28: 1641–1654.

103. Laffont G, Ferdinand P. Tilted short-period fiber-Bragg-grating induced coupling to cladding modes for accurate refractometry. *Measurement Science Technology* 2001; 12: 765–770.

104. Chen C, Xiong L, Caucheteur C, Mégret P, Albert J. Differential strain sensitivity of higher order cladding modes in weakly tilted fiber Bragg gratings. *Proceeding of SPIE* 2006; 6379: 63790E-1–63790E-7.

105. Froggatt M, Erdogan T. All-fiber wavemeter and Fourier-transform spectrometer. *Technical Digest of Optical Fiber Communication Conference*, 1999, San Diego, CA, paper PD21.

106. Holmes MJ, Kashyap R, Wyatt R. Physical properties of optical fiber sidetap grating filters: free-space model. *IEEE Journal of Selected Topics Quantum Electronics* 1999; 5: 1353–1365.

107. Li Y, Froggatt M, Erdogan T. Volume current method for analysis of tilted fiber gratings. *Journal of Lightwave Technology* 2001; 19: 1580–1591.

108. Li Y, Wielandy S, Carver GE, Durko HL, Westbrook PS. Influence of the longitudinal mode field in grating scattering from weakly guided optical fiber waveguides. *Optics Letters* 2004; 29: 691–693.

109. Walker RB, Mihailov SJ, Lu P, Grobnic D. Shaping the radiation field fiber Bragg gratings. *Journal of Optical Society America B* 2005; 22: 962–975.

110. Walker RB, Mihailov SJ, Grobnic D, Lu P, Bao XY. Direct evidence of tilted Bragg grating azimuthal radiation mode coupling mechanisms. *Optics Express* 2009; 17: 14075–14087.

111. Baek S, Jeong Y, Lee B. Characteristics of short-period blazed fiber Bragg gratings for use as macro-bending sensors. *Applied Optics* 2002; 41: 631–636.

112. Ouellette F, Gagnon D, Poirier M. Permanent photoinduced birefringence in a Ge-doped fiber. *Applied Physics Letters* 1991; 58: 1813–1815.

113. Bardal S, Kamal A, Russell PSJ. Photoinduced birefringence in optical fibers: a comparative study of low-birefringence and high-birefringence fibers. *Optics Letters* 1992; 17: 411–413.

114. Torres P, Valentec LCG. Spectral response of locally pressed fiber Bragg grating. *Optics Communications* 2002; 208: 285–291.

115. Gafsi R, El-Sherif MA. Analysis of induced-birefringence effects on fiber Bragg gratings. *Optical Fiber Technology* 2000; 6: 299–323.

116. de Matos CJS, Torres P, Valente LCG, Margulis W, Stubbe R. Fiber Bragg grating (FBG) characterization and shaping by local pressure. *Journal of Lightwave Technology* 2001; 19: 1206–1211.

117. Lee KS, Cho JY. Polarization-mode coupling in birefringent fiber gratings. *Journal of Optical Society America A* 2002; 19: 1621–1631.

118. Botero-Cadavid JF, Causado-Buelvas JD, Torres P. Spectral properties of locally pressed fiber Bragg gratings written in polarization maintaining fibers. *Journal of Lightwave Technology* 2010; 28: 1291–1297.

119. Müller MS, Hoffmann L, Sandmair A, Koch AW. Full strain tensor treatment of fiber Bragg grating sensors. *IEEE Journal of Quantum Electronics* 2009; 45: 547–553.

120. Müller MS, El-Khozondar HJ, Bernardini A, Koch AW. Transfer matrix approach to four mode coupling in fiber Bragg gratings. *IEEE Journal of Quantum Electronics* 2009; 45: 1142–1148.

121. Muller MS, El-Khozondar HJ, Buck TC, Koch AW. Analytical solution of four-mode coupling in shear strain loaded fiber Bragg grating sensors. *Optics Letters* 2009; 34: 2622–2624.

122. El-Khozondar HJ, Muller MS, Wellenhofer C, El-Khozondar RJ, Koch AW. Four-mode coupling in fiber Bragg grating sensors using full fields of the fundamental modes. *Fiber and Integrated Optics* 2010; 29: 420–430.

123. Liu Y, Feng X, Yuan S, Kai G, Dong X. Simultaneous four-wavelength lasing oscillations in an erbium-doped fiber laser with two high birefringence fiber Bragg gratings. *Optics Express* 2004; 12: 2057–2061.

124. Guan W, Marciante JR. Pump-induced, dual-frequency switching in a short-cavity, ytterbium-doped fiber laser. *Optics Express* 2007; 15: 14979–14992.

125. Lee BH, Nishii J. Self-interference of long-period fiber grating and its application as temperature sensor. *Electronics Letters* 1998; 34: 2059–2060.

126. Gu XJ. Wavelength-division multiplexing isolation fiber filter and light source using cascaded long-period fiber gratings. *Optics Letters* 1998; 23: 509–510.

127. Lee BH, Nishii J. Dependence of fringe spacing on the grating separation in a long-period fiber grating pair. *Applied Optics* 1999; 38: 3450–3459.

128. Allsop T, Neal R, Giannone D, Webb DJ, Mapps DJ, Bennion I. Sensing characteristics of a novel two-section long-period grating. *Applied Optics* 2003; 42: 3766–3771.

129. Guan ZG, Zhang AP, Liao R, He S. Wavelength detection of coherence-multiplexed fiber-optic sensors based on long-period grating pairs. *IEEE Sensor Journal* 2007; 7: 36–37.

130. Birks TA, Russell PSJ, Culverhouse DO. The acousto-optic effect in single-mode fiber tapers and couplers. *Journal of Lightwave Technology.* 1996; 14: 2519–2529.

131. Kim BY, Blake JN, Engan HE, Shaw HJ. All-fiber acousto-optic frequency shifter. *Optics Letters* 1986; 11: 389–391.

132. Birks TA, Farwell SG, Russell PSJ, Pannell CN. Four-port fiber frequency shifter with a null taper coupler. *Optics Letters* 1994; 19: 1964–1966.

133. Kim HS, Yun SH, Kwang IK, Kim BY. All-fiber acousto-optic tunable notch filter with electronically controllable spectral profile. *Optics Letters* 1997; 22: 1476–1478.

134. Jin T, Li Q, Zhao JH, Cheng K, Liu XM. Ultra-broad-band AOTF based on cladding etched single-mode fiber. *IEEE Photonics Technology Letters* 2002; 14: 1133–1135.

135. Dashti PZ, Li Q, Lin CH, Lee HP. Coherent acousto-optic mode coupling in dispersion compensating fiber by two acoustic gratings with orthogonal vibration directions. *Optics Letters* 2003; 28: 1403–1405.

136. Zhou Y, Gao K, Huang R, Geng J, Fang Z. Temperature and stress tuning characteristics of long-period gratings imprinted in Panda fiber. *IEEE Photonics Technology Letters* 2003; 15: 1728–1730.

137. Eggleton BJ, Ahuja A, Westbrook PS, Rogers JA, Kuo P, Nielsen TN, Mikkelsen B. Integrated tunable fiber gratings for dispersion management in high-bit rate systems. *Journal of Lightwave Technology* 2000; 18: 1418–1432.

138. Li L, Geng J, Zhao L, Chen G, Chen G, Fang Z, Lam CF. Response characteristics of thin-film-heated tunable fiber Bragg gratings. *IEEE Photonics Technology Letters* 2003; 15: 545–547.

139. Torres P, Valente LCG. Spectral response of locally pressed fiber Bragg grating. *Optics Communications* 2002; 208: 285–291.

140. Kersey AD, Davis MA, Patrick HJ, LeBlanc M, Koo KP, Askins CG, Putnam MA, Friebele EJ. Fiber grating sensors. *Journal of Lightwave Technology* 1997; 15: 1442–1463.

141. Zhao Y, Liao Y. Discrimination methods and demodulation techniques for fiber Bragg grating sensors. *Optics and Lasers in Engineering* 2004; 41: 1–18.

142. Jung J, Park N, Lee B. Simultaneous measurement of strain and temperature by use of a single fiber Bragg grating written in an erbium: ytterbium-doped fiber. *Applied Optics* 2000; 39: 1118–1120.

143. Patrick HJ, Williams GM, Kersey AD, Pedrazzani JR, Vengsarkar AM. Hybrid fiber Bragg grating/long period fiber grating sensor for strain/temperature discrimination. *IEEE Photonics Technology Letters* 1996; 8: 1223–1225.

144. Bhatia V, Campbell D, Claus RO. Simultaneous strain and temperature measurement with long-period gratings. *Optics Letters* 1997; 22: 648–650.

145. Yoffe GW, Krug PA, Ouellette F, Thorncraft DA. Passive temperature-compensating package for optical fiber gratings. *Applied Optics* 1995; 34: 6859–6861.

146. Haran FM, Rew JK, Foote PD. A strain-isolated fiber Bragg grating sensor for temperature compensation of fiber Bragg grating strain sensors. *Measurement Science and Technology* 1998; 9: 1163–1166.

147. Lin GC, Wang L, Yang CC, Shih MC, Chuang TJ. Thermal performance of metal-clad fiber Bragg grating sensors. *IEEE Photonics Technology Letters* 1998; 10: 406–408.

148. Zhang S, Lee SB, Fang X, Choi SS. In-fiber grating sensors. *Optics and Lasers in Engineering* 1999; 32: 405–418.

149. Spirin VV, Shlyagin MG, Miridonov SV, Jiménez FJM, Gutiérrez RML. Fiber Bragg grating sensor for petroleum hydrocarbon leak detection. *Optics and Lasers in Engineering.* 2000; 32: 479–503.

150. Yang S, Cai H, Geng J, Fang Z, Dan D, Sun L. Advanced fiber grating corrosion sensors for structural health monitoring. Proceeding of the Second International Conference on Structural Health Monitoring of Intelligent Infrastructure. Shenzhen, China, edited by Ou JP, Li H, and Duan ZD *Structural Health Monitoring and Intelligent Infrastructure.* London, Leiden, New York, Philadelphia, Singapore:Taylor & Francis, 2006, p. 441.

151. Lee B. Review of the present status of optical fiber sensors. *Optical Fiber Technology* 2003; 9: 57–79.

152. Todd MD, Johnson GA, Althouse BL. A novel Bragg grating sensor interrogation system utilizing a scanning filter, a Mach–Zehnder interferometer and a 3×3 coupler. *Measurement Sciences and Technology* 2001; 12: 771–777.

153. Kersey AD, Berkoff TA, Morey WW. Multiplexed fiber Bragg grating strain-sensor system with a fiber Fabry–Pérot wavelength filter. *Optics Letters* 1993; 18: 1370–1372.

154. Kersey AD, Berkoff TA, Morey WW. Fiber-optic Bragg grating strain sensor with drift-compensated high-resolution interferometric wavelength-shift detection. *Optics Letters* 1993; 18: 72–74.

155. Rao YJ, Jackson DA, Zhang L, Bennion I. Dual-cavity interferometric wavelength-shift detection for in-fiber Bragg grating sensors. *Optics Letters* 1996; 21: 1556–1558.

156. Shi WJ, Ning YN, Grattan KTV, Palmer AW. Novel hybrid interferometer stabilization scheme used in wavelength shift measurement for Bragg grating sensors. *Review of Scientific Instrument* 1998; 69: 1961–1965.

157. Huang R, Zhou Y, Cai H, Qu R, Fang Z. A fiber Bragg grating with triangular spectrum as wavelength readout in sensor systems. *Optics Communications* 2004; 229: 197–201.

158. Fallon RW, Zhang L, Everall LA, Williams JAR, Bennion I. All-fiber optical sensing system: Bragg grating sensor interrogated by a long-period grating. *Measurement Sciences and Technology* 1998; 9: 1969–1973.

159. Tsuda H. Fiber Bragg grating vibration-sensing system, insensitive to Bragg wavelength and employing fiber ring laser. *Optics Letters* 2010; 35: 2349–2351.

160. Kang MS, Yong JC, Kim BY. Suppression of the polarization dependence of fiber Bragg grating interrogation based on a wavelength-swept fiber laser. *Smart Materials and Structures* 2006; 15: 435–440.

161. Rao YJ, Ribeiro ABL, Jackson DA, Zhang L, Bennion I. Combined spatial- and time-division-multiplexing scheme for fiber grating sensors

with drift-compensated phase-sensitive detection. *Optics Letters* 1995; 20: 2149–2151.

162. Zhou B, Guan ZG, Yan CS, He S. Interrogation technique for a fiber Bragg grating sensing array based on a Sagnac interferometer and an acousto-optic modulator. *Optics Letters* 2008; 33: 2485–2487.

163. Cheng HC, Lo YL. Arbitrary strain distribution measurement using a genetic algorithm approach and two fiber Bragg grating intensity spectra. *Optics Communications* 2004; 239: 323–332.

164. Todd MD, Johnson GA, Vohra ST. Deployment of a fiber Bragg grating-based measurement system in a structural health monitoring application. *Smart Material and Structures* 2001; 10: 534–539.

165. Todd M, Seaver M, Wiener T, Trickey S. Structural monitoring using high-performance fiber optic measurement system. *Proceedings of SPIE* 2002; 4694: 149–161.

166. Udd E, Schulz W, Seim J, Haugse E, Trego A, Johnson P, Bennett TE, Nelson D, Makino A. Multidimensional strain field measurements using fiber optic grating sensors. www.bluerr.com.

CHAPTER 5

DISTRIBUTED OPTICAL
FIBER SENSORS

It is a unique feature that fiber itself can play the role of sensing. More-over, it provides the possibility of constructing a distributed sensor system. This chapter is devoted to distributed fiber sensors. The basic physical principles of optical scatterings, including elastic and inelastic scatterings, are introduced in Section 5.1. Section 5.2 describes the optical fiber time domain reflectometer (OTDR) and related distributed fiber sensors, which are based on the Rayleigh scattering effect. Section 5.3 introduces the distributed Raman temperature sensor system. Section 5.4 is engaged in distributed Brillouin sensors for both temperature and strain. The last section introduces interferometric distributed sensors.

5.1 OPTICAL SCATTERING IN FIBER

Early in the 1970s, soon after the breakdown of low-loss silica optical fiber, backscattering effect was investigated in detail [1–4]. The effect was used to characterize loss and imperfections of fiber [5]; and the special technology of OTDR was invented [6,7], which has been used widely in fiber communication technology. The fiber in OTDR is not

Fundamentals of Optical Fiber Sensors, First Edition.
Zujie Fang, Ken K. Chin, Ronghui Qu, and Haiwen Cai.
© 2012 John Wiley & Sons, Inc. Published 2012 by John Wiley & Sons, Inc.

only the medium for transporting optical probes and signals, but also the sensing element giving information about the status of the fiber. Shortly later, nonlinear optical scattering effects, including Raman scattering and Brillouin scattering, were also utilized to develop various distributed sensors. In this section, the basic physical mechanisms are introduced.

5.1.1 Elastic Optical Scattering

Optical scattering phenomena were investigated earlier in the 1800s. Lord Rayleigh gave an explanation of the colors in the sky based on optical scattering due to the thermal movement of molecules. Raman scattering and Brillouin scattering were discovered later in the 1920s. Physically, the scatterings are divided into two categories, elastic scattering, and inelastic scattering. The former is a linear collision process, which does not change the photon's energy, whereas the later is a nonlinear collision process, which changes the photon's energy. Generally, elastic scattering and inelastic scattering occur simultaneously in the medium, but with different strengths.

It is imaginable that an electromagnetic wave will be scattered by particles in a medium into different directions, with its wavelength unchanged. Mie and other researchers presented a theory giving an analytical solution of Maxwell equations for scatterings by spherical particles in the medium, which is regarded as a model of different scattering centers, such as real dust-like particles, suspended water drops, aerosols, and density fluctuations due to thermal movement of molecules. Mie's theory gives descriptions of spatial distribution of scattered light intensity, including their dependences on the wavelength of incident electromagnetic waves, and on its polarization [8]. Figure 5.1 gives polar diagrams for scattering of linearly polarized light by a spherical particle with the size parameter of $q = 2\pi a/\lambda$,

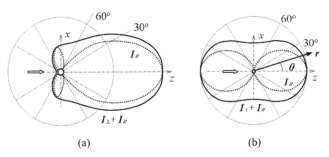

(a)   (b)

Figure 5.1 Polar diagram of (a) Mie scattering and (b) Rayleigh scattering.

where a is its radius. Figure 5.1(a) is for $q \approx 1$ of general Mie scattering as an example, and Figure 5.1(b) is for Rayleigh scattering, which is a special case of Mie scattering in limitation of $q \rightarrow 0$. The inner curves in doted lines are for $I_{\parallel}$, which stands for scattered wave polarized in x–y-plane; the outer curve in solid line is for $I_{\parallel} + I_{\perp}$, where $I_{\perp}$ stands for scattered wave polarized in y-direction. The figures give pictures of how the spatial distribution of scattering radiation depends on the size parameter q and the polarization. It depends also on the index of the scattering particle, including its conductivity, described as the imaginary part of the index. The formulas of Mie theory are expressed as infinitive series. The theory is useful and important in applications such as in analyses of weather conditions and air pollution, and in research of atmospheric turbulence. This book will not go further in detail; interested readers can find them in textbooks [8,9].

In the limit of $q \rightarrow 0$, that is, for Rayleigh scattering, the scattering intensity is given by the theory based on the electric bipolar radiation model, written as

$$I = \left(\frac{2\pi}{\lambda}\right)^4 \frac{a^6}{r^2} \left(\frac{\hat{n}^2 - 1}{\hat{n}^2 + 2}\right)^2 (\cos^2 \theta \cos^2 \varphi + \sin^2 \varphi), \qquad (5.1)$$

where r is the distance between the particle and observation point, $\hat{n}$ is the complex reflective index, φ is the angle of $\boldsymbol{r}$ in x–y-plane, and θ is in x–z-plane. The scattering intensities for $I_{\parallel}$ and $I_{\perp}$ are expressed as

$$I_{\parallel} = \left(\frac{2\pi}{\lambda}\right)^4 \frac{a^6}{r^2} \left(\frac{\hat{n}^2 - 1}{\hat{n}^2 + 2}\right)^2 \cos^2 \theta, \qquad (5.2a)$$

$$I_{\perp} = \left(\frac{2\pi}{\lambda}\right)^4 \frac{a^6}{r^2} \left(\frac{\hat{n}^2 - 1}{\hat{n}^2 + 2}\right)^2. \qquad (5.2b)$$

For a nonpolarized incident light, the scattering intensity is expressed as

$$I = \bar{I}_{\parallel} + \bar{I}_{\perp} = \left(\frac{2\pi}{\lambda}\right)^4 \frac{a^6}{2r^2} \left(\frac{\hat{n}^2 - 1}{\hat{n}^2 + 2}\right)^2 (1 + \cos^2 \theta). \qquad (5.3)$$

It can be seen from the figures and formulas that Rayleigh scattering intensity strongly depends on wavelength with relation of $\propto \lambda^{-4}$; moreover, the scattering shows wavelength-dependent spatial distribution, although it does not change the wavelength of the scattered photons. The scattering of shorter wavelength light is much stronger than that of longer wavelength, giving an explanation for the color of the sky.

It is also shown that the scattering shows a degree of polarization (DOP) that varies with the observation angles; a linearly polarized light is obtained in the observation direction perpendicular to the incident light beam, that is, at the angle of $\theta = \pi/2$. The characteristics are also found in observation of sky light.

Rayleigh scattering occurs in almost all mediums, including fibers, because thermal movement exists everywhere; it is thus utilized to develop distributed fiber sensors.

5.1.2 Inelastic Optical Scattering

In optical scattering, most photons are scattered elastically, meaning that only the radiation direction is changed, whereas the photon energy does not change. However, a fraction of the incident photons is scattered with energy change. Raman scattering and Brillouin scattering are the main inelastic scatterings with the photon's energy change. Figure 5.2 shows their spectral states schematically, where the red-shifted scattering is termed the Stokes band, whereas the blue-shifted one is the anti-Stokes band. The frequency shift of Raman scattering in silica is in the range of about 13 THz, whereas that of Brillouin scattering is about 10 GHz in the near infrared wavelength band [10]. More than two lines are depicted for Brillouin scattering, showing higher order scatterings. Higher order Raman scatterings may occur similarly; they are not depicted due to the limited scale of the figure.

Figure 5.3 gives an illustrative picture of the energy levels, showing that most electrons occupy the ground state, denoted by energy E_0, and some at the exited states $E_1 = E_0 + \Delta E$, which usually correspond to a vibration state intra molecule or between molecules, or a rotation state. When an incident photon with energy $h\nu$ is absorbed, the electron transits up to a virtual state with energy of $E_v = E_0 + h\nu$. The virtual

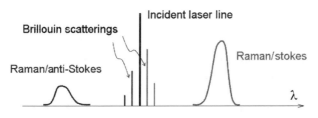

Figure 5.2   Spectral lines of nonlinear scatterings.

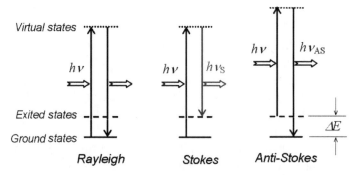

Figure 5.3 Energy level diagram of Stokes and anti-Stokes scatterings.

state is an unstable state, so that it transits down to the ground state or the exited state very quickly, and emit photons simultaneously for Rayleigh scattering or for Stokes scattering, respectively. Electrons at the exited states absorb the incident photon energy and transit from the virtual state down to the ground state, emit photons with energy of $hv_a = hv + \Delta E$, corresponding to anti-Stokes scattering.

Optical scattering is attributed to collisions between scatters and incident photons. In Mie–Rayleigh scattering, the movement of scatters is not taken into consideration. Actually, scatter's movement exists always; according to the Doppler effect, the movement will lead to frequency change of the secondary source radiation. In other words, some of the scattered photons acquire certain energy from the scattering particle, or lose energy to accelerate scatter movement. In view of the lattice dynamics of solid-state physics, or generally of condensed matter, the movement of molecules is quantized as phonons. The states of phonons are described in the space of energy versus wave vector ($E \sim K$), generally by two bands, that is, the optical frequency band and the acoustic frequency band, as shown schematically in Figure 5.4 [11]. K_1 and K_2 in the figure stand for wave vectors in different directions and with possible different cell period a_1 and a_2; the period is regarded as statistical and isotropic parameters in liquids, gases and noncrystalline matters, including fused silica.

The acoustic frequency phonons describe the waves caused by the force between molecules, whereas the optical frequency phonons are attributed to the relative motion of atoms inside the molecule, indicated by energy levels related to vibrations or rotations. Therefore three photon–phonon interactions may occur: the first is elastic interaction with the largest probability, that is, Mie–Rayleigh scattering; the second creates a phonon and causes energy loss of the photon (Stokes process); the third results in annihilation of a phonon and in energy

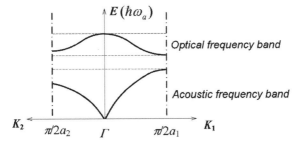

Figure 5.4 Schematic diagram of energy band structure of phonons.

increasing of the photon (anti-Stokes process). Raman scattering involves phonons in the optical frequency band with larger frequency shifts; whereas Brillouin scattering involves phonons in the acoustic frequency band with smaller frequency shifts.

To give a qualitative understanding of the scattering mechanism, the strength of the induced dipole moment is expressed by $\boldsymbol{P} = \chi \boldsymbol{E}$, where χ is the polarizability. Taking the molecule motion into consideration, the polarizability depends on the physical displacement dr of atoms from their equilibrium position:

$$\chi = \chi_0 + \frac{\partial \chi}{\partial r} dr = \chi_0 + \frac{\partial \chi}{\partial r} r_1 \cos(\omega_a t), \tag{5.4}$$

where r is the normal coordinate of atoms, r_1 is the maximum displacement from its equilibrium position, and ω_a stands for the frequency of a particular mode of phonon. The dipole moment is then written as

$$\boldsymbol{P} = \left[\chi_0 + \frac{\partial \chi}{\partial r} r_1 \cos(\omega_a t) \right] \boldsymbol{E}_0 \cos \omega t, \tag{5.5}$$

where the first term describes Mie–Rayleigh scattering, whereas the second term can be expressed as a sum of two harmonic motions, $\omega \pm \omega_a$, referring to Stokes and anti-Stokes processes, respectively, indicating the energy conservation:

$$\hbar \omega_S = \hbar \omega \pm \hbar \omega_a, \tag{5.6}$$

where ω, ω_S, and ω_a are the angular frequencies of the incident photon, the scattered photon, and the phonon, respectively. It is noticed that the scattering can take place for any frequency of the incident photons. This property makes the inelastic scattering different from the process

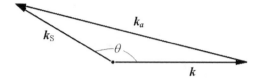

Figure 5.5 Momentum conservation in photon scattering.

of fluorescence, which involves transitions between a particular pair of energy levels.

Raman scattering and Brillouin scattering are attributed to the molecule motion; consequently, their characteristics depend on composition and structure of the material since different materials have different energy level structure of vibration and/or rotations, and different phonon energy band. Therefore, the effects have successfully been utilized to characterize materials. Spectral data of the effects have been measured for varieties of materials, including condensed matters, liquids and gases; and Raman spectroscopy is one of the important tools of material analysis. Characteristics of the three effects depend also on the status of the material, especially its temperature and strain, which lay the foundation of distributed fiber sensors.

In the process of particle collision, momentum must be conserved, expressed as

$$\hbar \mathbf{k}_S = \hbar \mathbf{k} \pm \hbar \mathbf{k}_a, \tag{5.7}$$

as shown in Figure 5.5. For Brillouin scattering the approximation of $|\mathbf{k}_S| \approx |\mathbf{k}|$ holds generally; the phonon frequency is deduced to be [10]

$$\omega_a = V_a |k_a| \approx 2V_a |k| \sin(\theta/2) = 4\pi \frac{nV_a}{\lambda} \sin(\theta/2), \tag{5.8}$$

where V_a is the sound velocity in material. In the fiber, only forward and backward scatterings need to be considered; it is seen in formula (5.8) that the largest frequency shift is obtained at $\theta = \pi$, that is for the backward scatterings. For forward scattering at $\theta = 0$, frequency shift is zero, meaning that forward Brillouin scattering will not occur. Therefore, the Brillouin frequency shift is $\nu_B = 2nV_a/\lambda$

The case of Raman scattering is different. The phonon wave vector for the forward and backward scatterings in fiber is expressed as

$$|k_a| = |\mathbf{k} \pm \mathbf{k}_S|. \tag{5.9}$$

It is shown in Figure 5.4 that the optical frequency band of phonon has a flat top in the $E \sim K$ diagram. It means that many phonons with the same energy can take part in Raman scattering with the momentum conservation satisfied; and both forward and backward scatterings are possible, without violation of energy and momentum conservations.

5.1.3 Stimulated Raman Scattering and Stimulated Brillouin Scattering

The electric field of incident light will accelerate the motion of electrons and ions, thus inducing dynamic strain of the material, called electrostriction. The effect reinforces the acoustic waves and enhances photon scatterings. When the intensity of incident light is high enough, this process will lead to amplification of the effects, termed the stimulated Raman scattering (SRS) and the stimulated Brillouin scattering (SBS). Such scatterings are coherent as a presupposition of the stimulated process, and can thus be used to amplify optical signals and to develop lasers. When the pump light is weak, the scattering is limited in spontaneous range. As the pump increases, the scattering increases with the pump, showing a behavior with threshold, above which the stimulated scattering increases in a much higher speed and is dominant over the spontaneous process.

In the SBS, molecules are pushed by electrostriction, and an acoustic wave is stimulated; such a process leads to formation of a dynamic grating, as shown schematically in Figure 5.6. It is regarded as a dynamic grating induced by the pump beam; and the grating enhances the reflected Brillouin scattering greatly.

When the SBS and SRS intensity is amplified high enough, the Stokes and anti-Stokes photons can also play a role of pumping, then the higher order scatterings occurs, resulting in multiple Stokes lines in the spectrum. The higher order scattering also causes changes in

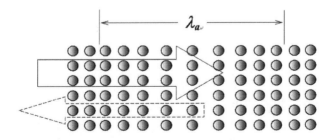

Figure 5.6 Photon-induced dynamic grating.

its propagation direction; detailed characteristics and theories can be found in related references [10,12].

Based on SRS and SBS, several important optical fiber devices are developed. One of them is the Raman lasers, in which the fiber plays the role of active media. An attractive advantage of the Raman laser is the possibility of opening up new ranges of lasing wavelength, no longer to be limited by transition between energy levels of the materials. SBS is also utilized to develop narrow line width lasers.

Another device is the distributed Raman fiber amplifier (DRFA), used in fiber communication systems. In DRFA, the fiber not only plays the role of transporting optical signals, but also acts as the active medium of optical amplifier. Compared with erbium-doped fiber amplifier (EDFA), the DRFA expands the working wavelength range widely, and remote pumping becomes possible. The measurement apparatus based on Raman spectroscopy and Brillouin scattering are also widely used in various areas of science and technology.

On the other hand, nonlinear optical effects have to be avoided in optical signal transportation, because they will impair the transported signals; especially when the low threshold SBS sets a limit of optical power handling capacity [2].

Distributed fiber sensors based on Rayleigh scattering, Raman scattering and Brillouin scattering are introduced in the following sections. Characteristics of SRS and SBS in optical fibers will be discussed further in Sections 5.3 and 5.4.

5.2   DISTRIBUTED SENSORS BASED ON RAYLEIGH SCATTERING

This section introduces several types of optical time domain reflectometers (OTDR), including the analog OTDR based on power detection of scattering intensity (Section 5.10), the polarization-dependent OTDR (Section 5.11), the coherent OTDR and phase-sensitive OTDR (Section 5.12), and the optical frequency domain reflectometer (Section 5.13); all of them are based on Rayleigh scattering.

5.2.1   Optical Time Domain Reflectometer

5.2.1.1 Basic Schemes of OTDR Based on Rayleigh scattering in fiber, the OTDR was invented earlier in the 1970s [3–5]. It has been developed as a commercially available apparatus, and used widely as an effective and important apparatus of investigating attenuation

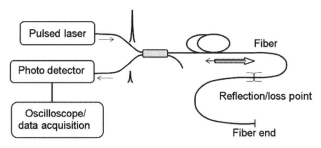

Figure 5.7   Schematic diagram of basic OTDR system.

characteristics of fiber loops, detecting faults in the loop, and as distributed sensors. Figure 5.7 shows schematically its basic structure.

In the OTDR, high-power optical pulses are launched into a fiber, which is being inspected, or is used as a distributed sensor, the pulses are scattered back along with propagation in the fiber, and their intensity decays due to the fiber loss. The return signal is thus a declined curve displayed in an oscilloscope, as shown in Figure 5.8. The fiber loss coefficient can be measured by the curve. Any lumped nonuniform points in the fiber, such as connections, fusions, faults, and extra losses, bring about additional reflections and losses, showing a sharp peak and dip in the return signal. It is noticed that OTDR can not only tell the existence and properties of loss points, but also give their locations by the time delay from the onset of input optical pulse.

External conditions, including temperature change, strains, and stresses, affect also the optical attenuation characteristics. These changes are reflected in Rayleigh back scattering and can then be sensed by the reflectometer. It is a typical distributed sensor, with a long fiber paved in the concerned area.

5.2.1.2 Basic Theory and Techniques of OTDR   Although the
principle of OTDR seems simple, its mechanism should be expressed

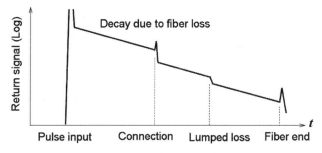

Figure 5.8   Schematic diagram of OTDR signals.

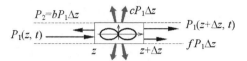

Figure 5.9   Scattered beams at a fiber section Δz.

by physical models. Reference [1,13–17] present basic equations to describe the variations of forward optical beam and the backward scattered signals, and give its solutions under different input optical pulses in analytical forms. References [18,19] discuss the characteristics of scattered waves for different fibers, including single-mode fibers, multimode fibers, and gradient index fibers. Apart from the intensity of a reflected wave, its phase shift and its polarization state are also parameters to carry the information of a fiber system; phase-sensitive OTDR and polarization-OTDR technologies are then developed. Reference [20] gives a review and comparison of the technologies.

A physical model of the scattering phenomena is shown in Figure 5.9 schematically, where b, f, and c stand for coefficients of backward scattering, forward scattering, and lateral scattering, including those in divergent angles. The coefficient of backward scattering b is proportional to a factor of fiber waveguide, proportional to $3(n_1^2 - n_2^2)/8n_1^2$ for step index fibers, which comes from the coupling between spatial distribution of scattering and fiber modes, and is related to the numerical aperture (NA) [18,19]. The attenuation in fibers by Rayleigh scattering is described by Beer–Lambert law as follows:

$$P(z) = P(z_1)\exp\left[-\int_{z_1}^{z}\alpha(z)\mathrm{d}z\right] = P(z_1)\exp[-\bar{\alpha}(z - z_1)], \quad (5.10)$$

where $\bar{\alpha} = \int_{z_1}^{z}\alpha(z)dz/(z - z_1)$ is the averaged attenuation coefficient. To take the variation of loss along the fiber into consideration, differential equations are used to describe the coupling between the input beam $P_1(z, t)$ and scattered beam $P_2(z, t)$, as depicted in Figure 5.9. For the elastic scattering the forward scattered beam is actually a major part of the input beam; and the lateral scattering and the backward scattering contribute to fiber loss α. A relation between the input pulse and the scattered pulse is then written as

$$P_1(z + \Delta z, t + \Delta t) = (1 - \alpha\Delta z)P_1(z, t) + b\Delta z P_2(z, t). \quad (5.11)$$

Taking the transmission time delay into account, a pair of equations is obtained:

$$\frac{\partial P_1}{\partial z} + \frac{1}{v_g}\frac{\partial P_1}{\partial t} = -\alpha P_1 + b P_2,$$

$$-\frac{\partial P_2}{\partial z} + \frac{1}{v_g}\frac{\partial P_2}{\partial t} = -\alpha P_2 + b P_1,$$

$$(5.12)$$

where v_g is the group velocity of the optical wave. The equation can be solved by Laplace transform analytically for different input optical pulse shapes [1]. By denoting new variables: $Q_1 = P_1 + P_2$ and $Q_2 = P_1 - P_2$, they obey an identical equation:

$$\frac{\partial^2 Q_{1,2}}{\partial t^2} + 2\alpha v_g \frac{\partial Q_{1,2}}{\partial t} + (\alpha^2 - b^2)v_g^2 Q_{1,2} - \frac{\partial^2 Q_{1,2}}{\partial \tau^2} = 0. \qquad (5.13)$$

Their general solutions are obtained by the method of variable separation with the following analytical forms:

$$Q_1(t,\tau) = Q_{10}(e^{-2\alpha v_g t} + A_1)(e^{-\sqrt{\alpha^2 - b^2} z} + B_1 e^{\sqrt{\alpha^2 - b^2} z}).$$

$$Q_2(t,\tau) = Q_{20}(e^{-2\alpha v_g t} + A_2)(e^{-\sqrt{\alpha^2 - b^2} z} + B_2 e^{\sqrt{\alpha^2 - b^2} z}).$$

$$(5.14)$$

With primary conditions of input optical pulse, the forward propagating beam $P_1(z, t)$ and backward scattered beam $P_2(z, t)$ can then be obtained. For a rectangular input pulse with pulse width of τ, the reflected signal can be expressed approximated as

$$P_2(t) = \begin{cases} P_{20}[1 - \exp(-2\alpha v_g t)] & (t \leq \tau) \\ P_{20}[1 - \exp(-2\alpha v_g \tau)]\exp[-2\alpha v_g (t - \tau)] & (t \geq \tau), \end{cases} \qquad (5.15)$$

where $P_{20} \propto b P_{in}$. Figure 5.10 shows schematically a normalized temporal variation of the backscattering signal. Other input pulse waveforms, such as the delta function, exponential function, and Gaussian, give similar behaviors. Different pulse waveforms bring about influence mainly in the range of $t \leq \tau$, which is not of concern in OTDR applications, especially in cases with narrow input pulses for high spatial resolution.

For nonuniform losses and lumped loss points, the variation along the fiber of attenuation coefficient $\alpha = \alpha(z)$ is just the information to be measured. Since the backscattering intensity is very weak, Equation

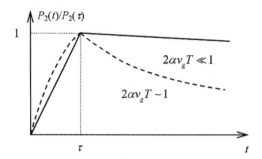

Figure 5.10   Normalized reflected signals of OTDR.

(5.12) can be simplified to $\partial P_2/\partial t \approx v_g b P_1$ and $\partial P_1/\partial z \approx -\alpha P_1$; and the solution is then expressed as:

$$P_2(t) \approx v_g b \int P_1(t)\mathrm{d}t = v_g b P_0 \int \exp[-\alpha(v_g t)v_g t]\mathrm{d}t. \qquad (5.16)$$

The Rayleigh scattering coefficient is very low, usually on the order of $10^{-5} \sim 10^6$ m^{-1}, estimated from the loss data [14]. It is essential in OTDR technology to extract the weak Rayleigh scattering signals. In addition, the scattering is a statistical process, resulting in strong noise. To increase the signal-to-noise ratio (SNR), averaging of multiple received traces is necessary. By integrating the backscattered signal over N pulses the SNR can be improved by a factor of $\sqrt{N}$. Correlation technique, photon counting, cooled detector, and proper data processing are also developed to increase SNR [21–24].

5.2.1.3 Performance of OTDR   As the characterization of fiber attenuation and the detection of faults and disturbances, main performances of OTDR include the following specifications [25]:

1. Length of the fiber span to be measured. It is obvious that a longer working length requires a higher intensity of the input optical pulse, and a higher SNR of the receiver and its electronics. Narrow pulse width and high-power pulsed lasers are necessary. However, the power level is limited by nonlinear optical effects [2] for the conventional OTDR. The input pulse intensity depends also on the coupling coefficient of laser source to fiber. The MMF is favorable to OTDR signal retrieving and used to build special distributed sensor systems.

Since Rayleigh scattering coefficient is inversely proportional to the fourth power of wavelength, the input pulse at shorter wavelengths, such as in 800 nm or visible range, is helpful to enhance the scattering signal. However, for silica fibers, the lowest loss is in 1,550 nm range, which corresponds not only to the longest working length, but also to the range people are concerned with mostly; it is thus used most frequently.

2. Resolution of localization. It depends mainly on two factors: the response of detection and the width of input pulse τ, similar to the principle of radar based on time of flight (TOF): $\delta L \propto 2v_g\tau$. For example, a laser pulse of 10 ns gives localization precision of 4 m. It should be noted that the pulse will be broadened in conventional fibers with certain dispersion, especially for very narrow pulses, which correspond to broader spectral widths and higher dispersion effects.

In the practical OTDR apparatus, some components must be used to couple the laser beam to the fiber, such as a directional fiber coupler. The reflected waves at the components and their connections will give a high signal to the receiver near the input end of the fiber, resulting in the problem of dead range, similar to the laser range finding system. To solve the problem an extra buffer fiber may be connected before the working fiber.

3. Time of the data readout. It includes the propagation time over the fiber length and the data storage and processing time. The repetition rate of optical source pulse should be set to $f \le v_g/2L$, where L is the fiber length, to avoid overlaps between returned signals of successive pulses. Therefore, the time of data readout is larger than $2NL/v_g$, where N is the pulse number in data averaging.

4. Sensitivity of signal variation and dynamical range. It is noted that there is a trade-off between sensitivity and data output time; and also a trade-off between sensitivity and the spatial resolution, since longer pulses give higher integrated scattering intensity. In order to enhance the sensitivity and other specifications, technologies different from the analog OTDR were developed: such as polarization-OTDR and phase-insensitive OTDR.

The OTDR instruments are now commercially available and play important roles in fiber technology engineering, and in research and development.

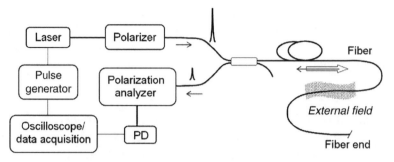

Figure 5.11 Schematic diagram of POTDR system.

5.2.2 Polarization OTDR

As introduced in Section 5.1, Rayleigh scattering has polarization dependence. It can be utilized to enhance the sensitivity of OTDR and to acquire more information on fiber characteristics. For this purpose, a polarized laser is taken as the probe, and a polarization analyzer is used in detection, as shown in Figure 5.11, which shows the basic configuration of polarization OTDR (POTDR) [26–28].

As analyzed in Chapter 3, the practical fiber possesses more or less birefringence. Apart from the internal factors in its fabrication processing, various external conditions, such as pressure, bending and twisting, electromagnetic fields and nonuniform temperature distributions, will induce birefringence. The polarization state of scattered beam includes the basic parameters: (1) ellipticity of the polarized field; (2) orientation of the polarization; (3) birefringence and PMD; and (4) the DOP. The POTDR is used to detect such effects and their evolution along the fiber.

As introduced in Section 3.4, the distributed birefringence can be modeled by a concatenated wave plate with a principal state of polarization [29–32]. Figure 5.12(a) gives an illustration of the concept. The backscattering waves from points between z and $z + dz$ are analyzed at position $z = 0$ by a polarization analyzer. Figure 5.12(b) shows schematically the polarization directions of input pulse, reflected pulses, and the analyzer azimuth in the lab coordinate.

The fiber section between 0 and z corresponds to an equivalent wave plate. The backscattering waves at points between z and $z + dz$ experience the same polarization variation in the fiber between 0 and z, giving outputs at the analyzer as [33]

$$P_{//,\perp}(t) = P_0 g_{//,\perp} \exp(-2\alpha v_g t) \\ \cdot [1 \pm \cos 2\theta \cos 2\gamma \pm \sin 2\theta \sin 2\gamma \cos(2\delta\beta v_g t)], \quad (5.17)$$

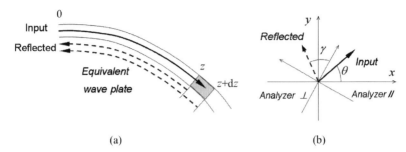

Figure 5.12 (a) Conceptional illustration of polarization interference and (b) Polarization directions and analyzer azimuth in the lab coordinate.

where $\delta\beta = \beta_e - \beta_o$ is the birefringence, γ is the polarization direction of the scattering; both of them may vary temporally and spatially. Factor g stands for an integral of the perturbation over fiber section dz and is weighted by mode field. The factor may be different for the two polarizations.

For a fiber section with short length and with stationary birefringence, a decayed oscillation POTDR curve is measured, as shown in Figure 5.13(a) schematically [27]. For longer fiber loops with randomly distributed birefringence, the distributed birefringence can no longer be regarded as a fixed equivalent wave plate. Furthermore, the multiple pickups and averaging of the data are necessary due to the weaker backscattering signal. The temporal variation of the distributed birefringence during the data processing makes the signal more random. Therefore, the POTDR trace looks more noisy, as shown in Figure 5.13(b), and a statistic analysis is necessary to retrieve the distributed birefringence.

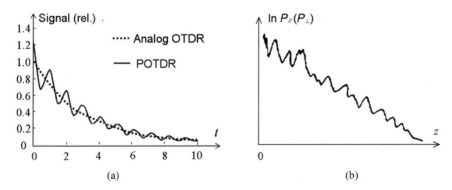

Figure 5.13 (a) Comparison of POTDR and analog OTDR signals and (b) with random variation.

The configuration of POTDR shown in Figure 5.11 is not the scheme that can mine the information of polarization states sufficiently. A further developed, fully POTDR uses a rotatable polarizer and rotatable quarter wave plate to change the polarization state of input laser pulse, instead of using a fixed polarizer [34,35]. In addition, a polarimeter, which gives four Stokes components of the detected beam, is used at the receiver port instead of the simple analyzer. By varying the polarization state of the input pulse in a range large enough on the Poincaré sphere, both the linear and circular birefringence distribution along the fiber can be evaluated. The principle of polarimeter is introduced in Appendix 3.

In the development of optical fiber communication systems, the polarization mode dispersion is an important subject. POTDR method is regarded as one of the effective tools in investigating and measuring PMD characteristics [35–37]. Based on the analyses of PMD properties, related algorithms are developed to evaluate the PMD parameters, including statistical results of beat length, polarization mode coupling coefficient, and variation of DOP. Essentials of PMD involve the spectral properties. The spectrum analysis of POTDR gives more information also for sensor applications [38,39].

5.2.3 Coherent OTDR and Phase Sensitive OTDR

In analog OTDR, the signal from the far end of a long fiber span may become too weak to be detected by direct detection (DD). An alternative method, the coherent detection, provides much higher sensitivity for the weak signal, which is termed coherent OTDR [40–42]. In the COTDR narrow linewidth, laser pulses are used as the probe, and the returned wave is mixed with the local oscillation (LO), giving beat signals.

In the analog OTDR, the optical phase is omitted, since low coherent sources are used, which is beneficial for reducing the interferometric noise. However, externally induced disturbance and intrinsic nonuniformity in the fiber causes not only changes of attenuation and polarization states, but also cause phase shift due to the optical path change. Actually, it is more sensitive than the attenuation change and the polarization change. To utilize the higher sensitivity of the phase, another type of OTDR, the phase-sensitive OTDR (ϕ-OTDR), is developed [40].

The configurations of COTDR and ϕ-OTDR systems are similar. Figure 5.14 shows a schematic diagram, where a narrow linewidth laser is used in continuous wave (cw) operation; and its output is chopped

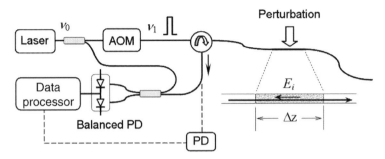

Figure 5.14   Schematic diagram of COTDR and ϕ-OTDR.

by an acoustic optical modulator (AOM) into a pulse train with narrow pulse width τ and repetition rate fitting to the fiber length L. Two detection schemes are shown in the figure. One is the direct detection: the backscattering signal power is detected by a PD, shown in the figure with dashed lines. The other is heterodyne detection. The probe beam is frequency shifted from v_0 to v_1, and a part of the laser beam is taken as the LO. The backscattered wave E_S and the local oscillator (LO) wave E_L are mixed at the fiber coupler. Two outputs of the coupler are proportional to $I_0 \pm (E_L E_S^* - E_S E_L^*) \cos[2\pi(v_1 - v_0)t]$, respectively, and detected by a balanced detector, canceling out their difference with direct current (DC) background.

For coherent detection the coherent length of the laser source should be longer than the entire length L to be measured, so that the source linewidth $\delta\lambda < \lambda^2/2L$ is required. Besides, the source power should be high enough; therefore EDFAs are often used. Fiber lasers with output power larger than that of conventional diode laser are also used in ϕ-OTDR [43].

The phase characteristics of the scattering and their effect on OTDR are analyzed in references [41,44–46]. The insert in Figure 5.14 shows a section of fiber corresponding to the pulse width $\Delta z = \tau/v_g$, and E_i stands for the scattering wave. It is recognized that the Rayleigh scattering occurs randomly both temporarily and spatially; the composite reflection is a sum of that from every scatter [46]:

$$r(t) = \sum_{i=1}^{M} r_i(t) \exp[j\phi_i(t)], \qquad (5.18)$$

where M is the scatter number, on the order of $\Delta z/\lambda$, distributing in section z to $(z + \Delta z)$ randomly. Assuming the scatterings are independent from each other, the probability density functions of the amplitude and phase of the reflection are expressed as

$$p(r) = (r/\sigma^2)\exp(-r^2/2\sigma^2) \quad (r > 0), \tag{5.19}$$

$$p(\phi) = 1/2\pi \quad (-\pi < \phi \le \pi), \tag{5.20}$$

where σ^2 is the reflectance with the largest probability, and is on the order of 10^{-7} for Rayleigh scattering [46]. The characteristics of interference in fiber are studied also in [47–49].

Due to the interferences of the large number of randomly distributed scattering light waves, the signal by direct detection shows large fluctuations, called fading noise, which conceals the information to be sensed. The statistical analysis of the fading noise indicates that it can be reduced by averaging over a large number of signal traces, and is inversely proportional to the square root of averaging number N: $\propto 1/\sqrt{N}$. A method of frequency shift averaging by sweeping the diode laser frequency was proposed. This plain averaging scheme has the shortcoming of low data processing speed and is not suitable for higher frequency vibrations. Nevertheless, the direct detection system is simple and incurs low cost. In addition band pass filters can be used to enhance the SNR. Due to the randomness of scattering, the detected trace is noisy; the information to be sensed is obtained from the difference of successively received traces, as described in [50–52].

The direct detection scheme has not revealed the information behind the phase sufficiently. The coherent detection helps to acquire information of interference amplitudes. If the composite reflection (5.18) is rewritten as $r(t) = |r_0 + \delta r(t)| \exp j[\phi_0 + \delta\phi(t)]$, where $\delta r(t)$ and $\delta\phi(t)$ are the variations to be sensed, the information in $\delta\phi(t)$ can be detected by heterodyne detection. In the scheme, mixing of E_S and E_L gives a beat signal written as

$$I(t) \propto E_L^2 + E_S^2 + 2E_L E_S \cos\theta \cos[2\pi(\nu_1 - \nu_0)t + \Delta\phi], \tag{5.21}$$

where $\Delta\phi = \phi_0 + \delta\phi(t) - \phi_L$ is the phase difference between E_S and E_L; $\theta(t)$ stands for the polarization deviated from LO, which also varies temporally and spatially. The heterodyne method plays a role of narrow band filter to reduce noise out of band, including DC background and low frequency noise; and also amplifies the beat signal by a high power LO, $\propto E_L E_S$. From the detected $I(t)$, the phase factor $\Delta\phi$ and the amplitude $2E_L E_S \cos\theta$ are retrieved. To avoid phase noise in successive sweeps, the phase factor has to be retrieved for every sweep,

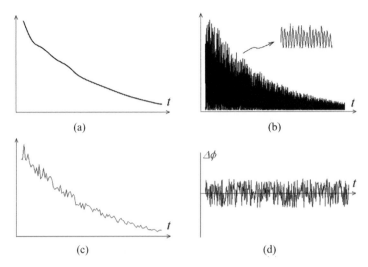

Figure 5.15 Illustrative comparisons of (a) analog OTDR; (b) heterodyne signal; (c) interference amplitude; and (d) phase variation.

and then averaged on multiple sweeps. Effective algorithms, such as fast Fourier transform (FFT) and autocorrelation, are needed for data processing. The system of ϕ-OTDR is almost the same as that of COTDR. It seems that the former is preferable for sensing dynamic signals, whereas the purpose of COTDR is for static weak signals.

Figure 5.15 shows comparisons among the detected traces of different technologies [18]. The curve in Figure 5.15(a) is the trace of analog OTDR; the signal is the averaged reflected power by direct detection. Figure 5.15(b) is the beat signal detected by heterodyne. The curve in Figure 5.15(c) is the amplitude of the interference, $\propto E_L E_S(t) \cos \theta(t)$, obtained by averaging over several beat periods and multiple sweeps. Figure 5.15(d) shows the phase shift variation retrieved from term $\cos[2\pi(\nu_1 - \nu_0)t + \Delta\phi(t)]$. It is seen that the last signal contains more details with higher sensitivity.

Reference [53] proposed and demonstrated a method of moving averaging and moving differential. The moving averaging allows using more data of adjacent traces; whereas moving differential is to calculate the difference between the averaged signal and a reference, giving dynamic information, such as very weak sound.

Combination of POTDR and ϕ-OTDR techniques shows attractive features for applications. The polarization divided detection by using a polarization beam splitter makes use of both polarization signals, and reduces also the polarization noise [52]. A tunable coherent laser and polarization analyzer are incorporated in OTDR to measured

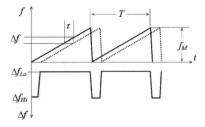

Figure 5.16 Schematic frequency-time diagram of OFDR.

differential group delay (DGD) and PMD distributions along optical fiber links [36].

5.2.4 Optical Frequency Domain Reflectometry

The above OTDR schemes resemble a pulsed radar system. They are used widely in telecommunication and fiber sensor technologies; but the disadvantage of pulsed OTDR is noticed. Its spatial resolution is limited by the pulse width of laser sources. To improve the resolution a narrower pulse is needed. But it leads to a low average power and low average SNRs. If a higher peak power source is used, the dispersion and nonlinear optical effects give other limitations. To overcome the difficulties another scheme was proposed and developed, that is the OFDR [54–57].

The OFDR uses the frequency-modulated continuous wave (FMCW) ranging technique, referring to the radar technology. In the system, a cw optical carrier with a periodic linear frequency sweep is employed. The detected optical echoes are with the same frequency sweep but with a delay, which is proportional to the propagation distance in fiber. The signal reflected back to the start point interferes with a reference signal to produce a correlation signal in the frequency domain, observed by a spectrum analyzer or a data-processing device. Figure 5.16 gives a schematic frequency-time diagram. It is shown in the figure that the frequency modulation amplitude is f_M and the sweep period is T; thus the slope of frequency sweep is $s = \mathrm{d}f/\mathrm{d}t = f_M/T$. The beat frequency of the reflected wave and reference signal depends on the delay time τ: $\Delta f = s\tau = 2sz/v_g$, where z is the distance of the scatter. Therefore, the positions of faults or external disturbances can be localized by the beat frequency.

References [58–61] give the basic theoretical analysis and experiment descriptions of the OFDR. The reference wave E_r and the scattering wave E_s within a modulation period are expressed as

$$E_r(t) = E_{r0} \exp j[\omega_0 t + 2\pi s t^2 + \phi(t)], \tag{5.22a}$$

$$E_s(t, \tau) = E_{s0} \exp j[\omega_0(t + \tau) + 2\pi s(t + \tau)^2 + \phi(t + \tau)]. \tag{5.22b}$$

At the receiver, their sum $E_T(t, \tau) = E_r(t) + E_s(t, \tau)$ is detected by the photodiode, giving

$$
\begin{aligned}
V(t, \tau) &= \eta \langle |E_T(t, \tau)|^2 \rangle \\
&= P_1 + P_2 + 2\sqrt{P_1 P_2} \xi_\phi \xi_T \cos(4\pi s \tau t + \omega_0 \tau + s \tau^2),
\end{aligned}
\tag{5.23}
$$

where $P_1 = \eta E_{r0}^2$, $P_2 = \eta E_{s0}^2$, and η is the responsivity of detection. ξ_ϕ is the phase correlation term, expressed as

$$\xi_\phi = \langle \exp j \Delta\phi(t, \tau) \rangle = \exp \frac{-\tau}{2\tau_{coh}} = \exp(-\pi \tau \delta\nu_{1/2}), \tag{5.24}$$

where τ_{coh} is the coherence time, and $\delta\nu_{1/2} = (2\pi \tau_{coh})^{-1}$ is the full width at half maximum (FWHM) linewidth of laser beam with Lorentzian line. Factor $\xi_T = \sin^2[(\omega - \omega_b)(T - \tau)/2]/[(\omega - \omega_b)(T - \tau)/2]^2$ comes from the periodic sweeping. The beat frequency $\omega_b = 4\pi s \tau = 8\pi f_M z / v_g T$ is then obtained from (5.23).

The frequency spectrum analyzer in the receiver or a data-processing device gives the frequency spectrum, which corresponds to the Fourier transform for one period of modulation near the beat frequency, written as [58]

$$V^2(\omega) \propto P_1 P_2 (T - \tau)^2 \frac{\sin^2[(\omega - \omega_b)(T - \tau)/2]}{[(\omega - \omega_b)(T - \tau)/2]^2} \exp \frac{-\tau}{\tau_{coh}}. \tag{5.25}$$

The noise characteristics are analyzed in reference [58], giving an expression of SNR in case of the phase noise dominant:

$$SNR = \frac{\exp(-\tau/\tau_{coh})}{1 - \exp(-\tau/\tau_{coh})}. \tag{5.26}$$

The maximum detectable delay time is set at $SNR = 1$, resulting in $\tau_{max} = \tau_{coh} \ln 2$. This means that the maximum detectable range is inversely proportional to the linewidth: $z_{max} = v_g \ln 2/(4\pi \delta\nu_{1/2})$.

It is shown in (5.25) that the spectral component at beating is proportional to the intensity of the scattered beam, which is expressed as [54]

$$P_2(z) \propto P_1 \left| \int_0^z \sigma(x) \exp\left[j2\beta(x, \omega_b) - \int_0^x \alpha(\zeta)\mathrm{d}\zeta \right] \mathrm{d}x \right|^2, \qquad (5.27)$$

where $\sigma(z)$ stands for the distributed scattering coefficient, $\alpha(z)$ is the distributed attenuation coefficient, and the spatial distribution and frequency dependence of propagation constant $\beta(x, \omega)$ is taken into consideration, including its dependence on temperature, strain and other physical conditions.

It is obvious that the spatial resolution depends also on the linewidth of the source and the effective receiver bandwidth, including the resolution of spectrum analyzer. The beat frequency spectrum (5.25) is in the form of the sinc function; its bandwidth depends on the modulation period T. As a result, the spatial resolution of OFDR is inversely proportional to the modulation frequency [60].

In practical OFDR systems, various laser sources are used, such as He–Ne laser, dye laser, and solid state laser. The most convenient and cost-effective source is the semiconductor laser (LD), whose lasing wavelength can be easily modulated by the injected current. However, the effect of the accompanied power modulation of the LD has to be solved. The fiber dispersion of $\beta = \beta(z, \omega)$ may influence the beat signal; but analysis indicates that it is usually negligible for the narrow linewidth source. The OFDR is also used to detect temperature change, based mainly on the thermal expansion of fiber material [62].

Polarization effects must be considered such as polarization fading and noise. By using polarization analyzers more information may be revealed [60], similar to POTDR; the distributed birefringence of SMF and characteristics of a polarization maintaining fiber can be investigated by OFDR [63,64].

Since the OFDR signal is obtained by the spectrum analyzer, or by digital data processing, the signal readout takes a longer time. The electronics and algorithm of data processing play important roles, not only for speeding up the output but also for improving precision.

5.3 DISTRIBUTED SENSORS BASED ON RAMAN SCATTERING

Raman scattering in optical fibers has been exploited to develop distributed sensors, especially for temperature sensing. In the literature, it is usually termed Raman distributed temperature sensor (RDTS), or

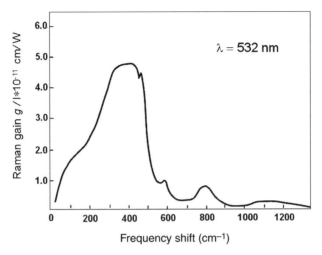

Figure 5.17 Raman gain spectrum for vitreous silica. (Reprinted with permission from reference [65]. © 1972 American Institute of Physics.)

distributed anti-Stokes Raman thermometry (DART). In this section, the basic characteristics of Raman scattering in optical fibers is presented, and the configurations of temperature sensors based on Raman scattering are introduced.

5.3.1 Raman Scattering in Fiber

As introduced in Section 5.1, Raman frequency shift is a good characterization of the composition and structure of materials. Raman spectroscopy plays an important role in physics and chemistry research and industrial applications. Since the invention and development of optical fibers, Raman spectroscopy has been used to analyze compositions and impurities in fibers; optical fiber devices based on Raman effect have also been developed [10].

References [65,66] show the measurements of the Raman oscillation in silica fibers, giving a Raman spectrum as shown in Figure 5.17. The peak wave number shift of Raman scattering of vitreous silica is 420 cm^{-1} measured by a laser beam at 532 nm, corresponding to frequency shift of 13 THz. The linewidth is quite large, about 9 THz; it is attributed to the molecular structure of vitreous silica like a random net. In the conventional fibers, some components are necessarily doped, such as GeO_2, B_2O_3, and P_2O_5. Reference [67] investigates the relative Raman cross sections of these dopants compared with the glass SiO_2 host. Table 5.1 gives the frequency shift and the relative peak

Table 5.1   Relative Peak Raman Cross Sections

Material	Refractive index	Frequency shift (THz)	Relative intensity	Relative cross section
SiO_2	1.46	13.2 (440 cm^{-1})	1	1
GeO_2	1.60	12.6 (420 cm^{-1})	7.4	9.2
B_2O_3	1.48	24.2 (808 cm^{-1})	4.6	4.7
P_2O_5	1.55	19.2 (640 cm^{-1})	4.9	5.7
–	–	41.7 (1390 cm^{-1})	3.0	3.5

Reprinted with permission from reference [67]. © 1978 American Institute of Physics.

Raman cross sections of the primary glass formers, measured by an argon ion laser at 514.5 nm. The data are helpful in characterizing fiber composition.

It is also shown that the effective Raman cross section depends on fiber structures [68]. It is mainly due to the different mode distributions between the step-index fiber and the gradient-index fiber, and between SMFs and MMFs, since the capture fraction of the scattering depends on the NA of the fiber [13,15,68]:

$$S = \frac{3}{2n^2(k_0w_0)^2} \approx \frac{3}{2n^2}(NA)^2, \tag{5.28}$$

where w_0 is the waist radius of Gaussian beam approximation of fiber near field. As a characterization tool, Raman spectroscopy is also used to investigate the composition of materials, such as the rate of OH uptake in fiber protection studies [69].

The properties of Raman scattering are also functions of some physical conditions, which is exploited to develop sensors, especially temperature sensors. It is shown that Stokes and anti-Stokes scatterings occur simultaneously in the materials, but with very different intensities. The basic factor for the difference is the populations, N_0 and N_1, at the ground state and the exited state, which obeys the Bose–Einstein distribution law, or its approximation, Maxwell–Boltzmann distribution law:

$$\frac{N_1}{N_0} \propto \exp\frac{-\Delta E}{k_B T}, \tag{5.29}$$

where k_B is Boltzmann's constant. The efficiency ratio of anti-Stokes over Stokes Raman scattering components is thus a function of temperature [70]:

$$R(T) = \frac{\eta_a}{\eta_s} = \frac{\lambda_s^4}{\lambda_a^4} \exp \frac{-\Delta E}{k_B T} = \left(\frac{\nu + \Delta \nu}{\nu - \Delta \nu}\right)^4 \exp \frac{-T_0}{T}. \tag{5.30}$$

Factor λ_s^4/λ_a^4 comes from the mechanism of the dipole model, similar to that in Rayleigh scattering [8]. In the formula, $\Delta \nu = \Delta E/h$, and $T_0 = \Delta E/k_B$, where h is Plank's constant, and ΔE is the energy gap between the excited state and ground state, which is a characteristic parameter for the materials. For vitreous SiO_2, ΔE is about 50 meV, corresponding to $\Delta \nu \approx 13$ THz; and $T_0 \approx 600$ K. It is seen that $R(T)$ increases with the temperature. The relative temperature sensitivity of the anti-Stokes intensity is reported to be $0.8\%°C^{-1}$ at room temperature [71]. The intensity of backward Raman scatterings is surely proportional to the probe laser power; the fiber loss leads to attenuations of both probe and Raman signal. However, the ratio of anti-Stokes and Stokes scattering intensities keeps a function of temperature, does not depend on the fiber loss; thus, the ratio is utilized to develop the distributed DART.

It is necessary to note that formula (5.30) holds for the spontaneous Stokes and anti-Stokes components, meaning that the incident beam (pump) should be controlled under the threshold of stimulated Stokes. For forward scattering in stationary state, the Stokes and pump intensities are described by rate equations of [10].

$$\frac{d}{dz} I_S = g_R I_p I_S - \alpha_S I_S, \tag{5.31a}$$

$$\frac{d}{dz} I_p = -\frac{\omega_p}{\omega_S} g_R I_p I_S - \alpha_p I_p, \tag{5.31b}$$

where the subscripts of S and p stand for the Stokes component and pump beam; g_R is the gain coefficient of SRS; $\alpha_{p,s}$ is their loss coefficients. The Raman scattering induced loss of the pump can be neglected below the threshold; the pump is expressed as $I_p = I_{p0} \exp(-\alpha_p z)$. Thus the Stokes is solved to be

$$I_S(z) = I_S(0) \exp(g_R I_{p0} z_{\text{eff}} - \alpha_S z), \tag{5.32}$$

where $z_{\text{eff}} = (1 - e^{-\alpha_p z})/\alpha_p$; the Stokes $I_S(0)$ at the input point is generated by spontaneous Raman scattering. It is shown that the Stokes ascends with distance z under condition of $g_R I_{p0} z_{\text{eff}} > \alpha_S z$, whereas the pump decays with z. The SRS threshold is defined as the input pump power at which the output Stokes power equals the pump power at $z = L$; that is, $P_S(L) = P_p(L) = P_{p0} \exp(-\alpha_p L)$, where powers are the integrals over the gain spectrum and summing up over all modes. By physical consideration, the critical pump power for Raman threshold is determined by [2,10]

$$g_R P_R^{\text{cr}} L_{\text{eff}} / A_{\text{eff}} \approx 16, \tag{5.33}$$

where $L_{\text{eff}} = (1 - e^{-\alpha_p L})/\alpha_p$, and A_{eff} is the effective core area. The gain coefficient g_R depends on the medium; for silica fibers the Raman threshold intensity I_R^{th} is about 10 MW/cm^2.

As a distributed sensor that requires high spatial resolution, short optical pulses are used as the probe in DART, the same as that in OTDR. For a pulse of 10ns, as typically used, its extension in fiber is just 2 m; moreover the backscattering Raman is mostly used, which propagates opposite to the probe. The amplification of the SRS in the short interaction length can reasonably be neglected. That is to say, the DART system is based on spontaneous Raman scattering; and the coherence and polarization dependence need not be considered.

5.3.2 Distributed Anti-Stokes Raman Thermometry

A distributed temperature sensor based on Raman scattering was first demonstrated in [69], where a conventional OTDR was used to detect expanded returned waveforms of both Stokes and anti-Stokes signals in time domain. The temperature resolution of 1K and the measurement range of $-50°$C to $100°$C were reported in [70,71].

The main difference between DART and conventional OTDR is that an optical spectrum analyzer, such as a grating spectrometer or by thin film filters, is installed in the apparatus to divide the Stokes and anti-Stokes components. A typical configuration of a DART system is shown in Figure 5.18, which consists of a pulsed probe laser, a 1:1 beam splitter to deliver optical pulses to a long sensing fiber, which is usually an MMF to transmit higher power optical pulses and to pick up the returned weak signal; two narrow band filters to allow the Stoke and anti-Stokes components detected separately and to block the Rayleigh scattering component. Although the Raman frequency shift of 13THz is quite large, about 100nm at 1550nm range, the side lobe suppression ratio of the filters must be high enough because the Raman scattering intensity is much lower than that of Rayleigh scattering, typically one thousandth lower. Moreover, the intensity of anti-Stokes

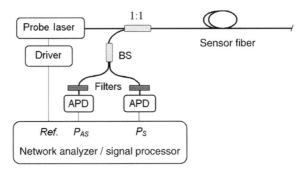

Figure 5.18 A typical configuration of DART.

components is much weaker than that of Stokes components; therefore, an asymmetric beam splitter is inserted in the fiber path, such as 90% for anti-Stokes channel and 10% for the other. The detected signals are analyzed by a network analyzer and data processor. The wavelength dependence of Raman scattering was analyzed in [72]. Long wavelength diode lasers (preferably in 1550 nm band) are beneficial for lower optical loss and longer working fiber, and for a higher factor of $(\lambda_s/\lambda_a)^4$ in formula (5.30).

Similar to the OTDR, several specifications are concerned for practical applications: working distance, precision of localization, precision and range of measured temperature, and readout time. The basic technical issue for better performance is to enhance the SNR. Higher SNR means longer working fiber length, fast response of data readout, higher spatial resolution and higher precision of temperature measurement. For the purpose, photomultiplier tubes were used at the earlier stage to detect signals by visible light sources. For the infrared range with much lower propagation loss, avalanche photodiodes (APD) with internal multiplication are usually used instead of the conventional photodiode. To enhance SNR, multiple data collection and averaging is necessary, and photon counting method is mostly adopted. The number of counts n_i obeys a Poissonian distribution; its relative accuracy is $\delta n_i/n_i = 1/\sqrt{n_i}$. The time-correlated single-photon counting technique and the multi-photon-timing technique show effectiveness in enhancing sensitivity [23,73]. It is noticed of cause that a longer measurement time is needed for the averaging.

One of the requirements for DART is a wide range of temperature measurements, especially for fire detection. From (5.29), the relative sensitivity drops with increasing temperature [73]:

$$\frac{1}{R}\frac{dR}{dT} = \frac{T_0}{T^2}, \tag{5.34}$$

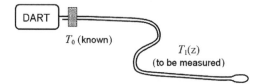

Figure 5.19   Calibration method with a loop arrangement.

implying that the resolution of temperature measurement decreases. Another technical issue is the protection of fiber in high-temperature environments. The polymeric jacket of the conventional fiber has to be replaced by heat-resistant materials. Reference [74] used gold coating and carbon coating to protect the fiber, making the fiber work at temperatures more than 500°C. However, the coating will induce additional losses due to thermal strains.

There are many factors in the scheme to pay attention to. The Rayleigh scattering leaked from the filters is a major source of the noise and background, which must be removed as much as possible. Different fiber losses for Stokes and anti-Stokes components and inhomogeneous loss distribution along the fiber also influence the precision of the temperature readout. DART signal is just for relative temperature change; calibrations are needed for the true temperature value. One method is to let the sensing fiber pass through some blocks with known temperature. Figure 5.19 shows a calibration method with loop arrangement [73,75].

The spatial resolution of DART is one of the main concerns. It is basically determined by the laser pulse width and the detector response, expressed as $\delta L = c\tau/2n_g$ if the detector response is fast enough. The first method for enhancing the spatial resolution is obviously to shorten the pulse width. However, a short pulse means fewer photons to be integrated and consequently longer data processing time. Some other methods are proposed within the limitations of laser pulse width and detector response. One of them is by using path delay multiplexing [76], as shown in Figure 5.20. In the system, the Raman scattering signal is integrated twice for data streams with time delay corresponding

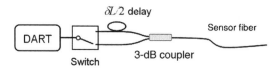

Figure 5.20   Spatial resolution enhanced DART with path delay multiplexing.

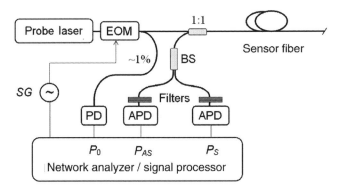

Figure 5.21 Basic scheme of the Raman optical frequency domain reflectometry.

to $\delta L/2$, showing that spatial resolution doubled experimentally. Accordingly the resolution can be further enhanced, so long as the SNR remains high enough.

5.3.3 Frequency Domain DART

As introduced in Section 5.2, the reflectometry based on Rayleigh scattering includes two basic methods: OTDR and OFDR. Similarly, Raman thermometry has two counterparts. Reference [77] proposes a Raman thermometry based on frequency domain analysis, as shown in Figure 5.21 schematically. The principle of Raman OFDR is similar to that of OFDR discussed in Section 5.2.4. The electro-optic modulator (EOM) is used to modulate the intensity of the probe laser with frequencies controlled by signal generator (SG).

Referring to formula (5.23), and taking the different coefficients of Raman scatterings at Stokes and anti-Stokes lines into consideration, their spectral components are written as

$$V^2(\omega_S) \propto \eta_S(T,z) P_r P_0 F_S e^{-\int (\alpha_S + \alpha_p/2)\mathrm{d}z}, \tag{5.35a}$$

$$V^2(\omega_{AS}) \propto \eta_{AS}(T,z) P_r P_0 F_{AS} e^{-\int (\alpha_{AS} + \alpha_p/2)\mathrm{d}z}, \tag{5.35b}$$

where the correlation factors are denoted as follows:

$$F_{S,AS} = (T_m - \tau_{S,AS})^2 \frac{\sin^2[(\omega - \omega_{S,AS})(T_m - \tau_{S,AS})/2]}{[(\omega - \omega_{S,AS})(T_m - \tau_{S,AS})/2]^2} \exp \frac{-\tau_{S,AS}}{\tau_{\mathrm{coh}}}. \tag{5.35c}$$

The efficiency ratio of Stokes over anti-Stokes components is then expressed as

$$R(T, z) = \frac{\eta_{AS}(T, z)}{\eta_S(T, z)} = \frac{V^2(\omega_{AS})}{V^2(\omega_S)} \frac{F_S}{F_{AS}} e^{\int (\alpha_{AS} - \alpha_S) dz}. \tag{5.36}$$

The term F_S/F_{AS} comes mainly from the index difference of Stokes and anti-Stokes lines and is basically near unity. The ratio has to be calibrated in practical systems, since there are other factors influencing the measured data, such as the transmissions of filters and differences in electronic circuits. As for the signal processing, discrete Fourier transform (DFT) and FFT algorithms are usually used; consequently, the probe source is modulated by sinusoidal waves with a series of equidistant modulation frequencies, as proposed in [76].

The DART systems are now technically mature and commercialized, widely applied as distributed temperature inspection systems and fire alarming facilities installed along tunnels, bridges, railways, electric cables, in power stations and in forests, and so on [74–79]. However, the sensitivity of Raman scattering to fiber strain is very low; therefore sensors based on Brillouin scattering are attractive, as introduced in Section 5.4.

5.4 DISTRIBUTED SENSORS BASED ON BRILLOUIN SCATTERING

As introduced in Section 5.1, Brillouin scattering is caused by interaction between photons and acoustic phonons, and is sensitive both to temperature and to strain. Sensors based on Brillouin scattering have attracted wide interest, being a trending subject in research and development.

5.4.1 Brillouin Scattering in Fiber

The basic properties and systematic theory of SBS are expounded in the literature [2,10,12,80]. It was found at the early stage of fiber technology that the threshold of input power for SBS is quite low [81–84] for narrow linewidth sources. The SBS brings about unfavorable effects for optical communication applications, such as propagation losses, crosstalk, and limitation of the signal power. To evaluate and discuss the effect, two coupled equations for variations of the backward Brillouin scattering and the pump are established:

$$\frac{d}{dz}I_S = -g_B I_p I_S + \alpha I_S, \tag{5.37a}$$

$$\frac{d}{dz}I_p = -g_B I_p I_S - \alpha I_p. \tag{5.37b}$$

Since the Brillouin frequency shift is very small, loss coefficients of the scattered beam and the pump are taken to be the same in the equations. Similar to Raman scattering, the intensity of Brillouin scattering below the threshold is solved as

$$I_S(z) = I_S(0)\exp(\alpha z - g_B I_{p0} z_{\text{eff}}). \tag{5.38}$$

It indicates that the intensity of backward Stokes increases in $-z$ direction if Brillouin gain $g_B I_{p0} z_{\text{eff}}/z$ is larger than α. The theoretical analysis indicates that the critical pump power for Brillouin threshold, P_B^{cr}, defined as equal to the SBS power at the input, is deduced to be

$$P_B^{\text{cr}} \simeq 21 A_{\text{eff}}/g_B L_{\text{eff}}. \tag{5.39}$$

The typical critical power for SMFs is measured experimentally to be 1 mW if the pump linewidth is near and narrower than the linewidth of Brillouin scattering [10], which is about 40 MHz at room temperature. The negative impact of SBS can be overcome by using sources with linewidth larger than Brillouin linewidth. It is not difficult to achieve in conventional communication systems with an intensity-modulated laser transmitter.

It is necessary to understand the spectral characteristics of Brillouin scattering, especially its line shape and linewidth. As introduced in Section 5.1, Brillouin scattering is attributed to the interaction of optical field and molecules, that is, the electrostriction. The sound wave is a strain wave in z direction; the stain-induced change of dielectric constant is expressed as $\delta\varepsilon = -\varepsilon^2 p_{12}e_z$ for the transverse field. The electrostriction is related with the coupled energy of the pump and the scattering field, $\varepsilon E_p \cdot E_S^*/2$. The equation of molecule's movement u is written as [12,80]

$$\frac{\partial^2 u}{\partial t^2} + \frac{1}{\tau_a}\frac{\partial u}{\partial t} - V_a^2\frac{\partial^2 u}{\partial z^2} = \frac{\varepsilon^2 p_{12}}{2\rho}\frac{\partial}{\partial z}(E_S^* E_p), \tag{5.40}$$

where τ_a is the decay time of the sound wave, that is, the lifetime of phonons; and the Stokes and pump are in the same direction of polarization. The amplitude of sound wave can then be solved from (5.40); and the nonlinear electric polarizations are obtained. The equations governing the pump and the Stokes are written as

$$\frac{\partial E_p}{\partial z} + \frac{1}{v_g}\frac{\partial E_p}{\partial t} + \frac{\alpha}{2}E_p = -\frac{g_B}{2}|E_S|^2 E_p, \qquad (5.41a)$$

$$-\frac{\partial E_S}{\partial z} + \frac{1}{v_g}\frac{\partial E_S}{\partial t} + \frac{\alpha}{2}E_S = \frac{g_B}{2}|E_p|^2 E_S. \qquad (5.41b)$$

The gain coefficient is deduced with a typical Lorenzian line shape, expressed as [85]

$$g_B(\nu) = \frac{g_0}{1 + (\nu - \nu_B)^2/(\Delta \nu_B/2)^2}, \qquad (5.42)$$

where ν is the deviation from the input optical frequency, $\Delta \nu_B = 1/(\pi \tau_a)$ is its FWHM linewidth. The Brillouin frequency shift ν_B is deduced to be $\nu_B = 2nV_a/\lambda$. The velocity of the longitudinal sound wave V_a in the fiber depends on the density and Young's modulus: $V_a = \sqrt{Y/\rho}$, which is measured to be $\sim 5{,}950$ m/s in silica fiber. Theoretical analyses and experiments indicate that the lateral deformation has little effect on the Brillouin frequency shift, implying that it is not needed to consider the transverse sound wave. It is then obtained that $\nu_B \sim 11$ GHz in 1,550 nm band. The phonon lifetime in silica fiber at room temperature is measured ≤ 10 ns, and $\Delta \nu_B$ is about $30 \sim 40$ MHz. The gain coefficient at the peak is deduced theoretically to be [10,80]

$$g_B(\nu_B) = \frac{2\pi n^7 p_{12}^2}{c\lambda_p^2 \rho V_a \Delta \nu_B}. \qquad (5.43)$$

Reference [85] studies Brillouin gain spectrum characteristics in detail for fibers with different GeO_2 compositions, and their temperature and strain dependencies.

The parameters Y, ρ and n are obviously functions of temperature and strains, and also the material properties. The sensitivity of ν_B to material composition is very high; relative differences in GeO_2 concentration as small as 0.01% can be detected [86,87]. Brillouin scattering is utilized to develop temperature sensors [86–89] and strain sensors

Table 5.2 Sensitivities of Brillouin Sensor in 1,550 nm Band

Parameter	$C_e \nu_B$	$C_T \nu_B$	D_e	D_T
–	48 kHz/$(\mu\varepsilon)$	1.1 MHz/K	$-0.077 \times 10^{-4}/(\mu\varepsilon)$	0.36/K

[90–92]. The relation of Brillouin frequency on strain and temperature is expressed as

$$\nu_B(e, T) = \nu_B(0, T_r)[1 + C_e e + C_T(T - T_r)], \qquad (5.44)$$

where T_r is the reference temperature. The coefficients are measured experimentally; the typical data are listed in Table 5.2.

One of the differences between Raman sensors and Brillouin sensors is that the parameter being detected is the amplitude of Stokes and anti-Stokes for the former, whereas it is the frequency shift of Stokes line for the later. Therefore, a pump laser source with narrow linewidth and the correlation detection with high resolution must be used to read out the changes of Stokes frequency in Brillouin sensors.

High spatial resolution is one of the important requirements for the distributed Brillouin sensor systems. The concept of OTDR is introduced to the Brillouin sensor, which is to detect spatially resolved reflected signals in time domain, called BOTDR. Its spatial resolution is mainly determined by the pulse width of a launched pump beam, expressed as $\delta L = c\delta t/2n_g$; for example, $\delta L = 10$ m for $\delta t = 100$ ns. It is obvious that the signal of BOTDR is attributed to the spontaneous Brillouin scattering in such a short length. Although the spatial resolution is proportional to the pulse width, it cannot be decreased infinitively. One of the limitations is the phonon lifetime τ_a. It is because pulse width shorter than the phonon lifetime means that the linewidth of the pump is broader than the linewidth of the Stokes line, resulting in a lower signal.

It is noticed that the signal induced by strain and that by temperature change are not distinguishable from the Brillouin frequency shift solely. Another parameter in Brillouin scattering sensitive to strain and temperature is needed with linearly independent coefficients. It is proved that a physical relation exists between Rayleigh scattering and Brillouin scattering: $I_R/I_B = (c_p - c_v)/c_v$, termed Landau–Placzek ratio, where c_p and c_v are the specific heats at constant pressure and constant volume, respectively [93]. It is deduced that the Brillouin scattering intensity is expressed as [94]

$$I_B = I_0 \frac{\pi^2 n^8 p_{12}^2}{\lambda^4} V (1 + \cos^2 \theta) \frac{k_B T}{\rho V_a^2}, \qquad (5.45)$$

where θ stands for the polarization direction and V is the volume. Therefore, the Brillouin power can be written as [95,96]

$$P_B = AT/v_B^2, \qquad (5.46)$$

giving its sensitivities in the expression of

$$P_B(e, T) = P_B(0, T_r)[1 + D_e e + D_T(T - T_r)]. \qquad (5.47)$$

The coefficients D_T and D_e are related to C_T and C_e as: $D_T = T^{-1} - 2C_T$ and $D_e = -2C_e$, where the weaker dependence of coefficient A on strain and temperature is not taken into account. The strain and temperature change are thus obtained from measured Brillouin frequency change δv_B and Brillouin power change δP_B. The latter is usually measured as a ratio to Rayleigh scattering power to cancel out any input power fluctuation. The experimentally obtained coefficients of power are listed in Table 5.2 [95], where $(\mu\varepsilon)$ is denoted for micro strain, that is, 10^{-6}.

The configuration of BOTDR and related technical issues will be discussed in Section 5.4.2. Another configuration of Brillouin sensor has two sources launching from two ends of the fiber simultaneously: one is a pulse source as a pump; the other is a cw source as a probe, termed Brillouin optical time domain analyzer (BOTDA), which will be discussed in Section 5.4.3.

5.4.2 Brillouin Optical Time Domain Reflectrometer

The main task of a Brillouin sensor is to acquire the frequency shift. Since the Landau–Placzek ratio in silica is about 30 at room temperature, it is very important to extract the Brillouin signal from the strong Rayleigh background. For this purpose a narrow linewidth pump source and a heterodyne detection are necessary in BOTDR; and an LO source is then needed for heterodyne detection. In practice, the local oscillator is just a beam divided from the pump and frequency shifted by an AOM. The backward Brillouin scattering wave is mixed with the LO, and a spectrum analyzer is used to measure their beat frequency. The method is called self-heterodyne, with distinction from that by using two solitary lasers.

Figure 5.22 shows basic configurations of a BOTDR system schematically [97–100]. In Figure 5.22(a), the source frequency v_0 is shifted to $v_0 + v_s$ with $v_s \approx v_B$, usually by AOMs. Since Brillouin frequency is around 11 GHz, much larger than a usual AOM frequency shift,

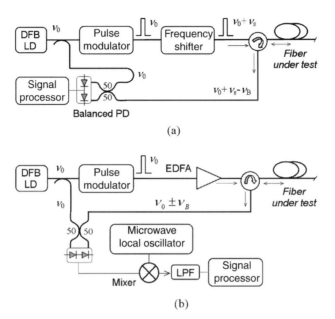

Figure 5.22 BOTDR with (a) frequency-converted source and (b) microwave heterodyne.

multiple shifting is needed by some recurring optical path [97,98]. The backscattered wave is mixed with the LO in a 3-dB coupler and detected by a balanced photodiode. The frequency variation to be detected is now contained in $(v_B - v_s)$. The heterodyne detection plays two roles: as an effective narrow bandwidth filter at v_B and signal amplification.

In Figure 5.22(b), the heterodyne occurs between the Brillouin wave and the LO with the same frequency as the probe, generating the beat signal at v_B; a high-speed PD (or a balanced PD preferably) and an RF spectrum analyzer have to be used [101,102]. To avoid detection and analysis of signals at such a high frequency, a microwave local oscillator at a fixed frequency of v_L is used to mix the signal again, as shown in the figure.

The DFB-LD is usually used as the single frequency source; and an electro-optic modulator (EOM) is used to chop the cw beam. EDFAs are often used to amplify the peak power of probe pulses, as depicted in Figure 5.22(b) and not shown in Figure 5.22(a) for simplicity. Apart from AOM, the electro-optic phase modulator (EOM) is also used to shift the frequency by using one of the side band harmonics with a filter [103].

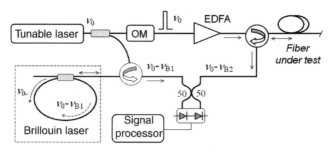

Figure 5.23 BOTDR with a Brillouin laser as the local oscillator.

A Brillouin laser is a frequency shifted source, and is used as the local oscillator [104], as illustrated schematically in Figure 5.23, where the Brillouin laser is composed by a fiber ring, as an example. The fiber in the ring may differ from the sensing fiber (fiber under test), giving two different Brillouin frequencies of v_{B1} and v_{B2}; and the heterodyne gives a signal at $v_{B2} - v_{B1}$. The pump laser has to be tuned to match one of the longitudinal modes of the ring; and the ring has to be stabilized to ensure the stability of Brillouin laser is good enough.

Polarization instability-caused signal fading takes place in BOTDR also, similar to that in OTDR. Reference [98] uses two Faraday cells as polarization scramblers to reduce the polarization noise. Optical filters, such as fiber Bragg gratings (FBGs), are often used to suppress the unwanted background; and electrical filters are also inserted in detection circuits as shown in Figure 5.22(b), denoted by LPF (low pass filter).

Basic performance concerned in the practical applications of BOTDR includes the following parameters: the spatial resolution, the frequency resolution, the measurement range, and the data output time. These parameters are generally in trade-off.

The detectable change of Brillouin frequency depends on the SNR. Due to the gain spectrum in Lorentzian form, the detected signal in voltage can be written as $V = V_0[1 - (v - v_B)^2/\Delta v^2]$ near the top. The peak is determined by setting its derivative to zero: $\delta V/V_0 = 2(v - v_B) \delta v/\Delta v^2 = 0$, which is written as $2\delta v^2/\Delta v^2$ at $v \approx v_B$. The left side of the expression can be regarded as $(SNR)^{-2}$ according to the definition of $SNR = P_{signal}/P_{noise} = (V_{signal}/V_{noise})^2$. Therefore, the minimum detectable change in v_B is expressed as [92]

$$\delta v = \frac{\Delta v}{\sqrt{2}(SNR)^{1/4}}. \tag{5.48}$$

The resolution for sensed strain and temperature change can accordingly be obtained.

The frequency width of the detected signal Δv is a composite parameter determined by Brillouin frequency width and the light source spectrum. The spectrum of a square waveform pulsed source is written as [105]

$$P_p(f) = P_0 \left[\frac{\sin \pi (f - f_0)\tau}{\pi (f - f_0)} \right]^2, \tag{5.49}$$

where f_0 is the central optical frequency; the composite spectrum is then expressed as

$$H(v) = H_0 \int_{-\infty}^{\infty} \frac{\sin^2[\pi(f - f_0)\tau]}{[\pi(f - f_0)]^2} \frac{(\Delta v_B/2)^2 df}{[(v - v_B) - (f - f_0)]^2 + (\Delta v_B/2)^2}. \tag{5.50}$$

An analytical result of the integral is given in formula (10) of [105]. Denoting $a = \pi \tau \Delta v_B$ the spectrum is approximated to

$$H(v) \approx \frac{H_0(a - 1 + e^{-a})}{\pi \Delta v_B} \left[1 - \frac{4(v - v_B)^2}{\Delta v^2} \right] \tag{5.51}$$

with

$$\Delta v^2 = (\Delta v_B)^2 \frac{a - 1 + e^{-a}}{a - 3 + (3 + 2a + a^2/2)e^{-a}}. \tag{5.51a}$$

It is shown that for a large pulse width with $a \gg 1$, $\Delta v \approx \Delta v_B$; for a narrow pulse width with $a \to 0$, $\Delta v \approx \sqrt{2}\Delta v_B/a = \sqrt{2}/\pi \tau$. It means that the uncertainty of frequency shift measurement is inversely proportional to the pump pulse width when it is near and less than phonon lifetime τ_a. Reference [106] presents simulations of pulse shape dependence on the spectrum of input light, with assumptions of square-like forms. Reference [107] shows a systematic experimental study on the dependence. If the waveform of a short pulse is regarded as a Gaussian (or a super-Gaussian), as it usually is, formula (5.49) should be replaced by a Gaussian, and (5.50) is replaced by a Voigt profile, which is the convolution of Gaussian and Lorentzian.

Similar to the OTDR based on Rayleigh scattering, the spatial resolution is determined by the pump pulse width τ: $\delta z = c\tau/2n$ if the speed of the detector is high enough. For the reason of spectral broadening analyzed above, the minimum spatial resolution is limited under

$\delta z_{\min} = c\tau_a/2n = c/2\pi n \Delta\nu_B$, which is near 1 m at room temperature. It is concluded that a trade-off exists between the spatial resolution and the precision of frequency measurement.

As shown in (5.48) the precision of frequency shift measurement depends on the SNR, which can be improved by increasing the pump power and also by multiple averaging. The power is limited by optical nonlinear effects, such as four-wave mixing (FWM), self-phase modulation (SPM), SRS, and SBS. The thresholds of the latter two effects are estimated by formulas (5.32) and (5.39). The data averaging needs higher accumulated detected energy with wider pulse width and higher pulse repetition rate. However, the repetition rate is limited by the working distance L: $f_r = c/2nL$. Denoting data output time as T_{out}, the times of data averaging is $N = f_r T_{\text{out}}$, and the SNR can generally be enhanced by $\sqrt{N}$.

The working distance depends obviously on the pump power and losses in the system, including fiber inherent loss, and losses in fiber couplers and filters. It depends also on the sensitivity of the balanced PD. Detailed analyses on the performances are given in [92,100]; further developments on data processing and system improvements can be found in references [108–111].

5.4.3 Brillouin Optical Time Domain Analyzer

This subsection is devoted to an introduction to BOTDA. Figure 5.24(a) shows its configuration schematically [113], in which a pulsed laser beam and a cw laser beam are injected into the sensing fiber from its two ends, respectively. The cw laser is tuned at $\nu_{\text{cw}} = \nu_p \pm \nu_B$, that is, Brillouin frequency shifted from that of the pulsed laser. In case of $\nu_{\text{cw}} = \nu_p + \nu_B$, the pulse will be amplified by SBS, whereas the cw beam experiences extra dynamic loss, which is usually called Brillouin loss. For $\nu_{\text{cw}} = \nu_p - \nu_B$, the pulsed energy will transfer to its Stokes component and amplify the cw power, called Brillouin gain. The evolutions of

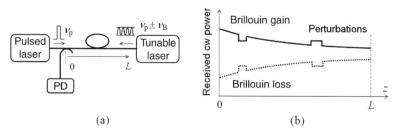

Figure 5.24   (a) Concept of BOTDA and (b) cw power variation in fiber.

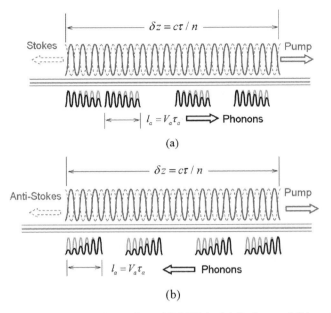

Figure 5.25 Conseptional illustration of BOTDA: (a) Stokes and (b) anti-Stokes.

cw power in the sensing fiber are illustrated in Figure 5.24(b) for the two cases, respectively. The evolution of cw power can be displayed in an oscilloscope or a computer, scanning with the pulses. The fiber attenuation is then measured by BOTDA; and external perturbations are detected accordingly [112,114–117].

Figure 5.25 shows conceptual illustrations of Stokes and anti-Stokes process in a pulse width τ. The acoustic wavelength is deduced from $v_B = 2nV_a/\lambda$ to be a half of the optical wavelength: $\lambda_a = V_a/v_a = \lambda/2n$. The phonons propagate codirectionally with the pump in Stokes process, whereas counterwise to the pump in anti-Stokes process. Therefore, more phonons are excited in Stokes process, implying a higher SBS gain; whereas annihilation of phonons occurs in anti-Stokes process, resulting in lower SBS gain.

The basic theory of BOTDA is presented in references [102,105]. The SBS couples the cw beam with the reversely propagating pulsed beam; the coupled wave equations (5.37a and 5.37b) are rewritten as

$$\frac{d}{dz}I_p = \pm gI_{cw}I_p - \alpha I_p, \tag{5.52a}$$

$$\frac{d}{dz}I_{cw} = \pm gI_{cw}I_p + \alpha I_{cw}. \tag{5.52b}$$

Here, the positive and negative signs are for the cases of Brillouin loss and Brillouin gain, respectively. To solve the equations the beams are obtained firstly with the Brillouin term neglected: $I_{cw}(z) = I_{cw}(L)e^{-\alpha(L-z)}$ and $I_p(z) = I_p(0)e^{-\alpha z}$. It is noted that the Brillouin term $gI_{cw}I_p = gI_{cw}(L)I_p(0)e^{-\alpha L}$ is a constant along the fiber length [112]. Since the pulse beam experiences Brillouin effect all the way in the propagation along the fiber, the second step is to investigate the evolution of pulsed beam induced by Brillouin effect:

$$I_p(z) = I_p(0) \exp[-\alpha z \pm \gamma(e^{\alpha z} - 1)], \qquad (5.53)$$

where $\gamma = (g/\alpha)I_{cw}(L)e^{-\alpha L}$. It is worth noting that in case of $v_{cw} = v_p + v_B$ the SBS will compensate the inherent fiber loss for the pulsed beam to a certain degree, and benefit extension of measurement distance. By substituting it into (5.52b), the increment of I_{cw} in the interacting range of z to $z + \delta z$ is written as [105,107]

$$\Delta I_{cw}(z) = I_{cw}(z + \delta z) - I_{cw}(z) = I_{cw}(z)\left\{ \exp\left[\int_z^{z+\delta z} (\alpha \pm gI_p)dz \right] - 1 \right\}$$

$$= I_{cw}(z)\left\{ \exp\left[\pm gI_p(0)e^{\pm\gamma+\alpha\delta z} \int_z^{z+\delta z} e^{-\alpha z} \exp(\mp\gamma e^{\alpha z})dz \right] - 1 \right\}.$$

$$(5.54)$$

For the I_p pulse width narrow enough with $\alpha\delta z \ll 1$, (5.54) can be approximated to

$$\Delta I_{cw}(z) \approx \pm gI_p(0)I_{cw}(L)\delta z \exp[\pm\gamma(e^{\alpha z} - 1)]. \qquad (5.55)$$

$\Delta I_{cw}(z)$ is just the signal measured in the BOTDA system. By $z = ct/2n$, the signal is displayed in an oscilloscope or recorded in a computer; and the loss variation along the fiber $\alpha(z)$ can be acquired. For SBS sensors, the variation of v_B with temperature and strains is obtained by detecting the peak of ΔI_{cw} by tuning the cw laser.

Figure 5.26 shows two basic configurations of BOTDA schematically. Configuration (a) uses two lasers for pump and probe, respectively. One of them is a slave laser, whose frequency is shifted by $\sim v_B$ from the master laser, and tuned by an optical phase locked loop (OPLL). Their frequencies are monitored by a spectrum analyzer. Configuration (b) uses one laser, whose output is divided into two beams: one is for the pump, the other is frequency shifted as the probe. In

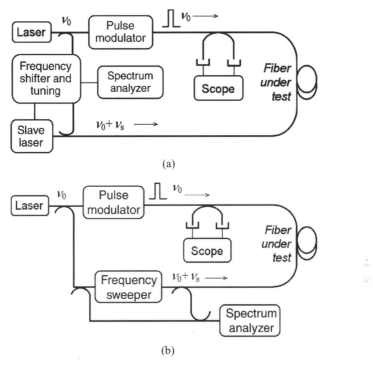

Figure 5.26 Basic configurations of BOTDA system.

experiments YAG lasers are often used for their better spectral characteristics; semiconductor DFB lasers are usually used in practical instruments. Since the SBS signal is contained in the forward propagation
cw beam, a PD or an APD is enough for detection; and an oscilloscope
or a computer-aided data processing is followed. Some components are
necessary in the system, such as filers, isolators, and polarization controllers, which are omitted in the figure for simplicity.

One of the advantages of BOTDA over BOTDR and DART is the
larger working distance; especially in case of $\nu_{cw} = \nu_p + \nu_B$, the amplified pulse can propagate much longer. A working distance up to
100 km was reported [118–120].

Similar to other time domain distributed sensors, it is necessary to
use shorter pulse width for higher spatial resolution. However, the
SNR will be reduced for shorter pulses as $\Delta I_{cw} \propto \delta z \propto \tau$. In addition, a short pulse causes spectrum broadening and lower precision of
frequency measurement, the same as that in BOTDR. Fortunately,
stimulated amplification will make the signal's spectrum narrowed in
principle. Denoting the ratio of frequency deviation over Brillouin

width as $q = \Delta v / \Delta v_B$, the gain is expressed as $g = g_0/(1 + q^2)$; and formula (5.55) is rewritten as

$$\Delta I_{cw}(v, z) = \frac{P_1}{1 + q^2} G^{1/(1+q^2)}, \qquad (5.56)$$

where $P_1 = \pm g_0 I_{cw}(L) I_p(0) \delta z$, $G = \exp[\pm g_0 I_{cw}(L) e^{-\alpha L} (e^{\alpha z} - 1)/\alpha]$. The half maximum is determined by $P(\Delta v_{1/2}, z) = P_1 G/2$, resulting in an equation for q:

$$G^{q^2/(1+q^2)} = 2/(1 + q^2), \qquad (5.57)$$

which gives reasonable approximations: $q_{1/2} \approx 1/\sqrt{G}$ for $G \sim 1$, and $\Delta v_{1/2} \approx \Delta v_B \sqrt{\ln 2 / \ln G}$ for a large G [139].

However, a higher amplification means higher cw background I_{cw}, which is limited by SBS threshold (5.39). It is found in experiments that DC leakage in between the pulses with moderate extinction ratio helps to reduce the spectral linewidth [121–125]. In this case, the waveform of pulse beam is regarded as the sum of a Gaussian short pulse and a low-amplitude square pulse with pulse width of repetition period $f_r^{-1} = 2nL/c$. The spectrum of detected signal is expressed as [121]

$$H(v) = \frac{(1 - c)g_0}{1 + 4(v - v_B)^2/\Delta v_B^2} + cV_t(\tau, \Delta v_B), \qquad (5.58)$$

where $V_t(\tau, \Delta v_B)$ is the Voigt convolution of Gaussian and Lorentzian, and c is the duty cycle. It is regarded that the first term plays an important role in retrieving the spectral information as $c \propto f_r \tau \ll 1$. The latter is basically the input Gaussian spectrum for pulse width much shorter than the phonon lifetime, and provides the temporal signal for high spatial resolution. Referring to such a mechanism, several methods are proposed to enhance the spatial resolution, such as the pre-pump [126], the double pulse [127], the dark pulse [128], and coded pulses [129]. These schemes enhance the spatial resolution effectively; resolutions down to centimeters have been achieved. There are other methods, which put emphases on data processing, for example, the second-order partial derivative and the differential pulse-width pair schemes [130–132]. The Brillouin scattering is also combined with other nonlinear optical effects to improve the performances of BOTDA, including the parameter amplification, Raman amplification, and the four wave mixing [133–137]. As a stimulated amplification, the polarization states must affect the SBS process. Reference [138] gives a theoretical analysis on the polarization properties of Brillouin scattering.

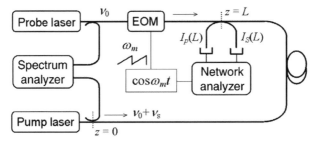

Figure 5.27   Typical BOFDA configuration.

It is needed to separate the effects of temperature and strains in sensors. In BOTDA, the signal of BOTDA experiences stimulated amplification (or loss), so that its power is affected by many factors. It is hard to use the power as a measured parameter as done in BOTDR. In practice, two parallel fibers are paved in such a way that one of them is strain relieved; or two fibers with different characteristics are used.

As the counterpart of BOTDA, the Brillouin optical frequency domain distributed sensor (BOFDA) attracts the attention of research and development also, with a similar principle to that of OFDR and frequency domain DART. Figure 5.27 shows a typical configuration, where the probe is modulated as $\cos 2\pi f_m t$, with a saw-tooth varied modulation frequency. The maximum tuning range Δf_m determines the spatial resolution: $\delta z = v_g / 2\Delta f_m$; and the working distance depends on the tuning step: $L = v_g / 2\delta f_m$. To give examples, 20 km fiber length requires frequency step of 5 kHz; and 1 m spatial resolution requires frequency-tuning range of 100 MHz.

In the BOFDA system, evolutions of the pump beam and the probe beam (Stokes component) are described [139] by equations of

$$\left(\frac{n}{c}\frac{\partial}{\partial t} + \frac{\partial}{\partial z}\right)I_p = (-\alpha - g_B I_S)I_p, \tag{5.59a}$$

$$\left(\frac{n}{c}\frac{\partial}{\partial t} - \frac{\partial}{\partial z}\right)I_S = (-\alpha + g_B I_p)I_S, \tag{5.59b}$$

with initial conditions of $I_p(0, t) = I_{p0}$ and $I_S(L, t) = I_{S0}(1 + \cos \omega_m t)$. The latter is for an ideal 100% modulated probe. The equations are solved approximately under small pump intensity:

$$I_S(z, t) = I_{S0}e^{\alpha_1(z-L)}\{1 + \cos[\omega_m t + k_m(z - L)]\}, \tag{5.60}$$

with $\alpha_1 = \alpha - g_B I_{p0}$ and $k_m = n\omega_m/c$. Then the change of pump intensity $I_p(z, t)$ caused by SBS is obtained by solving (5.59a) with (5.60) substituted. At the end of the fiber $I_p(L, t)$ and $I_S(L, t)$ are detected, and correlated in the network analyzer. The signals are processed by FFT to retrieve variations of the SBS effect along the fiber.

The effective measurement time for OTDA is limited by its duty cycle τ/T_r, where τ is the pulse width, T_r is the repetition period. In contrast, the whole period can be utilized in OFDA, and more averaging time is available for the same total measurement time, resulting in higher SNR. Apart from the linear scanning of modulation frequency, a composite modulation spectrum can be applied on the probe, which provides a capability of parallel operation and multiple folded measurements. On the other hand, the OFDA operation needs multiple data collection for different frequencies, whereas a single shot is needed for OTDA in principle. It means a longer measurement time for OFDA, and a stable state of the sensor fiber is required.

Based on the same mechanism, another scheme [140,141] uses sinusoidal modulated pump beam and probe beam, and interrogates the signal by their correlation, named Brillouin optical correlation domain analysis (BOCDA). Both the optical frequency modulation (FM) and the optical intensity modulation (IM) are used to scan the Brillouin gain spectrum; the spectrum of correlation signal contains information on fiber loss, strain, and temperature change.

Compared with OTDR, OTDA needs two beams launched into two ends of the fiber. To overcome this shortage, the Fresnel reflection from the far end of the fiber is utilized, as demonstrated in [142,143]. In the system, a short pulse train (pump) and a RF modulated pulse train (probe) are composed into one train; SBS occurs when the pump pulse meets the reflected probe pulse, and the probe meets the reflected pump as well. At the input end of the fiber, the reflected pulses are detected and analyzed. Many new techniques and schemes are being developed in the field of distributed fiber sensors based on Brillouin scattering.

5.5   DISTRIBUTED SENSORS BASED ON FIBER INTERFEROMETERS

The fiber interferometer is widely used for high-precision measurements and sensor technologies, including distributed and quasi-distributed sensor systems. Section 5.5.1 introduces sensor systems based on Mach–Zehnder interferometers (MZI), Michelson

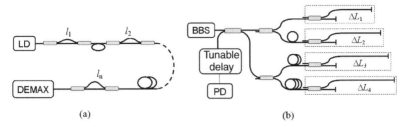

Figure 5.28   Quasi-distributed sensor systems based on (a) MZI and (b) MI.

interferometers (MI), Fabry–Perot interferometers (FPI), and Sagnac interferometers (SI), and discusses related technical problems. Section 5.5.2 is concentrated on the low coherence technology (LCT), which is a helpful method of signal localization. Section 5.5.3 introduces sensors based on multimode interference, especially the optical speckles.

5.5.1   Configuration and Characteristics of Interferometric Fiber Sensors

5.5.1.1   Structure of Quasi-Distributed Sensors and Signal Interrogation
A quasi-distributed sensor system is composed of multiple fiber optic interferometers, such as MZIs and FPIs, by connecting them with fiber sections, in series or parallel fashion, or a combination of both. The system usually covers a large area, while using a common source and common detection devices. The detected parameter in interferometers is the phase shift of

$$\delta\phi = \beta\delta L + L\delta\beta = \beta\delta L + k_0 L \delta n_{\text{eff}}. \tag{5.61}$$

The variation of physical conditions, including temperature and strain, can then be detected. Figure 5.28 shows two examples of quasi-distributed sensor systems, based on MZI [144,145] and MI [146,147], respectively.

A basic technical issue in the system is how to interrogate the signals, that is, separate each sensor's signal from the data stream. Reference [144] proposed multipoint unbalanced MZI systems with a short coherence length source, interrogated by matched MZI on the receiver end. Similar structures were proposed with a frequency-modulated source to scan the multiple MZI with different imbalances [145]. MIs are usually connected in parallel, as shown in Figure 5.28(b). A broadband source (BBS) with the coherence length shorter than the optical path differences (OPD) of MI is used; different sensors are distinguished in detection by a tunable delay device [147].

To shrink the size of fiber MZI and MI, more compact materials and designs are proposed and demonstrated. The fiber gratings are widely used in the quasi-distributed sensor systems, as introduced in Chapter 4. A fiber MI made of a two-core fiber is demonstrated [148]. The polarization maintaining fiber is used to build sensors based on polarization interference, as shown in [149,150] for example.

Several technical problems have to be solved before the fiber interferometric sensors are used in practical applications. One of them is to reduce and remove the impact of stray disturbances, since the fiber interferometer is very sensitive to the environmental conditions. Apart from careful and thoughtful packaging and installation of the sensor, reference [146] demonstrates a method of actively adjusting the OPD of MI with a piezo-electric transducer to eliminate the slowly varying disturbance and pick up the detected acoustic signals.

The fiber sensitivity to external disturbance causes not only the optical phase variation, but also polarization change, resulting in polarization direction deviation between the sensing wave and the reference wave. If the interference signal is expressed as $I = I_0(1 + V \cos \varphi)$ with φ to be detected, the visibility, or modulation depth, is expressed as [151,152]

$$V = \frac{I_{\max} - I_{\min}}{I_{\max} + I_{\min}} = \frac{2\sqrt{\alpha_s \alpha_r \kappa_1 \kappa_2 (1 - \kappa_1)(1 - \kappa_2)}}{\alpha_r \kappa_1 \kappa_2 + \alpha_s (1 - \kappa_1)(1 - \kappa_2)} \gamma(\tau) \cos \eta, \qquad (5.62)$$

where α_s and α_r are the losses in signal and reference paths; κ_1 and κ_2 are the beam split ratios of the two couplers; γ is the coherence factor, being a function of coherence time τ; η is the polarization deviation between signal and reference. It is rewritten as $V = V_0 \cos \eta$, where V_0 is generally insusceptible to external disturbances for a built interferometer. Ideally, for $\alpha_r = \alpha_s$, $\kappa_1 = \kappa_2 = 0.5$, and for a high coherence source used, $V_0 = 1$ is obtained. It is seen that the polarization deviation brings direct impact on the interference signal. In the worst case, the signal vanishes at $\eta = \pi/2$. The effect is termed polarization fading. Furthermore, it fluctuates randomly, inducing polarization noise. It is more critical for a distributed system with remote sensors installed in a large area.

Several methods are proposed to solve the effect. A direct compensation method is to actively adjust the polarization state by a polarization controller. The method relies on the detection of the polarization deviation and a feedback circuit, leading to higher cost, especially for multisensor systems. Another effective method is to build the interferometer with the polarization maintaining fiber, which is much less sensitive to the external disturbance than conventional fiber. On the other hand, a light source with polarization scrambled and with its coherence kept high is beneficial to smoothen the polarization effect; though the interferometer's sensitivity may decrease somewhat, but the stability is enhanced greatly.

Faraday rotation mirrors (FRMs) are widely used in fiber MIs, as introduced in Section 3.4. By using FRM, the light wave experiences a round trip way with orthogonal polarization changes, so that the externally induced disturbance is compensated. However, polarization-dependent components, such as polarization-dependent fiber couplers, will degrade FRM's effectiveness, as analyzed in [153].

As discussed in Section 3.3, interferometers composed by 2×2 couplers have inherent shortcomings, that is, the null sensitivity at points with phase of $m\pi$, whereas interferometers with 3×3 couplers can overcome the problem. It is shown in reference [154] that $N \times N$ topology interferometers allow for instantaneous extraction of magnitude and phase information, and the Doppler frequency shift consequently. Detailed analyses on optical fiber sensor technologies are given in [155].

5.5.1.2 Distributed Sensor Loops Based on the Sagnac Interferometer

The SI possesses a unique feature: reciprocal phase variations in the loop are cancelled out in its output, which makes it insusceptible to external disturbances and suitable for gyros and for magnetic sensors. It is noted that the cancelled effect are static factors. The phase factors of CW and CCW paths, induced by temporarily varied disturbances, are not necessarily cancelled out, especially for a long loop. Such a property is utilized as a distributed sensor.

Figure 5.29 illustrates the nonreciprocity in time domain [156]. If dynamic disturbances, such as vibrations or impulsive forces, occur at some point of the Sagnac loop, the induced dynamic phase changes propagate in CW and CCW ways to the couplers with phase difference expressed as

$$\Delta\Phi = \Phi_{CW} - \Phi_{CCW} = \Delta\Phi_0 + \varphi(t - \tau_1) - \varphi(t - \tau_2), \qquad (5.63)$$

where $\Delta\Phi_0$ is the background nonreciprocal phase, $\tau_{1,2} = L_{1,2}/v_g$ is the propagation time in CW and CCW ways. For a sustained vibration, $\varphi(t) = \varphi_0 \sin \omega_s t$, the phase factor is deduced as

$$\Delta\Phi \propto \varphi_0 \sin \omega_s \Delta\tau \cos[\omega_s(t - \bar{\tau})], \qquad (5.64)$$

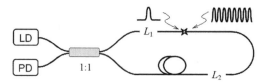

Figure 5.29   Nonreciprocity of purturbation occuring in Sagnac loop.

where $\bar{\tau} = (\tau_1 + \tau_2)/2$ and $\Delta\tau = (\tau_1 - \tau_2)/2$. It is shown that the phase difference is not cancelled out unless $\Delta\tau = 0$, that is, $L_2 = L_1$, or $\omega_s \Delta\tau = m\pi$. The vibration is then detected by the interference signal $I \propto \cos\Delta\Phi$. For an impulsive disturbance with $\varphi(t) = \varphi_0 e^{-t/T}$ assumed, the interference phase is

$$\Delta\Phi \propto [e^{-(t-\tau_1)/T} - e^{-(t-\tau_2)/T}], \tag{5.65}$$

meaning that two sequent signals will be detected with time interval of $(\tau_1 - \tau_2)$, if the sustained time of disturbance is short enough. Generally, (5.63) can be rewritten as [157]

$$\Delta\Phi \approx \frac{n}{c}(L_2 - L_1)\frac{d\varphi}{dt} = \frac{2nz}{c}\frac{d\varphi}{dt}, \tag{5.66}$$

where z is the distance of the disturbance from the midpoint of the fiber loop, $d\varphi/dt$ is the phase change rate induced by the disturbance to be detected; mostly it is attributed to the strain e induced by external forces:

$$\frac{d\varphi}{dt} = \frac{2\pi}{\lambda}\left(\frac{\partial n}{\partial e}l + \frac{\partial l}{\partial e}n\right)\frac{\partial e}{\partial t}. \tag{5.66a}$$

Expressions (5.64-5.66) provide a possibility of localizing the disturbance site. But a duality with $\pm\Delta\tau$ in the interference signal of $\propto \cos\Delta\Phi$ has to be removed; or otherwise only one half of the loop can be taken as the sensor fiber, leaving the other half idle. To localize the disturbance site, many schemes were proposed [156–161]. Reference [156] utilized Fourier analysis to determine the position of disturbance, based on the fact that a real disturbance is not a single frequency vibration, but with different frequency components.

Another idea is to build a composite Sagnac loop with a common fiber section for the distributed sensor and additional loops with different lengths or a variable length. Figure 5.30 shows two examples. In Figure 5.30(a), four Sagnac loops are contained with whole lengths of L, $L + l_1$, $L + l_2$, and $L + l_1 + l_2$. In Figure 5.30(b), two loops with lengths

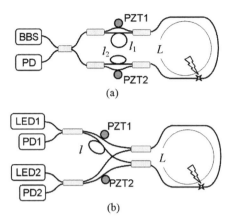

Figure 5.30   Composite Sagnac loop distributed sensors.

of L and $L + l$ are formed, incorporated with two sources and two detectors.

To distinguish the different loops and different sources, optical phase modulators driven by PZT are inserted in the loop, as shown in the figure, so that by spectrum analyzer the detected signal can be decomposed spectral components to identify the different paths and to determine the disturbance site.

The SI composed of high birefringence fiber is often used, as shown schematically in Figure 5.31 [162]. Two polarization modes forms two Sagnac loops with time delay of $\delta\tau = (n_x - n_y)(L_1 - L_2)/c$. To enhance the spatial resolution, the FMCW technique is also used, similar to that of OFDR in Section 5.2.

5.5.2   Low Coherence Technology in a Distributed Sensor System

5.5.2.1   *Principle of LCT*   As well known, low coherence light waves can interfere only in a short range, where two waves have passed equal optical paths. This property has been developed as a special method, called the LCT, or the white light interferometer (WLI). The technology is used to localize the position of measured parameters. It is also

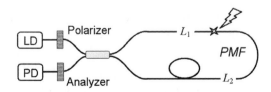

Figure 5.31   Sagnac interferometer composed of PMF.

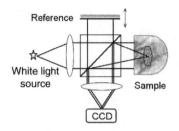

Figure 5.32   Typical OCT setup.

used to pick up the reflection and scattering from an object buried inside a diffusive medium, to have its image retrieved, called optical coherence tomography (OCT) [163]. A typical setup of OCT is an MI, as shown in Figure 5.32.

The sample and the surrounding medium are characterized by their refractive index, absorption coefficient, and scattering coefficient; all of these parameters are functions of poisson z and wavelength λ: $n(k, z)$, $\alpha(k, z)$, and $\sigma(k, z)$. In the interferometer, the two beams scattered from the sample and reflected from the reference mirror back to the beam splitter (BS) are written respectively as

$$E_1 = t_1 \sigma(z) E_0 \exp\left[-\int_{z_1}^{z} \alpha(s) ds\right] e^{i2n(z)k(z-z_1)+i2k(z_1-z_0)}, \quad (5.67a)$$

$$E_2 = t_2 r\, E_0 e^{i2k(z_2-z_0)}, \quad (5.67b)$$

where z_0 is the position of BS, z_1 is the surface position of the diffusive medium, t_1 and t_2 are the field split ratio of BS, and r is the reflectivity of the reference mirror. The scattering coefficient is assumed with a phase factor as $\sigma = |\sigma| \exp i\phi_\sigma$. The interference intensity is expressed as

$$I = |E_1 + E_2|^2 = E_0^2 [t_1^2 |\sigma|^2 A^2 + t_2^2 r^2 + 2t_1 t_2 r |\sigma| A \cos \Phi], \quad (5.68)$$

where $\Phi = 2k[n(z - z_1) + (z_1 - z_2)] + \phi_\sigma = k\Delta L$ is the phase shift with the equivalent optical path difference ΔL; $A = \exp[-2\int_{z_1}^{z} \alpha(s) ds]$ is the round trip attenuation. This expression is for a single frequency optical wave. If a broadband source is used, its spectral properties must be taken into account. Assuming the spectral profile is Gaussian, written as

$$g(k) = (g_0/\delta k) \exp[-(k - k_0)^2/(\delta k)^2], \quad (5.69)$$

with $1/e$ line width of $\delta\omega = c\delta k$, the measured optical signal is then obtained to be

$$\bar{I} = I_0[t_1^2 |\sigma|^2 A^2 + t_2^2 r^2 + 2t_1 t_2 r |\sigma| A \cos(k_0 \Delta L) e^{-(\delta k \Delta L)^2}], \qquad (5.70)$$

where the spectral dependences of n, α, and σ are neglected for simplicity. It is indicated that the interference signal reaches the maximum when $\Delta L = 0$; the signal declines with ΔL increasing in a rate dependent on the linewidth δk: the larger the line width, the faster the signal declines with ΔL. The spatial resolution is thus determined by the coherence length of the source; for example, if a light source at 1,300 nm range with line width of 20 nm is used, the spatial resolution is near $\delta l = \lambda^2/(2\delta\lambda) \approx 40\mu m$.

By varying the reference mirror's position and detecting the interference maximum, the scattering and/or reflection of the sample $\sigma(z)$ can be measured layer by layer. If the focus scans transversely on the sample by a certain device, the sample's image in the three-dimensional space can be acquired. The influence of the surrounding diffusive medium to the visibility is diminished greatly, because the interference between the wave reflected from the mirror and the scattering wave by medium gives very weak intensity, if the optical path difference is beyond the coherence length. The OCT technology has been used widely in biological research and medicine, and in optical waveguide characterization as well [164,165].

5.5.2.2 Applications of LCT in Distributed Sensing
By using its localization function, LCT is useful for interrogation of distributed fiber sensor systems. References [166–168] propose and demonstrate its effectiveness, where the fiber sensor unit is basically a butt connection of fiber facets. Sensor units are distributed along the fiber and connected in a loop or as one of the arms of an MI, as shown in Figure 5.33. The reflected beams from each facet will interfere with the beam reflected from the reference arm at the fiber coupler; it is thus sensitive to

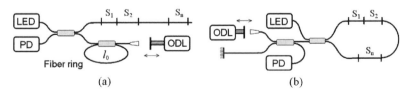

Figure 5.33 Fiber sensor systems with LCT interrogation in configurations of (a) MI and (b) SI.

variations of the fiber section lengths, induced by strain or temperature change, for example. The sensor system can be regarded as a series of low finesse Fabry–Perot cavities with resonances at different wavelengths.

In the system, a broadband light emitting diode (LED) is used as a low coherence source. Multiple reflected waves interfere with the optical wave reflected from the optical delay line (ODL). Only the two beams with equal optical path give high interference intensity. Two interrogation configurations are shown in the figure: (a) an adjustable ODL is set in the reference arm of the MI and (b) an MI is built at the receiving port of the Sagnac loop. By adjusting ODL, the sensor location corresponding to the detected interference signal can be identified. To expand the delay range, a fiber ring is inserted, as shown in Figure 5.33(a), in which the multiple recurring roundtrips, combined with the ODL provide a wide usable range of optical path. The decayed amplitude with the multiple recurring causes SNR decrease, but its localization function maintains in a certain range.

In the configuration with the Sagnac loop, light waves propagating in CW and CCW ways reflected from different sensor elements interfere at the coupler. The corresponding OPD for different sensor units are different from each other, making their localization possible. Another interrogation configuration is a Mach–Zehnder interferometer with an adjustable ODL in one of its beams [167]. Since multiple interferences exist, the received signals are functions of complicated factors; detailed analyses and data processing are necessary to interrogate the signals.

The LCT is also useful in fiber grating characterization and FBG sensors. References [169,170] use an MI with a long broadband fiber grating, such as the chirped FBG (CFBG), inserted in one of its arms, and with an adjustable optical delay line connected in the reference arm, and a broadband LED was used as the low coherence source, as shown in Figure 5.34. The set-up can be used to measure the Bragg wavelength variation along the CFBG. The adjustable ODL used in

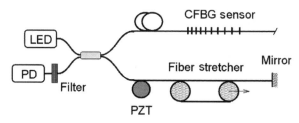

Figure 5.34   CFBG sensor and its interrogation.

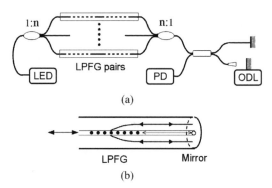

(a)

(b)

Figure 5.35 (a) Low-coherent interrogation scheme for multiplexed sensors based on LPFG pairs and (b) MI structure composed of LPFG and fiber facet.

the experiment is a fiber stretcher with a long fiber wrapped several circles on two translating posts; and a PZT phase modulator is inserted to make the interference signal modulated and to suppress the DC background.

If the CFBG in the above configuration is replaced by a long period fiber-grating sensor, the LCT is also beneficial for the interrogation. References [171,172] present systems of multiplex LPFG pair sensors with low coherence sources and a Michelson interferometric interrogation, as shown in Figure 5.35(a). The LPFG pair is actually a Mach–Zehnder interferometer with fiber core and cladding as its two arms, as analyzed in Section 4.3.5 (see Figure 4.43). The spectrum of LPFG pair shows multiple interference peaks with the interval determined by the spacing between the two LPFG. The peak interval can be taken as the signature of each LPFG pair, and corresponding signal is picked up based on LCT by using a scanned ODL in the MI. Figure 5.35(b) shows another sensor with an LPFG as the beam splitter and a highly reflective fiber facet to form a Michelson structure [173]. Similar to the LPFG pair, it is used to sense changes of length between grating and mirror, the grating period change, and the index change induced by strain and temperature variation. The interrogation based on LCT can distinguish the sensed signal from different sensors.

5.5.3 Sensors Based on Speckle Effect and Mode Coupling in Multimode Fiber

When a laser beam illuminates a piece of white paper (or any object with diffused reflection), speckles are often observed. It is attributed to random fluctuation of medium index and the phase fluctuation caused

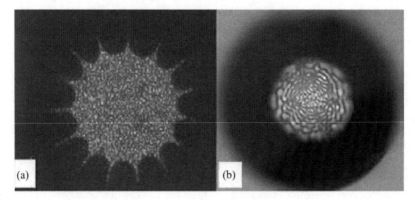

Figure 5.36   Speckle patterns observed in (a) a grapefruit fiber and (b) an MMF. (Reprinted with permission from reference [175].)

by rough reflective surface [174]. Speckles are also observed in the near field at MMF output facets. Figure 5.36 shows speckle patterns observed at the near field of a grapefruit fiber and a conventional MMF with a He–Ne laser [175]. It is found further that the pattern of speckles changes with the fiber deformation. This phenomenon is used as a distributed sensor, called fiber specklegram sensor. Reference [176] developes a statistical model to analyze the specklegram sensor and demonstrates signal readout methods by a spatial filter. References [177,178] use a charge-coupled device (CCD) to record the speckle pattern and to calculate its inner product and mean intensity variation in their speckle sensors. CCD can provide higher sensitivity and adaptability than ordinary detection by a single detector.

A typical configuration of a specklegram sensor is shown in Figure 5.37, where the laser beam is launched into a MMF through the SMF pigtail, and the output end of MMF is picked up with another SMF. Such a structure is usually called the SMS structure. It is found experimentally that the speckle pattern has a spatially statistical distribution with some characteristic size. A higher sensitivity is obtained when the receiving aperture is matching with the characteristic size. Therefore, a small gap is set and adjusted between the MMF and the SMF2 to enhance the sensitivity to the speckle pattern variation.

The mechanism of speckle in MMF was attributed to the interference of the transverse modes [179]. However, a different argument was presented in [180]: usually the modes propagate adiabatically without intermixing. Reference [181] found that the sensitivity in the sensor with a multiple longitudinal mode laser as its source is much higher

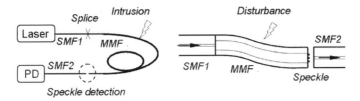

Figure 5.37   Schematic diagram of a fiber specklegram sensor.

than that with a single longitudinal-mode laser. It was also found that the spot size of MMF output keeps nearly the same as that at the facet of SMF1, even if the MMF has a length up to 200 m, and is wound onto a drum with diameter of 150 mm or smaller, as shown in Figure 5.38(a). In comparison, when a gap between SMF1 and MMF exists, the spot is seen to fill the core, as shown in Figure 5.38(b).

The observations indicate that the interference between longitudinal modes plays an important role in specklegram sensors; and the coupling between the transverse modes in the fiber is very weak; the fundamental mode in SMF1 does not excite many modes in MMF when it is launched at the center of MMF. The property coincides with the result of modal delay measurement in Reference [182], demonstrating that the pulse of different modes propagate independently without coupling with each other.

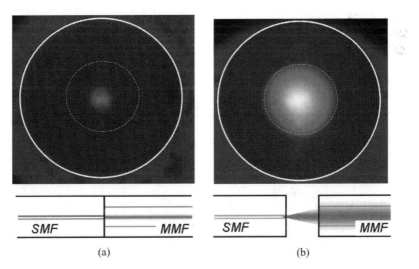

Figure 5.38   Near-field patterns of a long MMF by two input ways: (a) tight connection and (b) loose connection with a small gap of ~1 mm. (Reprinted with permission from reference [181].)

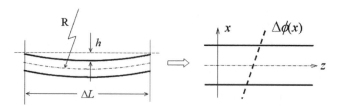

Figure 5.39 Illustration of additive phase caused by bending. (Reprinted with permission from reference [181].)

Based on the above experimental results, the mechanism of a specklegram is described as follows. Since the inter-modal coupling is negligibly small, the output at the end of the MMF can be expressed as

$$E_{\text{out}} = \sum_l (a_l + \Delta a_l) \exp[j\delta_l(r, \theta)][g_{0l}(r, \theta)e^{j\beta_{0l}L}]$$
$$\approx \sum_l a_l \exp[j\delta_l(r, \theta)][g_{0l}(r, \theta)e^{j\beta_{0l}L}]. \tag{5.71}$$

The modes with $m > 0$ are omitted due to the orthogonality between modes, leaving only $(0, l)$ mode of the MMF in the expansion, denoted by g_{0l}. a_l is the expansion coefficient of the input LP_{01} mode of the SMF, Δa_l is the externally induced amplitude change in the L-long fiber, which can be neglected since the coupling between modes are weak. $\delta_l(r, \theta)$ is the phase variation, which is believed to give main contributions to the speckles. As a typical case, the phase variation induced by bending is considered, as shown in Figure 5.39, where an equivalent phase front change is caused by bending:

$$\delta_l(x, y) = \beta_{0l}\frac{\Delta L}{R}x = \beta_{0l}\frac{8hx}{\Delta L}. \tag{5.72}$$

Taking the multiple bending along the fiber into account, the additive phase is expressed as

$$\delta_l(x, y) = \beta_{0l} \sum_i \frac{\Delta L_i}{R_i}(x \cos \varphi_i + y \sin \varphi_i), \tag{5.73}$$

where φ_i stands for the bending direction. Thus the phase distribution in the transverse cross section induced by bending, and other deformations, generates the speckle pattern. When a SMF is used to pick up a part of the near field, the output intensity is expressed as

$$I \propto \sum_{l,l'} a_l^* a_{l'} \int_{\text{SMF}} g_{0l}^* g_{0l'} \cos\left[(\beta_l - \beta_{l'})\frac{x \Delta L}{R}\right] dx dy. \qquad (5.74)$$

Different wavelengths correspond to different phases of (5.73); beating of two longitudinal modes, λ_p and λ_q, gives a signal proportional to the phase factor of

$$\delta_l(\lambda_p) - \delta_l(\lambda_q) = 2\pi n_{0l}\left(\frac{1}{\lambda_p} - \frac{1}{\lambda_q}\right)\frac{x \Delta L}{R} \approx \beta_{0l}\frac{x \Delta L \Delta \lambda}{\lambda R}. \qquad (5.75)$$

The output signal picked up by the SMF is then written as

$$I \propto \sum_{l,p,q} a_l^*(\lambda_p)a_l(\lambda_q) \int_{\text{SMF}} |g_{0l}|^2 \cos\left(\beta_{0l}\frac{x \Delta L \Delta \lambda}{\lambda R}\right) dx dy. \qquad (5.76)$$

The effects of longitudinal mode beating and transverse mode interference are compared by two factors: one is the mode field product $|g_{0l}|^2$ versus ($g_{0l}^* g_{0l'}$); the other is the phase factor: $\beta_{0l} x \Delta \lambda / \lambda$ versus ($\beta_l - \beta_{l'})x$. Both ratios are much larger than unity. The experimental results of higher sensitivity with multiple longitudinal mode sources are thus explained by the model.

Taking the temperature dependence of index and fiber length into consideration, the SMS structure is also sensitive to temperature variation. Reference [183] gives demonstrations of the SMS structure with gradient index MMF as a temperature and strain sensor. Transmission characteristics of the SMS structure in a broad wavelength range was studied in [184], showing that it can also be used as a spectral filter. Instead of MMF, few-mode fibers, mostly two-mode fibers, are also used to build a SMS structure sensor with similar characteristics and applications [185–187].

PROBLEMS

5.1 What is Rayleigh scattering? How do you explain the color of the sky in sunny weather?

5.2 What is Raman scattering and Brillouin scattering? Why are they called inelastic scattering? What kind of transitions are they related to? How are the scatterings related to the phonons?

5.3 What are the fundamental conservation conditions of the scatterings? Describe the angular dependence of Brillouin scattering. Why does the forward Brillouin scattering not occur?

5.4 What is the difference between Stokes component and anti-Stokes component of the two inelastic scatterings?

5.5 Describe the basic characteristics of the three scatterings. What factors determine the frequency shift of Brillouin scattering? What factors determine the bandwidth of Brillouin scattering? How much is it in silica fiber?

5.6 What are the stimulated Raman and Brillouin scatterings? What are the higher order Raman and Brillouin scatterings? Describe the propagation direction of the higher order scatterings.

5.7 Describe the basic concept of OTDR; describe the waveform of OTDR signal. What characteristics will show for the fiber intrinsic loss, for connection loss, and for break down of fiber?

5.8 What parameters determine the spatial resolution of OTDR? What parameters determine the fiber span of OTDR and sensitivity to the disturbances?

5.9 Describe the features of POTDR and COTDR, the differences between them and the ordinary OTDR.

5.10 How does OFDR work? How does it localize the position of disturbance?

5.11 What characteristics of Raman scattering can be used to measure the temperature? Write the basic equations of the temperature dependence.

5.12 Describe the basic DART configuration. What are the differences between DART and OTDR systems?

5.13 What are the physical mechanisms of strain dependence and temperature dependence of Brillouin scattering characteristics?

5.14 By what characteristics of Brillouin scattering can the strain signal and the temperature signal be demodulated?

5.15 What is the difference between the detection method of BOTDR and ordinary OTDR? How do you interrogate the signals in BOTDR? Why is an LO needed?

5.16 What is the limit of spatial resolution in BOTDR?

5.17 Describe the principles of BOTDA. What are the main differences between BOTDR and BOTDA? What advantages and disadvantages do they have, why?

5.18 Describe the systems of BOTDA based on Brillouin gain and on Brillouin loss. What features do they have?

5.19 How do you interrogate the signal in BOTDA? Why can the limit of spatial resolution of BOTDR be broke though in BOTDA?

5.20 What is the temporal nonreciprocity of the Sagnac loop? How do you make use of it to localize the external disturbance in an intrusion sensor based on a Sagnac loop?

5.21 What is the LCT? How do you make use of it to establish a system of OCT? What parameters determine the spatial resolution of LCT?

5.22 What effects cause fading of interferometric sensor signals? How do you avoid or remove the effects?

5.23 What is the fiber speckle? How does external disturbance affect the speckle? Describe the mode coupling between the fundamental mode of an SMF and the modes of an MMF.

REFERENCES

1. Kapron FP, Teter MP, Maurer RD. Theory of backscattering effects in waveguides. *Applied Optics* 1972; 11: 1352–1356.
2. Smith RG. Optical power handling capacity of low loss optical fibers as determined by stimulated Raman and Brillouin-scattering. *Applied Optics* 1972; 11: 2489–2494.
3. Lines ME. Scattering losses in optic fiber materials. 1. A new parameterization. *Journal of Applied Physics* 1984; 55: 4052–4057.
4. Lines ME. Scattering losses in optic fiber materials. 2. Numerical estimates. *Journal of Applied Physics* 1984; 55: 4058–4063.
5. Barnoski MK, Jensen SM. Fiber waveguides – novel technique for investigating attenuation characteristics. *Applied Optics* 1976; 15: 2112–2115.
6. Barnoski MK, Rourke MD, Jensen SM, Melville RT. Optical time domain reflectometer. *Applied Optics* 1977; 16: 2375–2379.
7. Personick SD. Photon probe – optical fiber time domain reflectometer. *Bell System Technical Journal* 1977; 56: 355–366.

8. Born M, Wolf E. *Principles of Optics*. Seventh Edition. Cambridge: Cambridge University Press, 1999.

9. Bohren CF, Huffmann DR. *Absorption and Scattering of Light by Small Particles*. Hoboken: Wiley-Interscience, 1983.

10. Agrawal GP. *Nonlinear Fiber Optics*. Singapore: Elsevier Science, 2004.

11. Kittel C. *Introduction to Solid State Physics*. Sixth Edition. Hoboken: John Wiley & Sons, Inc., 1986.

12. Shen YR. *The Principles of Nonlinear Optics*. Hoboken: John Wiley & Sons, Inc., 2003.

13. Brinkmeyer E. Analysis of the backscattering method for single-mode optical fibers. *Journal of the Optical Society of America* 1980; 70: 1010–1012.

14. Aoyama KI, Nakagawa K, Itoh T. Optical time domain reflectometry in a single-mode fiber. *IEEE Journal of Quantum Electronics* 1981; 17: 862–868.

15. Philen DL, White IA, Kuhl JF, Mettler SC. Single-mode fiber OTDR - experiment and theory. *IEEE Journal of Quantum Electronics* 1982; 18: 1499–1508.

16. Mickelson AR, Eriksrud M. Theory of the backscattering process in multimode optical fibers. *Applied Optics* 1982; 21: 1898–1909.

17. Hartog AH, Gold MP. On the theory of backscattering in single-mode optical fibers. *Journal of Lightwave Technology* 1984; 2: 76–82.

18. Healey P. Review of long wavelength single-mode optical fiber reflectometry techniques. *Journal of Lightwave Technology* 1985; 3: 876–886.

19. So VCY, Jiang JW, Cargill JA, Vella PJ. Automation of an optical time domain reflectometer to measure loss and return loss. *Journal of Lightwave Technology* 1990; 8: 1078–1083.

20. Bethea CG, Levine BF, Cova S, Ripamonti G. High-resolution and high-sensitivity optical-time-domain reflectometer. *Optics Letters* 1988; 13: 233–235.

21. Rogers A. Distributed optical-fiber sensing. *Measurement Science and Technology* 1999; 10: R75–R99.

22. Healey P, Hensel P. Optical time domain reflectometry by photon-counting. *Electronics Letters* 1980; 16: 631–633.

23. Stierlin R, Ricka J, Zysset B, Battig R, Weber HP, Binkert T, Borer WJ. Distributed fiberoptic temperature sensor using single photon-counting detection. *Applied Optics* 1987; 26: 1368–1370.

24. Legre M, Thew R, Zbinden H, Gisin N. High resolution optical time domain reflectometer based on 1.55 μm up-conversion photon-counting module. *Optics Express* 2007; 15: 8237–8242.

25. Healey P. Instrumentation principles for optical time domain reflectometry. *Journal of Physics E-Scientific Instruments* 1986; 19: 334–341.

26. Rogers AJ. Polarization optical time domain reflectometry. *Electronics Letters* 1980; 16: 489–490.

27. Kim BY, Choi SS. Backscattering measurement of bending-induced birefringence in single-mode fibers. *Electronics Letters* 1981; 17: 193–194.

28. Rogers AJ. Polarization-optical time domain reflectometry - a technique for the measurement of field distributions. *Applied Optics* 1981; 20: 1060–1074.

29. Vandeventer MO. Polarization properties of Rayleigh backscattering in single-mode fibers. *Journal of Lightwave Technology* 1993; 11: 1895–1899.

30. Poole CD, Wagner RE. Phenomenological approach to polarization dispersion in long single-mode fibers. *Electronics Letters* 1986; 22: 1029–1030.

31. Wuilpart M, Rogers AJ, Megret P, Blondel M. Fully-distributed polarization properties of an optical fiber using the backscattering technique. *Proceeding of SPIE* 2000: 4087: 396–404.

32. Wuilpart M, Megret P, Blondel M, Rogers AJ, Defosse Y. Measurement of the spatial distribution of birefringence in optical fibers. *IEEE Photonics Technology Letters* 2001; 13: 836–838.

33. Nakazawa M, Horiguchi T, Tokuda M, Uchida N. Measurement and analysis on polarization properties of backward Rayleigh-scattering for single-mode optical fibers. *IEEE Journal of Quantum Electronics* 1981; 17: 2326–2334.

34. Ellison JG, Siddiqui AS. A fully polarimetric optical time-domain reflectometer. *IEEE Photonics Technology Letters* 1998; 10: 246–248.

35. Corsi F, Galtarossa A, Palmieri L. Polarization mode dispersion characterization of single-mode optical fiber using backscattering technique. *Journal of Lightwave Technology* 1998; 16: 1832–1843.

36. Huttner B, Gisin B, Gisin N. Distributed PMD measurement with a polarization-OTDR in optical fibers. *Journal of Lightwave Technology* 1999; 17: 1843–1848.

37. Wuilpart M, Ravet G, Megret P, Blondel M. Polarization mode dispersion mapping in optical fibers with a polarization-OTDR. *IEEE Photonics Technology Letters* 2002; 14: 1716–1718.

38. Sunnerud H, Olsson BE, Andrekson PA. Technique for characterization of polarization mode dispersion accumulation along optical fibers. *Electronics Letters* 1998; 34: 397–398.

39. Zhang Z, Bao X. Distributed optical fiber vibration sensor based on spectrum analysis of Polarization-OTDR system. *Optics Express* 2008; 16: 10240–10247.

40. Healey P, Malyon DJ. OTDR in single-mode fiber at 1.5 μm using heterodyne detection. *Electronics Letters* 1982; 18: 862–863.

41. Healey P. Fading in heterodyne OTDR. *Electronics Letters* 1984; 20: 30–32.

42. King JP, Smith DF, Richards K, Timson P, Epworth RE, Wright S. Development of a coherent OTDR instrument. *Journal of Lightwave Technology* 1987; 5: 616–624.

43. Choi KN, Taylor HF. Spectrally stable Er-fiber laser for application in phase-sensitive optical time-domain reflectometry. *IEEE Photonics Technology Letters* 2003; 15: 386–388.

44. Izumita H, Furukawa S, Koyamada Y, Sankawa I. Fading noise-reduction in coherent OTDR. *IEEE Photonics Technology Letters* 1992; 4: 201–203.

45. Shimizu K, Horiguchi T, Koyamada Y. Characteristics and reduction of coherent fading noise in Rayleigh backscattering measurement for optical fibers and components. *Journal of Lightwave Technology* 1992; 10: 982–988.

46. Park JH, Lee WK, Taylor HF. A fiber optic intrusion sensor with the configuration of an optical time domain reflectometer using coherent interference of Rayleigh backscattering. *Optical and Fiber Optic Sensor Systems* 1998; 3555: 49–56.

47. Juskaitis R, Mamedov AM, Potapov VT, Shatalin SV. Interferometry with Rayleigh backscattering in a single-mode optical fiber. *Optics Letters* 1994; 19: 225–227.

48. Healey P. Statistics of Rayleigh backscatter from a single-mode optical fiber. *Electronics Letters* 1985; 21: 227–228.

49. Shatalin SV, Treschikov VN, Rogers AJ. Interferometric optical time domain reflectometry for distributed optical fiber sensing. *Applied Optics* 1998; 37: 5600–5604.

50. Choi KN, Juarez JC, Taylor HF. Distributed fiber-optic pressure/seismic sensor for low-cost monitoring of long perimeters. *Proceeding of SPIE* 2003; 5090: 134–141.

51. Juarez JC, Maier EW, Choi KN, Taylor HF. Distributed fiber-optic intrusion sensor system. *Journal of Lightwave Technology* 2005; 23: 2081–2087.

52. Juarez JC, Taylor HF. Polarization discrimination in a phase-sensitive optical time-domain reflectometer intrusion-sensor system. *Optics Letters* 2005; 30: 3284–3286.

53. Lu Y, Zhu T, Chen L, Bao X. Distributed vibration sensor based on coherent detection of phase-OTDR. *Journal of Lightwave Technology* 2010; 28: 3243–3249.

54. Eickhoff W, Ulrich R. Optical frequency domain reflectometry in single-mode fiber. *Applied Physics Letters* 1981; 39: 693–695.

55. Macdonald RI. Frequency domain optical reflectometer. *Applied Optics* 1981; 20: 1840–1844.

56. Giles IP, Uttam D, Culshaw B, Davies DEN. Coherent optical-fibre sensors with modulated laser sources. *Electronics Letters* 1983; 19: 14–15.

57. Kingsley SA, Davies DEN. OFDR diagnostics for fiber and integrated-optic systems. *Electronics Letters* 1985; 21: 434–435.

58. Uttam D, Culshaw B. Precision time domain reflectometry in optical fiber systems using a frequency modulated continuous wave ranging technique. *Journal of Lightwave Technology* 1985; 3: 971–977.

59. Ghafoorishiraz H, Okoshi T. Fault location in optical fibers using optical frequency domain reflectometry. *Journal of Lightwave Technology* 1986; 4: 316–322.

60. Barfuss H, Brinkmeyer E. Modified optical frequency domain reflectometry with high spatial-resolution for components of integrated optic systems. *Journal of Lightwave Technology* 1989; 7: 3–10.

61. Juskaitis R, Mamedov AM, Potapov VT, Shatalin SV. Distributed interferometric fiber sensor system. *Optics Letters* 1992; 17: 1623–1625.

62. Rathod R, Pechstedt RD, Jackson DA, Webb DJ. Distributed temperature-change sensor based on Rayleigh backscattering in an optical fiber. *Optics Letters* 1994; 19: 593–595.

63. Huttner B, Reecht J, Gisin N, Passy R, von der Weid JP. Local birefringence measurements in single-mode fibers with coherent optical frequency-domain reflectometry. *IEEE Photonics Technology Letters* 1998; 10: 1458–1460.

64. Froggatt ME, Gifford DK, Kreger S, Wolfe M, Soller BJ. Characterization of polarization-maintaining fiber using high-sensitivity optical-frequency-domain reflectometry. *Journal of Lightwave Technology* 2006; 24: 4149–4154.

65. Stolen RH, Tynes AR, Ippen EP. Raman oscillation in glass optical waveguide. *Applied Physics Letters* 1972; 20: 62–64.

66. Stolen RH, Ippen EP. Raman gain in glass optical waveguides. *Applied Physics Letters* 1973; 22: 276–278.

67. Galeener FL, Mikkelsen JC, Geils RH, Mosby WJ. Relative Raman cross-sections of vitreous SiO_2, GeO_2, B_2O_3, and P_2O_5. *Applied Physics Letters* 1978; 32: 34–36.

68. Whitbread TW, Wassef WS, Allen PM, Chu PL. Profile dependence and measurement of absolute Raman-scattering cross-section in optical fibers. *Electronics Letters* 1989; 25: 1502–1503.

69. Walrafen GE, Krishnan PN, Hardison DR. Raman investigation of the rate of OH uptake in stressed and unstressed optical fibers. *Journal of Lightwave Technology* 1984; 2: 646–649.

70. Dakin JP, Pratt DJ, Bibby GW, Ross JN. Distributed optical fiber Raman temperature sensor using a semiconductor light-source and detector. *Electronics Letters* 1985; 21: 569–570.

71. Hartog AH, Leach AP, Gold MP. Distributed temperature sensing in solid-core fibers. *Electronics Letters* 1985; 21: 1061–1062.

72. Samson PJ. Analysis of the wavelength dependence of Raman backscatter in optical fiber thermometry. *Electronics Letters* 1990; 26: 163–165.

73. Hobel M, Ricka J, Wuthrich M, Binkert T. High-resolution distributed temperature sensing with the multiphoton-timing technique. *Applied Optics* 1995; 34: 2955–2967.

74. Feced R, Farhadiroushan M, Handerek VA, Rogers AJ. A high spatial resolution distributed optical fiber sensor for high-temperature measurements. *Review of Scientific Instruments* 1997; 68: 3772–3776.

75. Kimura A, Takada E, Fujita K, Nakazawa M, Takahashi H, Ichige S. Application of a Raman distributed temperature sensor to the experimental fast reactor JOYO with correction techniques. *Measurement Science and Technology* 2001; 12: 966–973.

76. Pandian C, Kasinathan M, Sosamma S, Rao CB, Murali N, Jayakumar T, Raj B. One-dimensional temperature reconstruction for Raman distributed temperature sensor using path delay multiplexing. *Journal of Optical Society of America B* 2009; 26: 2423–2426.

77. Farahani MA, Gogolla T. Spontaneous Raman scattering in optical fibers with modulated probe light for distributed temperature Raman remote sensing. *Journal of Lightwave Technology* 1999; 17: 1379–1391.

78. Hartog A. Distributed fiber optic temperature sensors - technology and applications in the power Industry. *Power Engineering Journal* 1995; 9: 114–120.

79. Yilmaz G, Karlik SE. A distributed optical fiber sensor for temperature detection in power cables. *Sensors and Actuators A* 2006; 125: 148–155.

80. Tang CL. Saturation and spectral characteristics of Stokes emission in stimulated Brillouin process. *Journal of Applied Physics* 1966; 37: 2945–2955.

81. Ippen EP, Stolen RH. Stimulated Brillouin-scattering in optical fibers. *Applied Physics Letters* 1972; 21: 539–541.

82. Uesugi N, Ikeda M, Sasaki Y. Maximum single frequency input power in a long optical fiber determined by stimulated Brillouin scattering. *Electronics Letters* 1981; 17: 379–380.

83. Cotter D. Observation of stimulated Brillouin scattering in low-loss silica fiber at 1.3 μm. *Electronics Letters* 1982; 18: 495–496.

84. Waarts RG, Braun RP. Crosstalk due to stimulated Brillouin scattering in monomode fiber. *Electronics Letters* 1985; 21: 1114–1115.

85. Nikles M, Thevenaz L, Robert PA. Brillouin gain spectrum characterization in single-mode optical fibers. *Journal of Lightwave Technology* 1997; 15: 1842–1851.

86. Tkach RW, Chraplyvy AR, Derosier RM. Spontaneous Brillouin scattering for single-mode optical-fiber characterization. *Electronics Letters* 1986; 22: 1011–1013.

87. Shibata N, Waarts RG, Braun RP. Brillouin gain spectra for single-mode fibers having pure-silica, GeO_2-doped, and P_2O_5-doped cores. *Optics Letters* 1987; 12: 269–271.

88. Culverhouse D, Farahi F, Pannell CN, Jackson DA. Potential of stimulated Brillouin scattering as sensing mechanism for distributed temperature sensors. *Electronics Letters* 1989; 25: 913–915.

89. Culverhouse D, Farahi F, Pannell CN, Jackson DA. Stimulated Brillouin scattering - a means to realize tunable microwave generator or distributed temperature sensor. *Electronics Letters* 1989; 25: 915–916.

90. Horiguchi T, Kurashima T, Tateda M. Tensile strain dependence of Brillouin frequency shift in silica optical fibers. *IEEE Photonics Technology Letters* 1989; 1: 107–108.

91. Kurashima T, Horiguchi T, Tateda M. Thermal effects on the Brillouin frequency-shift in jacketed optical silica fibers. *Applied Optics* 1990; 29: 2219–2222.

92. Horiguchi T, Shimizu K, Kurashima T, Tateda M, Koyamada Y. Development of a distributed sensing technique using Brillouin scattering. *Journal of Lightwave Technology* 1995; 13: 1296–1302.

93. Wait PC, Newson TP. Landau Placzek ratio applied to distributed fiber sensing. *Optics Communications* 1996; 122: 141–146.

94. Schroede J, Mohr R, Macedo PB, Montrose CJ. Rayleigh and Brillouin scattering in K_2O-SiO_2 glasses. *Journal of the American Ceramic Society* 1973; 56: 510–514.

95. Parker TR, Farhadiroushan M, Handerek VA, Rogers AJ. Temperature and strain dependence of the power level and frequency of spontaneous Brillouin scattering in optical fibers. *Optics Letters* 1997; 22: 787–789.

96. Parker TR, Farhadiroushan M, Handerek VA, Rogers AJ. A fully distributed simultaneous strain and temperature sensor using spontaneous Brillouin backscatter. *IEEE Photonics Technology Letters* 1997; 9: 979–981.

97. Shimizu K, Horiguchi T, Koyamada Y, Kurashima T. Coherent self-heterodyne Brillouin OTDR for measurement of Brillouin frequency-shift distribution in optical fibers. *Journal of Lightwave Technology* 1994; 12: 730–736.

98. Kurashima T, Tateda M, Horiguchi T, Koyamada Y. Performance improvement of a combined OTDR for distributed strain and loss measurement by randomizing the reference light polarization state. *IEEE Photonics Technology Letters* 1997; 9: 360–362.

99. Ohno H, Naruse H, Kihara M, Shimada A. Industrial applications of the BOTDR optical fiber strain sensor. *Optical Fiber Technology* 2001; 7: 45–64.

100. Ohno H, Naruse H, Yasue N, Miyajima Y, Uchiyama H, Sakairi Y, Li ZX. Development of highly stable BOTDR strain sensor employing microwave heterodyne detection and tunable electric oscillator. *Proceeding of SPIE* 2001; 4596: 74–85.

101. Maughan SM, Kee HH, Newson TP. 57-km single-ended spontaneous Brillouin-based distributed fiber temperature sensor using microwave coherent detection. *Optics Letters* 2001; 26: 331–333.

102. Alahbabi MN, Cho YT, Newson TP. 100 km distributed temperature sensor based on coherent detection of spontaneous Brillouin backscatter. *Measurement Science and Technology* 2004; 15: 1544–1547.

103. Izumita H, Sato T, Tateda M, Koyamada Y. Brillouin OTDR employing optical frequency shifter using side-band generation technique with high-speed LN phase-modulator. *IEEE Photonics Technology Letters* 1996; 8: 1674–1676.

104. Geng JH, Staines S, Blake M, Jiang SB. Distributed fiber temperature and strain sensor using coherent radio-frequency detection of spontaneous Brillouin scattering. *Applied Optics* 2007; 46: 5928–5932.

105. Naruse H, Tateda M. Trade-off between the spatial and the frequency resolutions in measuring the power spectrum of the Brillouin backscattered light in an optical fiber. *Applied Optics* 1999; 38: 6516–6521.

106. Maruse H, Tateda M. Launched pulse-shape dependence of the power spectrum of the spontaneous Brillouin backscattered light in an optical fiber. *Applied Optics* 2000; 39: 6376–6384.

107. Cho SB, Kim YG, Heo JS, Lee JJ. Pulse width dependence of Brillouin frequency in single mode optical fibers. *Optics Express* 2005; 13: 9472–9479.

108. Maughan SM, Kee HH, Newson TP. Simultaneous distributed fiber temperature and strain sensor using microwave coherent detection of spontaneous Brillouin backscatter. *Measurement Science and Technology* 2001; 12: 834–842.

109. Inaudi D, Glisic B. Integration of distributed strain and temperature sensors in composite coiled tubing. *Proceeding of SPIE* 2006; 6167: 16717-1–16717-10.

110. Koyamada Y, Sakairi Y, Takeuchi N, Adachi S. Novel technique to improve spatial resolution in Brillouin optical time-domain reflectometry. *IEEE Photonics Technology Letters* 2007; 19: 1910–1912.

111. Wang F, Zhang X, Lu Y, Dou R, Bao X. Spatial resolution analysis for discrete Fourier transform-based Brillouin optical time domain reflectometry. *Measurement Science and Technology* 2009; 20: 025202-1–025202-10.

112. Horiguchi T, Tateda M. Optical-fiber-attenuation investigation using stimulated Brillouin scattering between a pulse and a continuous wave. *Optics Letters* 1989; 14: 408–410.

113. Horiguchi T, Tateda M. BOTDA - nondestructive measurement of single-mode optical fiber attenuation characteristics using Brillouin interaction: theory. *Journal of Lightwave Technology* 1989; 7: 1170–1176.

114. Horiguchi T, Kurashima T, Tateda M. A technique to measure distributed strain in optical fibers. *IEEE Photonics Technology Letters* 1990; 2: 352–354.

115. Kurashima T, Horiguchi T, Tateda M. Distributed-temperature sensing using stimulated Brillouin scattering in optical silica fibers. *Optics Letters* 1990; 15: 1038–1040.

116. Bao X, Webb DJ, Jackson DA. 32-km distributed temperature sensor based on Brillouin loss in an optical fiber. *Optics Letters* 1993; 18: 1561–1563.

117. Bao X, Dhliwayo J, Heron N, Webb DJ, Jackson DA. Experimental and theoretical studies on a distributed temperature sensor based on Brillouin scattering. *Journal of Lightwave Technology* 1995; 13: 1340–1348.

118. Bao X, Webb DJ, Jackson DA. Recent progress in distributed fiber optic sensors based upon Brillouin scattering. *Proceeding of SPIE* 1995; 2507: 175–185.

119. Kurashima T, Horiguchi T, Yoshizawa N, Tada H, Tateda M. Measurement of distributed strain due to laying and recovery of submarine optical fiber cable. *Applied Optics* 1991; 30: 334–337.

120. DeMerchant M, Brown A, Bao X, Bremner T. Structural monitoring by use of a Brillouin distributed sensor. *Applied Optics* 1999; 38: 2755–2759.

121. Bao X, Brown A, DeMerchant M, Smith J. Characterization of the Brillouin-loss spectrum of single-mode fibers by use of very short (< 10-ns) pulses. *Optics Letters* 1999; 24: 510–512.

122. Lecoeuche V, Webb DJ, Pannell CN, Jackson DA. Transient response in high-resolution Brillouin-based distributed sensing using probe pulses shorter than the acoustic relaxation time. *Optics Letters* 2000; 25: 156–158.

123. Afshar S, Ferrier GA, Bao X, Chen L. Effect of the finite extinction ratio of an electro-optic modulator on the performance of distributed probe-pump Brillouin sensor systems. *Optics Letters* 2003; 28: 1418–1420.

124. Zou L, Bao X, Wan Y, Chen L. Coherent probe-pump-based Brillouin sensor for centimeter-crack detection. *Optics Letters* 2005; 30: 370–372.

125. Kalosha VP, Ponomarev EA, Chen L, Bao X. How to obtain high spectral resolution of SBS-based distributed sensing by using nanosecond pulses. *Optics Express* 2006; 14: 2071–2078.

126. Kishida K, Li CH, Nishiguchi K. Pulse pre-pump method for cm-order spatial resolution of BOTDA. *Proceeding of SPIE* 2005; 5855: 559–562.

127. Cho SB, Lee JJ. Strain event detection using a double-pulse technique of a Brillouin scattering-based distributed optical fiber sensor. *Optics Express* 2004; 12: 4339–4346.

128. Brown AW, Colpitts BG, Brown K. Dark-pulse Brillouin optical time-domain sensor with 20-mm spatial resolution. *Journal of Lightwave Technology* 2007; 25: 381–386.

129. Soto MA, Bolognini G, Pasquale FD. Analysis of pulse modulation format in coded BOTDA sensors. *Optics Express* 2010; 18: 14878–14892.

130. Yu Q, Bao X, Ravet F, Chen L. Simple method to identify the spatial location better than the pulse length with high strain accuracy. *Optics Letters* 2005; 30: 2215–2217.

131. Li W, Bao X, Li Y, Chen L. Differential pulse-width pair BOTDA for high spatial resolution sensing. *Optics Express* 2008; 16: 21616–21625.

132. Dong Y, Bao X, Li W. Differential Brillouin gain for improving the temperature accuracy and spatial resolution in a long-distance distributed fiber sensor. *Applied Optics* 2009; 48: 4297–4301.

133. Li Y, Bao X, Dong Y, Chen L. A novel distributed Brillouin sensor based on optical differential parametric amplification. *Journal of Lightwave Technology* 2010; 28: 2621–2626.

134. Bao X, Chen L. Recent progress in optical fiber sensors based on Brillouin scattering at University of Ottawa. *Photonic Sensors* 2011; 1: 102–117.

135. Alahbabi MN, Cho YT, Newson TP. Long-range distributed temperature and strain optical fibre sensor based on the coherent detection of spontaneous Brillouin scattering with in-line Raman amplification. *Measurement Science and Technology* 2006; 17: 1082–1090.

136. Martin-Lopez S, Alcon-Camas M, Rodriguez F, Corredera P, Ania-Castanon JD, Thevenaz L, Gonzalez-Herraez M. Brillouin optical time-domain analysis assisted by second-order Raman amplification. *Optics Express* 2010; 18: 18769–18778.

137. Herraez MG, Thevenaz L, Robert P. Distributed measurement of chromatic dispersion by four-wave mixing and Brillouin optical-time-domain analysis. *Optics Letters* 2003; 28: 2210–2212.

138. Vandeventer MO, Boot AJ. Polarization properties of stimulated Brillouin scattering in single-mode fibers. *Journal of Lightwave Technology* 1994; 12: 585–590.

139. Garus D, Gogolla T, Krebber K, Schliep F. Brillouin optical-fiber frequency-domain analysis for distributed temperature and strain measurements. *Journal of Lightwave Technology* 1997; 15: 654–662.

140. Hotate K, Ong SSL. Distributed dynamic strain measurement using a correlation-based brillouin sensing system. *IEEE Photonics Technology Letters* 2003; 15: 272–274.

141. Song KY, He Z, Hotate K. Effects of intensity modulation of light source on Brillouin optical correlation domain analysis. *Journal of Lightwave Technology* 2007; 25: 1238–1246.

142. Nikles M, Thevenaz L, Robert PA. Simple distributed fiber sensor based on Brillouin gain spectrum analysis. *Optics Letters* 1996; 21: 758–760.

143. Thevenaz L, Nikles M, Fellay A, Facchini M, Robert P. Application of distributed Brillouin fiber sensing. *Proceeding of SPIE* 1998; 3407: 374–381.

144. Brooks JL, Wentworth RH, Youngquist RC, Tur M, Kim BY, Shaw HJ. Coherence multiplexing of fiber-optic interferometric sensors. *Journal of Lightwave Technology* 1985; 3: 1062–1072.

145. Blotekjaer K, Wentworth R, Shaw HJ. Choosing relative optical-path delays in series-topology interferometric sensor arrays. *Journal of Lightwave Technology* 1987; 5: 229–235.

146. Liu K, Ferguson SM, Measures RM. Fiber optic interferometric sensor for the detection of acoustic emission within composite-materials. *Optics Letters* 1990; 15: 1255–1257.

147. Zhou X, Iiyama K, Hayashi K. Detection scheme of coherence-multiplexed sensor signals using an optical loop with a frequency shifter: sensitivity enhancement. *IEEE Photonics Technology Letters* 1994; 6: 767–769.

148. Yuan L, Yang J, Liu Z, Sun J. In-fiber integrated Michelson interferometer. *Optics Letters* 2006; 31: 2692–2694.

149. Taylor RM, Webb DJ, Jones JDC, Jackson DA. Extended-range fiber polarimetric strain sensor. *Optics Letters* 1987; 12: 744–746.

150. Donlagic D, Lesic M. All-fiber quasi-distributed polarimetric temperature sensor. *Optics Express* 2006; 14: 10245–10254.

151. Wanser KH, Safar NH. Remote polarization control for fiberoptic interferometers. *Optics Letters* 1987; 12: 217–219.

152. Kersey AD, Marrone MJ, Dandridge A, Tveten AB. Optimization and stabilization of visibility in interferometric fiber-optic sensors using input-polarization control. *Journal of Lightwave Technology* 1988; 6: 1599–1609.

153. Ferreira LA, Santos JL, Farahi F. Polarization-induced noise in a fiber optic Michelson interferometer: influence of the coupler. *Proceeding of SPIE* 1994; 2360: 351–354.

154. Choma MA, Yang C, Izatt JA. Instantaneous quadrature low-coherence interferometry with 3×3 fiber-optic couplers. *Optics Letters* 2003; 28: 2162–2164.

155. Culshaw B. Optical fiber sensor technologies: opportunities and -perhaps- pitfalls. *Journal of Lightwave Technology* 2004; 22: 39–50.

156. Hoffman PR, Kuzyk MG. Position determination of an acoustic burst along a Sagnac Interferometer. *Journal of Lightwave Technology* 2004; 22: 494–498.

157. Chtcherbakov AA, Swart PL, Spammer SJ. Dual wavelength Sagnac-Michelson distributed optical fiber sensor. *Proceeding of SPIE* 1996; 2838: 301–307.

158. Udd E. Sagnac distributed sensor concept. *Proceeding of SPIE.* 1991; 1586: 46–52.

159. Russell SJ, Brady KRC, Dakin JP. Real-time location of multiple time-varying strain disturbances, acting over a 40-km fiber section, using a novel dual-Sagnac interferometer. *Journal of Lightwave Technology* 2001; 19: 205–213.

160. Fang XJ. Fiber-optic distributed sensing by a two-loop Sagnac interferometer. *Optics Letters* 1996; 21: 444–446.

161. Fang X. A variable-loop sagnac interferometer for distributed impact sensing. *Journal of Lightwave Technology* 1996; 14: 2250–2254.

162. Campbell M, Zheng G, Wallace PA, HolmesSmith AS. A distributed stress sensor based on a birefringent fiber Sagnac ring. *Proceeding of SPIE* 1996; 2838: 138–142.

163. Huang D, Swanson EA, Lin CP, Schuman JS, Stinson WG, Chang W, Hee MR, Flotte T, Gregory K, Puliafito CA, Fujimoto JG. Optical coherence tomography. *Science* 1991; 254: 1178–1181.

164. Schmitt JM, Knuttel A, Bonner RF. Measurement of optical-properties of biological tissues by low-coherence reflectometry. *Applied Optics* 1993; 32: 6032–6042.

165. Takada K, Takato N, Noda J, Uchida N. Interferometric optical-time-domain reflectometer to determine backscattering characterization of silica-based glass waveguides. *Journal of Optical Society America A* 1990; 7: 857–867. ˙

166. Yuan L, Zhou L, Jin W. Quasi-distributed strain sensing with white-light interferometry: a novel approach. *Optics Letters* 2000; 25: 1074–1076.

167. Yuan L, Zhou L, Jin W, Yang J. Low-coherence fiber-optic sensor ring network based on a Mach-Zehnder interrogator. *Optics Letters* 2002; 27: 894–896.

168. Yuan L, Jin W, Zhou L, Hoo YL, Demokan AS. Enhancement of multiplexing capability of low-coherence interferometric fiber sensor array by use of a loop topology. *Journal of Lightwave Technology* 2003; 21: 1313–1319.

169. Volanthen M, Geiger H, Cole MJ, Laming RI, Dakin JP. Low coherence technique to characterize reflectivity and time delay as a function of wavelength within a long fiber grating. *Electronics Letters* 1996; 32: 757–758.

170. Volanthen M, Geiger H, Dakin JP. Distributed grating sensors using low-coherence reflectometry. *Journal of Lightwave Technology* 1997; 15: 2076–2082.

171. Zhang AP, Guan ZG, He S. Optical low-coherence reflectometry based on long-period grating Mach-Zehnder interferometers. *Applied Optics* 2006; 45: 5733–5739.

172. Guan ZG, Zhang AP, Liao R, He S. Wavelength detection of coherence-multiplexed fiber-optic sensors based on long-period grating pairs. *IEEE Sensors Journal* 2007; 7: 36–37.

173. Liu W, Guan ZG, Liu G, Yan C, He S. Optical low-coherence reflectometry for a distributed sensor array of fiber Bragg gratings. *Sensors and Actuators A* 2008; 144: 64–68.

174. Dainty JC. *Laser Speckle and Related Phenomena.* Berlin:Springer-Verlag, 1975.

175. Wang Y, Cai H, Qu R, Fang Z, Marin E, Meunier JP. Specklegram in a grapefruit fiber and its response to external mechanical disturbance in a single-multiple-single mode fiber structure. *Applied Optics* 2008; 47: 3543–3548.

176. Spillman WB, Kline BR, Maurice LB, Fuhr PL. Statistical-mode sensor for fiber optic vibration sensing uses. *Applied Optics* 1989; 28: 3166–3176.

177. Yu FTS, Wen M, Yin S, Uang CM. Submicrometer displacement sensing using inner-product multimode fiber speckle fields. *Applied Optics* 1993; 32: 4685–4689.

178. Pan K, Uang CM, Cheng F, Yu FTS. Multimode fiber sensing by using mean-absolute speckle-intensity variation. *Applied Optics* 1994; 33: 2095–2098.

179. Yu FTS, Zhang J, Yin S, Ruffin PB. Analysis of a fiber specklegram sensor by using coupled-mode theory. *Applied Optics* 1995; 34: 3018–3023.

180. Anderson DZ, Bolshtyansky MA, Zeldovich BY. Stabilization of the speckle pattern of a multimode fiber undergoing bending. *Optics Letters* 1996; 21: 785–787.

181. Li J, Cai H, Geng J, Qu R, Fang Z. Specklegram in a multiple-mode fiber and its dependence on longitudinal modes of the laser source. *Applied Optics* 2007; 46: 3572–3578.

182. Ahn TJ, Kim DY. High-resolution differential mode delay measurement for a multimode optical fiber using a modified optical frequency domain reflectometer. *Optics Express* 2005; 13: 8256–8262.

183. Liu Y, Wei L. Low-cost high-sensitivity strain and temperature sensing using graded-index multimode fibers. *Applied Optics* 2007; 46: 2516–2519.

184. Kumar A, Varshney RK, Antony S, Sharma P. Transmission characteristics of SMS fiber optic sensor structures. *Optics Communications* 2003; 219: 215–219.

185. Vengsarkar AM, Michie WC, Jankovic L, Culshaw B, Claus RO. Fiberoptic dual-technique sensor for simultaneous measurement of strain and temperature. *Journal of Lightwave Technology* 1994; 12: 170–177.

186. Bohnert K, Dewit GC, Nehring J. Coherence-tuned interrogation of a remote elliptic-core, dual-mode fiber strain sensor. *Journal of Lightwave Technology* 1995; 13: 94–103.

187. Arbore MA, Digonnet MJF, Pantell RH. Analysis of the insertion loss and extinction ratio of two-mode fiber interferometric devices. *Optical Fiber Technology* 1996; 2: 400–407.

CHAPTER 6

FIBER SENSORS WITH SPECIAL APPLICATIONS

Among the various fiber sensors, the fiber optic gyroscope (FOG), fiber optic hydrophone, fiber Faraday sensor, and fiber sensors based on surface plasmon are worthy of special attention, not only for their special applications but also for their technical features. Since some monographs and a great number of papers have been published, this chapter gives just a brief introduction.

6.1 FIBER OPTIC GYROSCOPE

The FOG, as an excellent rotation sensor, is regarded as one of the most successful and most precise devices of fiber sensor technology. It is widely used in many important application areas, such as compasses and navigational devices for various vehicles, including automobiles, boats, airplanes, missiles, and submarines, well and tunnel logging, and latitude position sensors [1–5]. Compared with the mechanical gyros, the FOG shows high sensitivities, low cost and good robustness with no moving parts. The physical principle, theory, characteristics and related mechanisms, and fabrication techniques of the FOG are expounded in hundreds of references [6–8] and textbooks [2,9] systematically in

Fundamentals of Optical Fiber Sensors, First Edition.
Zujie Fang, Ken K. Chin, Ronghui Qu, and Haiwen Cai.
© 2012 John Wiley & Sons, Inc. Published 2012 by John Wiley & Sons, Inc.

detail. Roughly, three types of FOG are developed: interferometric FOG (IFOG), resonant FOG (RFOG), and Brillouin FOG (BFOG). This section gives a brief introduction to the FOGs with emphasis on IFOG.

6.1.1 Interferometric FOG

The basic concept of the Sagnac effect is introduced in Section 3.3.1. For a practical fiber optic gyro, it is necessary to understand its physical mechanisms in details and to solve technical issues to meet the requirement of applications.

6.1.1.1 Phase Bias   The rotation-induced phase difference between CW and CCW waves in a Sagnac loop is expressed as (3.107). The signal of a plain Sagnac loop is in the form of $\cos\phi_R = \cos(4\omega N \boldsymbol{A} \cdot \boldsymbol{\Omega}/c^2)$, which is insensitive in a region near $\phi_R \sim 0$, and is unable to distinguish the direction of rotation. As discussed in Section 3.3.4, the 3×3 coupler gives three output signals with mutual phase shift of $\pi/3$, so that the problem is avoided [12]. Another method is to insert a phase modulator (PM) in the loop [8–11], as denoted by PM in Figure 6.1. A conventional PM is composed of a fiber coil wound on a piezoelectric transducer (PZT) cylinder. The phase modulation can not only move the working point to the quadrature point to avoid the problem of dead region, but also benefit noise reduction.

The additive phase difference between CW and CCW waves by the phase modulation comes from the temporal non-reciprocity, as discussed in Section 5.5. The phase factor is now a sum of the rotation-induced phase and the modulated phase:

$$\Phi = \phi_R + \phi_m(t_1) - \phi_m(t_2) \approx \phi_R + \Delta\phi_m(\Delta t), \tag{6.1}$$

where $t_{1,2} = nL_{1,2}/c$ is the propagation time from the phase modulator to the loop coupler with L_{12} being the respective fiber lengths. The modulator is usually inserted near the coupler, i.e. $L_1 \approx 0$ and

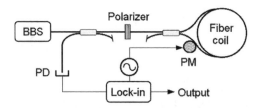

Figure 6.1   Basic structure of IFOG with a PM.

$L_2 \approx L$. To move the working point to the quadrature the phase is modulated sinusoidal with frequency of ω_m; the phase factor is then written as $\Phi = \phi_R + \phi_0[\cos \omega_m(t - \tau) - \cos \omega_m t]$, where $\tau = nL/c$ is the phase delay of the loop. It can be rewritten as $\Phi = \phi_R + \phi_m \sin \omega_m t$ with $\phi_m = 2\phi_0 \sin(\omega_m \tau/2)$ by moving the time coordinate. The interference signal is expanded into Fourier series as

$$\cos \Phi = \cos \phi_R \cos(\phi_m \sin \omega_m t) - \sin \phi_R \sin(\phi_m \sin \omega_m t)$$
$$= \left[J_0(\phi_m) + 2 \sum_{n=1}^{\infty} J_{2n}(\phi_m) \cos 2n\omega_m t \right] \cos \phi_R$$
$$- 2 \left[\sum_{n=1}^{\infty} J_{2n-1}(\phi_m) \sin(2n - 1)\omega_m t \right] \sin \phi_R.$$

To suppress the noise and drift, the signal is retrieved by correlation with the modulation wave [5]:

$$I \propto \langle \cos \Phi \cdot \sin \omega_m t \rangle \propto J_1[2\phi_0 \sin(\omega_m \tau/2)] \sin \phi_R. \qquad (6.2)$$

The working point is thus moved to the quadrature.

The modulation frequency is optimized to $\omega_m \tau = \pi$ for the largest argument of $J_1(2\phi_{m0})$ and J_1 reaches its maximum at $2\phi_{m0} = 1.83$. It is shown [8] that the optimized modulation frequency $f_m = 1/2\tau = c/2nL$ is also helpful for eliminating the spurious signals due to an imperfect phase modulation.

Besides the PZT PM, the integrated optic chip (IOC) is also developed, which has both functions of PM and optical coupler, as shown in Figure 6.2. The IOC is usually made of lithium niobate (LiNbO$_3$) waveguide, which generates two opposite phase at the two end parts of the loop, with more flexible waveforms and higher modulation frequency than PZT PM.

The scheme of Figure 6.1 is termed an open loop structure. It is noted that the signal is not linearly proportional to the angle velocity in the open loop structure, especially for higher speed rotations. Therefore, a close loop structure is developed [13,14], as shown in Figure 6.2.

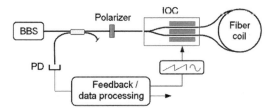

Figure 6.2 Close-loop scheme of IFOG with an IOC PM.

In the scheme, the phase bias is adjusted by feedback electronics to nullify the interferometric signal, and the angular velocity is calculated from the voltage applied on the phase modulator to balance the phase factor to be measured. For the purpose, the phase is usually modulated sinusoidal and scanned linearly by a serrated wave; digital function generators and data processors are often used [14]. A combination of PZT PM and a feedback element can serve the purpose. However, it is more convenient, flexible and reliable to use the IOC device, which is also suitable for mass production. The close-loop configuration not only counts the fringes of interferometric signal, greatly increases the dynamic range, but also has the merit of immunity to light intensity fluctuation and detects electronics instability [9].

6.1.1.2 *Parasitic Reflections and Rayleigh Scattering*

For a high-sensitive Sagnac rotation sensor any parasitic nonreciprocal effect will cause serious errors and spurious signals. An ideal Sagnac loop is considered without any reflections inside. In actuality, reflections exist in the loop inevitably, such as at butt joints between the fiber facet and the optical component, at an imperfect loop coupler with split ratio deviating from the exact 3dB, and at fiber splices. All these reflections will cause parasitic interferences. To reduce the effects careful designs and packaging are necessary, such as with tilted facets of fiber and IOC components [2].

The parasitic reflection composes an additional Michelson interferometer [8], as shown in Figure 6.3. With notations of $\varphi_i = \beta l_i$ and $\varphi = \varphi_1 + \varphi_2$, the CW and CCW waves are written as

$$
\begin{aligned}
E_{CW} &= A + b = A_0 e^{-\alpha l/2} e^{j(\varphi + \phi_R/2)} - rB_0 e^{-\alpha l_2} e^{j2\varphi_2}, \\
E_{CCW} &= B + a = B_0 e^{-\alpha l/2} e^{j(\varphi - \phi_R/2)} + rA_0 e^{-\alpha l_1} e^{j2\varphi_1},
\end{aligned}
\tag{6.3}
$$

where α is the fiber loss; and the opposite signs for the field reflection in two directions are expressed. The interference signal is obtained as

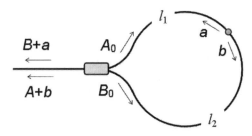

Figure 6.3   Interferences between primary wave and scattered waves.

$\Delta I \propto (E_{CW}E_{CCW}^* + c.c.)$, which is deduced to be

$$\Delta I \propto 2\sqrt{I_A I_B}e^{-2\alpha l}\cos\phi_R + 2re^{-\alpha l}\left(I_A e^{-2\alpha l_1} - I_B e^{-2\alpha l_2}\right)$$
$$\times \cos(\phi_R/2 + \varphi_2 - \varphi_1), \qquad (6.4)$$

where $I_A = |A_0|^2$ and $I_B = |B_0|^2$; and the interferences between the parasitic reflections are neglected since they are usually very weak. The formulas are deduced in the case where there is no phase modulation for simplicity.

It is seen that the parasitic interferences give serious influence to the gyro signals. An ideal 3dB loop coupler is important for suppressing the parasitic effect, especially from those in the middle part of the loop. It has been pointed out that paired parasitic reflections, symmetric to the loop middle, have a tendency to cancel each other out [8]. By using (6.3) on the paired reflections the interference gives parasitic signal of $\propto (I_A - I_B)(e^{-2\alpha l_1} + e^{-2\alpha l_2})$. It means that if 3dB beam split ratio of the coupler is ensured, the parasitic effect is eliminated.

It is worth noting that a low coherence source is critical for reducing the parasitic signals. Similar to the discussion on the low coherence technology in Section 5.5, for a Gaussian spectrum source, the integral over the spectrum gives $\Delta I \sim \cos[\beta(l_2 - l_1)]\exp[-(\delta\beta)^2(l_2 - l_1)^2]$, where $\delta\beta$ is the wave vector width, corresponding to the linewidth. It is noticed that the reflection at the middle point of the loop gives the largest influence. On the other hand, the 3dB beam split ratio of the coupler makes the parasitic effect term with form of $\propto (e^{-2\alpha l_1} - e^{-2\alpha l_2})$, which is diminished at the middle with an approximation of $\approx 2\alpha e^{-\alpha l}(l_2 - l_1)$.

Obviously, the Rayleigh backscattering induces the same effect [15–17]. The back reflection a and b in Figure 6.3 is now considered continuously distributed. The interference can surely be reduced by using a low coherence source; and the scattering occurring near the loop center gives the largest contributions due to the smaller OPD and equal intensities. In practice, superluminescence diodes (SLDs) with typical coherence length $L_c = 20 \sim 30$ µm, or amplified spontaneous emitting fiber sources (ASE) of erbium-doped fiber amplifier, are used widely [18,19]. The Rayleigh scattering occurs randomly, as discussed in Chapter 5, leading to a noisy signal. Therefore the phase modulation and correlation detection is needed to suppress the noise; and a proper modulation frequency of $f_m = 1/2\tau$ gives the best effect for reducing the interference of scatterings near the loop center.

The power imbalance between CW and CCW waves is another error source, as can be seen from (6.3). It is found that a power difference as small as 10 nW creates an intolerable error [8]. Apart from the above-discussed parasitic reflections, Kerr effect induces also such an imbalance [20]. The effect comes from the third-order electric polarization: $P \propto \chi^{[3]} |E|^2 E$, leading to a nonlinear refractive index, expressed as $n_{NL} = n_{II} |E|^2$. Here, the optical intensity includes both CW and CCW waves. The nonlinear indexes for the two waves are given as [2,21]

$$\Delta n_1 = n_{II}(|E_1|^2 + 2|E_2|^2),$$
$$\Delta n_2 = n_{II}(|E_2|^2 + 2|E_1|^2),$$

(6.5)

and the nonreciprocal index difference is $\delta n_{NL} = n_{II}(|E_2|^2 - |E_1|^2)$. In an ideal case that the split ratio of loop coupler is exactly 1:1, and the fiber loss is precisely uniform, the Kerr effect can be neglected. In reality, however, deviations of the split ratio are inevitable; and the fiber loss and its spatial and temporal variation are not symmetric to the loop center. For the random varying optical power $P(t)$, the error of rotation signal is expressed as [8]

$$\delta \Omega \propto (1 - 2K) \frac{\langle P^2(t) \rangle - 2\langle P(t) \rangle^2}{\langle P(t) \rangle},$$

(6.6)

where K is the split ratio. Although the Kerr index is estimated to be very small, its effect may accumulate in the long loop fiber, especially in cases where the light is highly coherent. Since the CW and CCW wave form a standing wave in the loop, the Kerr effect creates an index grating, which reflects the incident optical wave with corresponding wavelength, being one of the mechanisms of nonreciprocity. When a low coherence source is used the index grating induced by Kerr effect is limited near the loop center in a distance of the coherence length. It is therefore concluded that a low coherence source is with critical importance for reducing the effects of parasitic reflections, Rayleigh backscattering, and Kerr effect in IFOG.

6.1.1.3 Polarization Dependence

Among the various parasitic effects, polarization dependence is one of the most encountered effects. As analyzed theoretically, the Sagnac effect is independent of medium properties [2,7]; both polarization modes will give the same signal of rotation. However, the cross-coupling between the two modes will lead to parasitic interferences [2,5,8]. The polarization dependence of the Sagnac loop comes from several factors. The loop fiber is usually wound many turns on a cylindrical spool with a typical radius of several centimeters, resulting in bending-induced birefringence, and accumulated phase retardation up to hundreds of radians. A totally polarization-independent coupler may also be impractical. The PM is usually with polarization dependence, especially the IOC device, because it is usually based on a planar waveguide with asymmetric geometry.

Referring to the concept of the principal state of polarization (PSP), if the input polarization is adjusted to coincide with PSP of the loop, the polarization dependence is eliminated. It is the function of polarizers in Figures 6.1 and 6.2, which makes the input polarization in the PSP direction and ensures that the detected signal has the same polarization. However, the polarization extinction of the polarizer is finite; moreover, the polarization dependence varies temporally, due to varying temperature and strains, leading to signal drift. Therefore, one fixed polarizer is not enough. Two measures are thus adopted: one is with a polarization maintaining fiber (PMF) coil; the other is with depolarized source [2,9,22–25].

Figure 6.3 can serve as an illustration of polarization dependences, if waves a and b are regarded as the cross-coupled polarization waves. The interferences between them and the primary waves A and B depend on the coherence of the source. The depolarization length is introduced to describe the spectral dependence of polarization effects: $L_D = l_c/B$, where $l_c = \lambda^2/2\delta\lambda$ is the spectral coherence length, B is the birefringence. The polarization-induced spurious phase is expressed as [2,5,8,23]

$$\Delta\phi \sim \rho h \sqrt{L L_D}, \qquad (6.7)$$

where $\rho = I_y/(I_x + I_y)$ stands for the polarization extinction ratio of the polarizer, L is the loop length, h is a measure of polarization crosstalk per unit length of fiber, as discussed in Section 3.4. $\Delta\phi$ includes two compositions: one comes from the phase difference between the primary wave and the cross-coupled wave, the other from their intensity difference. It is shown that a broadband source is also necessary to reduce the polarization-induced phase error.

The PMF has much higher immunity to external disturbances than the ordinary single mode fiber; however the cross-coupling exists inevitably. Reference [23] demonstrates a scheme by using birefringence modulators (B-mods) to suppress the effect, as shown in Figure 6.4(a). The second scheme is depolarized FOG (DFOG) with depolarizers inserted, as shown in Figure 6.4(b) [9]. If the light is fully depolarized by DP1, the polarization dependence will be eliminated.

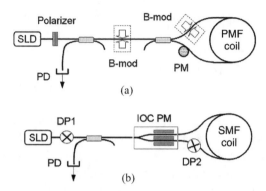

Figure 6.4   (a) IFOG with B-mods and (b) IFOG with depolarizers (DP).

However, some optical components used in FOG will polarize the traversed wave; it is, therefore, necessary to insert the second depolarizer (DP2) inside the Sagnac loop. The depolarizer is not ideal in practice; reference [25] analyzes the effect of residual DOP, and proposes a scheme with two depolarizers on the opposite sides of the gyro loop to reduce errors related to the nonreciprocal birefringence.

6.1.1.4 Thermally Induced Nonreciprocity   It is stated that the reciprocity applies only to time invariant systems. The temporally varied strain and temperature states will lead to nonreciprocity. In practice, the fiber coil is wound tightly onto a cylindrical spool. The temperature distributions in the coil and the strains of fiber are always varying temporally and spatially along the fiber more or less. References [9,26] analyze the time-varying thermal perturbations across the fiber coil, called the Shupe effect, which is thought the largest error source in IFOG with other factors optimized. The disturbances of mechanical stress, including vibration and acoustic waves, must be removed by careful packaging and sheltering. The thermally induced nonreciprocity due to temperature fluctuation needs more attention, especially in the starting period of the gyroscope operation. In case a temperature-gradient distribution exists in the fiber coil, the induced phase error by the thermal-optic effect is attributed to the phase difference between waves from two symmetric points to the loop center, at l and at $L - l$ with propagating time of $\Delta t = n(2l - L)/c$; the sum is expressed as [27]

$$\delta\phi = Q \int_0^{L/2} [\Delta\dot{T}(l) - \Delta\dot{T}(L - l)](2l - L)\mathrm{d}l, \qquad (6.8)$$

where $Q = (n\beta/c)(n^{-1}\mathrm{d}n/\mathrm{d}T + \alpha)$, α is the thermal expansion coefficient of silica. It is seen that the phase error can be canceled out if the two halves of the loop experience the same temperature fluctuation and symmetrical distributions to the center, that is, $\Delta \dot{T}(l) = \Delta \dot{T}(L-l)$. Therefore, it is important to make the variation symmetric to the loop center; in other words, the fiber points with equidistant lengths to the loop center should be located at the same position in the fiber coil spool, where the fiber sections encounter the same temperature variation. For the purpose, the fiber sections equidistant from the center should be placed together as close as possible in fiber winding; several methods are thus developed [9,27–30].

Loop fiber is usually wound in the way of fiber by fiber and layer by layer. The fiber forms a gentle helix with a pitch of fiber diameter (with its jacket) with a certain twisting for the long fiber length, inducing cross-coupling between the two polarization modes. For the requirement of symmetric arrangement, the equidistant fiber section layers should be interleaved. Figure 6.5(a) shows a fiber loop coil spool with temperature distributions in the axial direction and radial direction. Figure 6.5(b) gives a scheme of typical winding pattern, called quadrupolar winding. Compared with simple winding that has one end placed in the innermost layer and the other end placed at the outer layer, the quadrupolar scheme reduces the Shupe effect to about $1/N^2$, where N is the number of layers [9]. The heat conduction in the bulk and the transient temperature evolution are analyzed for different fibers [28,31]. Special fibers for FOG are developed with a smaller diameter and thinner jacket to increase the effective area of fiber coil and improve its thermal performance. NB, Figure 6.5(b) gives just a conceptional illustration; the fiber should be wound in the same direction to prevent the Sagnac effect from being cancelled with each other between circles.

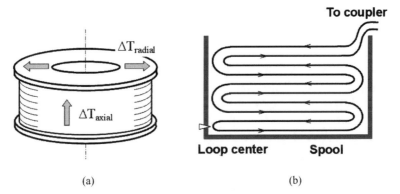

Figure 6.5   (a) Fiber coil spool with temperature fluctuation and (b) quadrupolar-winding pattern of fiber coil.

6.1.1.5 Faraday Effect and Earth Rotation   The polarization state is changed by Faraday rotation along the loop fiber in the external magnetic field, but its effect will be cancelled out by the line integral over the loop: $\varphi_F = V \oint B \cdot dl = 0$ [2,32,33], which is the same both for CW and CCW propagations, unless the magnetic is generated by current traversing the loop. However, if cross-coupling between the two polarization modes exists, the effect of Faraday rotation accumulated along the loop will give different results for the counter propagating waves. The analyses indicate that the fiber twisting gives a major contribution, since the twisting causes polarization plane rotation, added to the Faraday rotation. Moreover, the fiber itself often has residual shear strains distributed randomly, impossible with symmetric distributions to the loop center. All these factors cause the nonreciprocal effect of Faraday rotation. It is indicated that the PMF is beneficial in suppressing the influence due to its short depolarization length L_D.

On the other hand, the Sagnac sensor can be utilized as a magnetic field sensor, as analyzed in [33]. The earth magnetic field exists everywhere, which brings about a bias uncertainty of about 10 deg /h [32]. To eliminate the effect, careful packaging with magnetic shielding is necessary.

Another environmental effect is the rotation of the earth. At the two poles the background rotation is $\Omega_E = 15$ deg /h. At different latitudes θ, the rotation depends on gyro's orientation: $\Omega_{\text{vert}} = \Omega_E \sin \theta$ for a vertically placed gyro, $\Omega_{\text{NS}} = \Omega_E \cos \theta$ for the north–south direction, and $\Omega_{\text{EW}} = 0$ for the east–west direction. Therefore, the gyro can be directly used as a compass [5].

6.1.1.6 Scale Factor and Fundamental Limit   Scale factor is one of the important specifications of a gyroscope. It is defined as the ratio of phase change over angular velocity variation:

$$d\phi/d\Omega = 4\pi LR/c\lambda. \tag{6.9}$$

Gyro applications, such as navigation equipment, require a scale factor with high precision, high stability, and high linearity, because a real-time orientation angle is needed during the navigation, whereas small errors will be accumulated in a continuous long period, giving a serious deviation of integrated angle. It is estimated that the minimum detectable phase of less than 10^{-7} rad should be acquired; a resolution down to 10^{-4} deg /h is required.

The close-loop scheme is the best method to give maximum and linear sensitivity. The algorithm and related electronics are investigated and developed, especially by the digital technology. Detailed analyses can be found in references [2,5,13,14,34]. It is seen from (6.9) that the precision of the scale factor depends on the stability of fiber coil area and fiber length, and on the stability of source wavelength. The latter includes the stability of its spectrum. Since the Sagnac interferometer (SI) works generally at the near equal optical paths for detecting rotations with low rates, a broadband source can be used, giving the signal without ambiguities. However, the scale factor is related to an effective wavelength, averaged with its spectrum. It needs careful attention and technical measures since the spectra of widely used SLD and ASE sources will vary with pump current and temperature. Sources in 1,550 nm band are used mostly because of the lowest propagation loss, though a shorter wavelength is beneficial to higher scale factor. The ASE source is made of erbium doped fiber with about 30 nm spectral width; it can be spliced directly with single mode loop fiber by fusion, and thus with better reliability, becoming the most preferable choice.

Under ideal conditions with the harmful and unwanted nonreciprocities and various instabilities compensated, the last limitation is noise, including photon noise, shot noise, and thermal noise of detectors and electronic circuits. Such white noise determine the finite minimal detectable signal, signal drifting, and randomly walking errors. For a photon flow, its shot noise behaves in Poisson distribution; its standard deviation is written as

$$\sigma_N = \sqrt{2N\,\Delta f}, \tag{6.10}$$

where Δf is the detection frequency bandwidth. The photon flow rate N is about 7.5×10^{12} s^{-1} at 1500 nm wavelength and optical power $P = 1$ μW. The noise-equivalent phase error can then be obtained [8]:

$$\Delta\phi_{\text{rms}} = \frac{\sigma_N}{N} = \sqrt{\frac{2\Delta f}{N}} \propto \sqrt{\frac{\Delta f}{P}}. \tag{6.11}$$

References [2,4,5,8] give estimations of the theoretical limitations and comparisons of IFOG with other technologies. Figure 6.6 shows a comparison of gyroscope technologies and applications [4], including microelectromechanical system technology (MEMS), dynamically tuned gyroscope (DTG, mechanical type), ring laser gyroscope (RLG), and attitude heading reference systems (AHRS).

Schemes other than the above introduced are also proposed for IFOG, such as by using frequency-modulated continuous-wave (FMCW) technology [35].

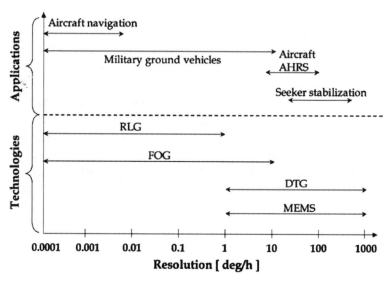

Figure 6.6   Applications and comparison of different gyro technologies. (Reprinted with permission from reference [4].)

6.1.2   Brillouin Laser Gyro and Resonance Fiber Optic Gyroscope

If a ring laser rotates in the inertial coordinate system, its two opposite output beams will show different phases by the Sagnac effect; the RLG has thus been developed earlier. The Brillouin laser is attractive due to its low threshold and it not needing any active medium. The Brillouin scattering forms a standing wave in the fiber ring when equal pumps are injected from both sides of the coupler. When the fiber ring rotates, Doppler shifts with opposite signs occur, giving frequency difference between the two outputs. From the resonance condition $\beta L = 2m\pi$, the frequency difference is deduced as

$$\omega_\Omega = \frac{-R\Omega_A}{nc}\omega_0, \tag{6.12}$$

where R is the fiber coil radius, $\Omega_A = (A \cdot \Omega)/|A|$ is the projection of angular velocity vector on the normal of the fiber ring, ω_0 is the resonance frequency in the static case.

Figure 6.7 shows a schematic structure of Brillouin ring laser gyro [36]. It can be regarded as a Sagnac loop with a fiber ring inserted. The fiber ring is pumped from the two ports of the coupler, which act also as the two output ports of the laser. A laser source with narrower

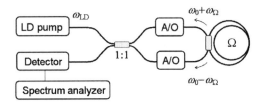

Figure 6.7 Schematic diagram of Brillouin laser gyro.

linewidth, such as a laser diode, is used as the pump. The pump is divided into two beams, injecting into the ring from the coupler. When the system rotates with angular velocity Ω, the two output beams of the Brillouin laser have different lasing frequencies, to be detected by correlation detection or by a spectrum analyzer. Two acousto-optic modulators are inserted into the loop so that the pump has frequency shifted properly for the correlation and signal processing.

It is seen that the resonance frequency shift does not depend on the fiber length of the ring, but a longer fiber is preferable since the resonance peak will be sharper and beneficial to higher detection resolution. Its sharpness also depends on the split ratio of the ring coupler, and reaches the maximum at the critical coupling (see Section 3.3).

A dead region at low angular velocity exists in simple RLG structures due to optical frequency lock-in induced by competition of longitudinal modes. Some methods were proposed to solve the problem. It is indicated theoretically and experimentally that the lock-in effect exists also in Brillouin RLGs. In addition, most parasitic effects existing in IFOG remain in BFOG, careful investigation and technical measures are needed [37].

Resonance fiber optic gyro (RFOG) makes use of the same effect in the fiber ring, but in passive operation mode. A typical close-loop scheme is shown in Figure 6.8, where the frequency of narrow linewidth LD beam is tuned $\delta\omega$ by its pumping current, the LD beam is divided into two and injected into the fiber ring from its two ports. One of them is frequency shifted $\Delta\omega$ by an acousto-optic modulator (AOM).

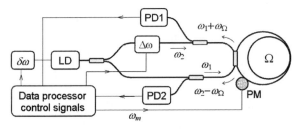

Figure 6.8 Schematic diagram of a typical RFOG.

A PM is inserted in the ring to set the working bias to the quadrature point. Two photo detectors are used to receive the outputs of the fiber ring from its two ports. One of the detected signals is used to feedback control the laser frequency coincident with the ring resonance. The other is taken as the deviation induced by the rotation, to feedback control AOM to nullify the deviation signal, and then acquire the rotation velocity by the voltage on AOM. Thus the close-loop operation is fulfilled.

It is noted that the fiber ring is no longer an all-pass filter since a long fiber is usually used. Moreover, the Brillouin backscattering occurring in the fiber ring is a new and important loss factor. By neglecting the coupler loss, the transmission of the ring is expressed as

$$\frac{I_1}{I_0} = 1 - \frac{1 - A_{re}}{1 + F \sin^2(\beta L/2)}, \tag{6.13}$$

where $F = 4t\gamma/(1 - t\gamma)^2$ is the finesse, $A_{re} = (t - \gamma)^2/(1 - t\gamma)^2$ is the attenuation at resonances, and a loss factor $\gamma = \exp(-\alpha L/2)$ is denoted. Based on the formulas, requirements of system designing, such as fiber length in the ring and beam split ratio of the coupler, are presented.

The nonreciprocal factors and parasitic effects in IFOG affect also the performance of RFOG. The birefringence generates two sets of resonances; whereas the depolarized broadband source is not suitable to RFOG, and thus a polarization scrambled narrow line source is needed [24]. The influence of Rayleigh backscattering is more serious for the high coherence wave. Kerr effect becomes larger in the ring with higher finesse due to higher intensity inside the ring. More schemes and methods are being studied and developed.

6.2 FIBER OPTIC HYDROPHONE

The acoustic sensor is extremely important for all aspects of social life, from civil activities to military events. A famous example is the piezoelectric sonar. Since the first fiber hydrophone was reported in the 1970s [38–41], a variety of fiber optic hydrophones and geophones have been demonstrated, and related technologies are developing continuously. The fiber hydrophones are now available commercially and used widely in sea fishery, marine biology, seismic wave detection, oil well drilling, ultrasonic medicine, coast guarding, ship and submarine navigation, and so on. This section gives a brief introduction to fiber hydrophone technology, describes its basic structure and main technical problems.

6.2.1 Basic Structures

The purpose of a hydrophone is to detect sound waves in water, and further to localize and distinguish the sound source. It is actually a dynamic pressure sensor; and multiple sensors form an array, or a system. The fiber hydrophones are categorized into active and passive sensors. The passive sensor includes generally two types: one is based on optical intensity variation [42,43], which has lower sensitivities; the other is the interferometric sensor based on the phase change. The active sensor makes use of the fiber laser. The hydrophone is basically composed of three main parts: (1) the sensor head, which transfers the pressure to fiber strains; (2) the interferometer, and the multiplex system; or the configuration of fiber laser; and (3) the signal interrogation and processing unit.

6.2.1.1 Sensor Head   Two typical structures of the sensor head are usually adopted [44–47]. Figure 6.9 shows a hollow mandrel with a fiber coil wound and fixed on it. The diameter of the mandrel will be deformed under the sound pressure, and an axial strain of the fiber occurs. To enhance its sensitivity, a polymer material with lower Young's Modulus may be placed underneath the fiber coil, as shown in Figure 6.9(b). Another structure is a fiber section coated with a thick and/or multiple layers, usually polymer materials with lower Young's Modulus and larger Poisson ratio, as shown in Figure 6.9(c). The polymer coating with a large cross section will expand axially under the pressure, and pull the fiber strained axially, if the bonding between the fiber and the coating is tight enough. The thick-coated fiber can also be used in the mandrel structure for enhanced sensitivity. The sensor unit may

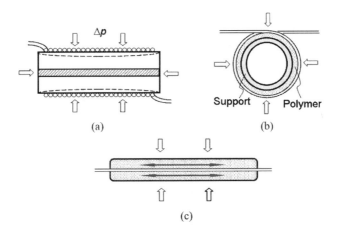

Figure 6.9   Transducers of acoustic pressure.

be further packaged into an acoustic resonant cavity to improve its frequency response [48].

It is no doubt necessary to design the mechanisms carefully; reference [45] and others give detailed analyses of the strains of some typical structures, based on the dynamic flexural plate theory. The phase change of a propagating light wave, induced by acoustic pressure, is expressed as

$$\Delta\phi = \left[\left(1 - \frac{1}{2}n^2 p_{12}\right)e_z - \frac{1}{2}n^2(p_{11} + p_{12})e_r\right]nk_0 L, \qquad (6.14)$$

where the axial strain e_z and the radial strain e_r are dependent on the structure of the pressure transduction mechanism and its materials, and proportional to the pressure. It is seen that a longer sensing fiber L is obviously advantageous to a higher sensitivity. The sensitivity of the hydrophone to the acoustic pressure p is generally defined as

$$M = \frac{\partial\phi}{\partial p}\bigg|_{n,\lambda}, \qquad (6.15)$$

in units of $rad/\mu Pa$; and the normalized sensitivity $M_n = \phi^{-1}(\partial\phi/\partial p)_{n,\lambda}$ is used to characterize the sensitivity per optical path length.

In active hydrophones, the most often used configuration is a fiber laser incorporated with a fiber Bragg grating, whose reflection peak wavelength is sensitive to the acoustic pressure, expressed as [49–52]

$$\frac{\Delta\lambda}{\lambda} = -\frac{1 - 2v}{Y}\left[1 - \frac{n^2}{2}(p_{11} + 2p_{12})\right]\Delta p. \qquad (6.16)$$

The gratings are usually imprinted in the active medium, such as erbium-doped fiber, to form a distributed feedback laser or a distributed Bragg reflection laser. A proper coating, as shown in Figure 6.9(c) will enhance the sensitivity of FBG to the acoustic pressure.

6.2.1.2 *Interferometer and Interrogation* Hydrophones incorporated with different interferometers have been studied experimentally, including MZI, MI, SI, and FPI; the fiber gratings are also used for acoustic sensing [53]. Figure 6.10 shows schematically the interferometers. The sensor and the reference fiber coil are connected into the two beams of the interferometer, respectively. The reference fiber coil is

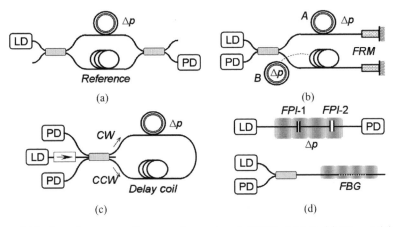

Figure 6.10 Interferometers for acoustic sensing: (a) MZI; (b) MI; (c) SI; and (d) FPI and FBG.

used to set a proper biased optical path difference (OPD $= n\Delta L$). The output of the interferometer is written as

$$I \propto \cos\frac{2\pi v n \Delta L}{c} = \cos(\phi_0 + M\,\Delta p). \qquad (6.17)$$

It is seen that a larger OPD gives a higher sensitivity, but a smaller linear dynamic range.

The SI is also sensitive to the sound wave [54] based on the temporal nonreciprocity, as analyzed in Section 5.5.1. To make use of the effect, a delay coil is inserted in the loop, making different time delays of the acoustic signal between CW and CCW propagations. A 3×3 coupler is used in the scheme to avoid quadrature point drift [55]. The Sagnac hydrophone has unique features: it is immune to the external disturbances of the acoustic frequency band detected; as a common-mode interferometer, a broadband source can be used, and some negative effects induced by narrow linewidth sources are removed; it eliminates drifts of working point, which occurs often in other interferometers.

A key technical issue is the interrogation of the required acoustic signals from large-scale sensor arrays and the retrieval of information from the noisy background and random drift. Some existing negative factors relate to the working principles of interferometer, such as polarization fading, temperature fluctuation, drift of working point, and phase noise. It is seen from (6.17) that the sensitivity depends on the phase bias; the highest sensitivity is obtained at $\phi_0 = (m + 1/2)\pi$. Several techniques are used to ensure the quadrature point; one of them is to add a PZT PM [56], with its driving voltage feedback adjusted to make the phase actively tracked. The scheme is called active homodyne, widely used in interferometric sensors. However, it is not convenient to put PZT components under water.

Another scheme is the phase-generated carrier method (PGC) [57–59], by which a phase modulation is added in the interferometer with the modulation frequency higher than the acoustic frequency of interest, and with the modulation amplitude large enough to generate a larger side band. The phase modulated signal detected form the interferometer is expressed as

$$I = I_0\{1 + V \cos[\varphi \cos \omega t + \phi(t)]\}, \tag{6.18}$$

where V is the visibility of interference fringes, determined by the difference between the optical intensities of two beams; φ is the phase modulation amplitude, ω is the modulation frequency; $\phi(t)$ is the detected signal, written as $\phi(t) = \phi_0 \cos \Omega t + \psi(t)$, with the acoustic frequency Ω and the external disturbance $\psi(t)$. The signal can be expanded as

$$I/I_0 = 1 + V \left\{ \left[J_0(\varphi) + 2 \sum_{n=1}^{\infty} (-1)^n J_{2n}(\varphi) \cos 2n\omega t \right] \cos \phi(t) \right.$$

$$\left. - 2 \left[\sum_{n=1}^{\infty} (-1)^{n+1} J_{2n-1}(\varphi) \cos(2n-1)\omega t \right] \sin \phi(t) \right\}. \tag{6.19}$$

By mixing with the local oscillation of ω and 2ω, the low order harmonic components are obtained:

$$I_{0\omega} = \alpha_0 I_0 [1 + V J_0(\varphi) \cos \phi(t)], \tag{6.20a}$$

$$I_{1\omega} = -2\alpha_1 I_0 V J_1(\varphi) \sin \phi(t), \tag{6.20b}$$

$$I_{2\omega} = -2\alpha_2 I_0 V J_2(\varphi) \cos \phi(t), \tag{6.20c}$$

where α_0, α_1, and α_2 stand for mixing and filtering efficiencies for the three harmonics, which may not be necessarily the same, but can be calibrated. Formulas hold under the condition that the modulation frequency ω is higher than the band of detected acoustic signal.

The acoustic signal to be detected can be calculated based on (6.19); however, the externally induced drift may result in signal fading and large errors. To remove the fading, time derivatives and cross-multiplying are made, resulting in

$$I_{2\omega}\dot{I}_{1\omega} - I_{1\omega}\dot{I}_{2\omega} = 2\alpha_1\alpha_2 I_0^2 V^2 J_1(\varphi)J_2(\varphi)\dot{\phi}(t). \tag{6.21}$$

The detected signal can then be acquired by integrating (6.21) in data processing, which extends the linear dynamic range greatly. If the externally induced drift is much slower than the acoustic wave, the derivative of the phase can be approximated as

$$\dot{\phi}(t) = -\Omega\phi_0 \sin \Omega t + \dot{\psi}(t) \approx -\Omega\phi_0 \sin \Omega t. \tag{6.22}$$

The acoustic wave does not necessarily have a single frequency; therefore, the frequency spectrum analyzer is needed in data processing. With a certain frequency Ω, term $\cos \phi$ is expanded as

$$\cos \phi = \left[J_0(\phi_0) + 2 \sum_{n=1}^{\infty} (-1)^n J_{2n}(\phi_0) \cos 2n\Omega t \right] \cos \psi(t)$$

$$- 2 \left[\sum_{n=1}^{\infty} (-1)^{n+1} J_{2n-1}(\phi_0) \cos(2n-1)\Omega t \right] \sin \psi(t) \tag{6.23}$$

and a similar expression for $\sin \phi$. It is seen that the variation of $\psi(t)$ will change the detected acoustic frequency spectrum. Other errors and noise sources, such as fluctuations of source power and visibility, should also be paid attention to in practice.

Two methods are usually used to realize the phase modulation. One is by a piezo-electric modulator, which is not suitable for underwater applications. The other is with a modulated laser, externally by an optical waveguide phase modulator, for example, or internally by its driving current, if a laser diode is used. The latter is convenient in practical applications, and used widely. It is known that the lasing frequency of the laser diode can be modulated by the current, expressed as

$$v = v_0 + \frac{\partial v}{\partial i} \Delta i \cos \omega t = v_0 + \Delta v \cos \omega t. \tag{6.24}$$

The signal of interferometer depends on its OPD ΔL:

$$I \propto \cos \frac{2\pi v n \Delta L}{c} = \cos \left(\frac{2\pi v_0 n \Delta L}{c} + \frac{2\pi n \Delta v \Delta L}{c} \cos \omega t \right), \tag{6.25}$$

where ΔL is the length difference between sensing fiber and reference fiber. Correspondingly the parameters in (6.18) are written as $\varphi = 2\pi n \Delta v \Delta L / c$ and $\phi(t) = 2\pi v_0 n \Delta L / c$. It is indicated that phase modulation can be realized by modulation of source frequency. The amplitude of equivalent phase modulation φ depends on the OPD and the modulation depth of laser frequency; therefore an unbalanced interferometer with a proper OPD is needed. The laser diode can be modulated easily and with high-speed response, making the system design freely. In addition, sensors with different OPD can be discriminated by scanning the laser frequency. It should be noted, however, that the output of the laser diode is modulated by the current simultaneously; this accompanied intensity modulation should be compensated. The PGC method has advantages of larger dynamic range, higher linearity, lower additive phase noise, and simpler electronics.

The interferometers are used also in the source unit and detection unit of the hydrophone system. Reference [60] describes a hydrophone with double Mach–Zehnder interferometers: one is used in the sensor head (at the "wet" end), and the other is used in the source unit (at the "dry" end). By actively modulating the latter to match the former, the signal fading is reduced. In the scheme, an acousto-optic modulator is used to generate two separate wavelengths. Such a configuration shows advantages: a laser source with a relatively short coherence length can be used; the OPD fluctuation of the sensor MZI can be compensated by using two probe wavelengths.

Polarization fluctuation is one of the major error sources. PMF is effective if the cost is not considered. The Faraday rotation mirror (FRM) is widely used to obviate the problem, as depicted in Figure 6.10(b); and depolarized or polarization scrambled optical beams are also helpful.

The hydrophone is often sensitive to the acceleration since it acts as an inertial force causing the fiber coil mandrel to deform, which should be distinguished from the acoustic signal. An effective method is to use paired sensors [45], as shown in Figure 6.10(b), where fiber coil B will be inserted as the reference. Two coils have the same sensitivity to the acceleration, but opposite sensitivities to the acoustic pressure. For example, the pressure is applied inside coil B as shown in the figure.

6.2.2 Sensor Arrays and Multiplexing

The practically used hydrophone and geophone must form a sensor array to detect the acoustic signal over a large area, for example, in several kilometers or more, whereas the spacing between the sensors may be only a few meters. On the other hand, sensor arrays packaged in a board are needed to detect the phase front of acoustic wave, in order to localize the sound source and to distinguish the target. It is an important subject to design and to build a large-scale sensor array and a multiplexing system [54,61–64], which should consider the requirements:

1. a great sensor number, usually up to 500;
2. a large covering area, such as several tens of kilometers;
3. low optical power with efficient and low loss network configurations;
4. low cross-talk between sensors;
5. suitable for application conditions, for example, deep underwater with acceptable performance of every sensor.

The typical configurations of a hydrophone array system are based on the parallel connection and series connection, and their combinations. Figure 6.11 shows several typical configurations: an MI array in parallel connection (a), an MI with multiple sensors in series connection (b), and a parallel connected SIs (c). MZI may be used instead of MI in configuration (a). Sensor arrays up to several tens of units are built by the configurations as one array module.

Large scale hydrophone networks must employ certain multiplexing technologies, basically the time division multiplex (TDM) and wavelength division multiplex (WDM) [62]. In TDM, some fiber delay lines are inserted in the path to distinguish and locate the sensor positions; in WDM, it is needed to use multiple optical add-drop multiplexers (OADM), as shown in Figures 6.12(a) and (b). For such large fiber systems, the loss of fiber and all optical components must be compensated by optical amplifiers (EDFA), shown in the figures. The characteristics

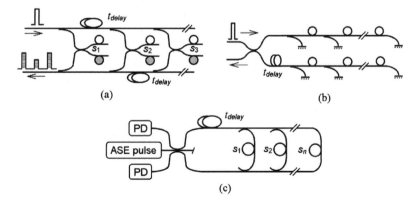

(a)

(b)

(c)

Figure 6.11 Typical hydrophone sensor arrays.

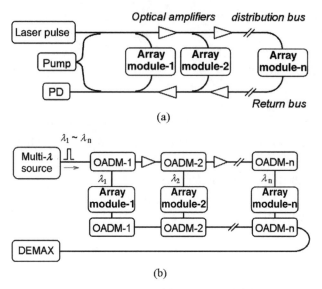

Figure 6.12 Large scale arrays with (a) TDM and (b) WDM technologies.

and performances of sensor arrays and multiplexing systems are analyzed theoretically and simulated in reference [65]. This book will not go into detail here.

6.2.3 Low Noise Laser Source

The applications require a hydrophone with high performance, such as high phase resolution (≤ 1 μrad), large linear dynamic range (10^7), small volume of sensor head, large-scale array covering large area, and low power consumption. The performance is finally limited by the noise, such as the shot noise and the thermal noise in detection and A/D converter, the thermal noise of interferometer fiber, the noise of optical amplifiers, and the noise of laser sources. The average power detected from the interferometer is typically around 100 μW; therefore, the laser noise is believed to give the greater contribution than others. The noise of laser source includes the relative intensity noise (RIN) and the phase noise. As an interferometric sensor, the effect of source phase noise on the detected signal is

$$\delta\phi = \frac{2\pi n \Delta L}{c}\delta v, \tag{6.26}$$

where $\delta\nu$ is the frequency deviation induced by the phase noise. It is seen that the lowest detectable phase change is proportional to linewidth of light source. It is estimated from (6.26) that for resolution of ≤ 1 µrad, the phase noise (the frequency fluctuation) of ~ 10 Hz/$\sqrt{\text{Hz}}$ is required for the interferometer with OPD of ~ 1 m. The intensity noise induced phase noise is given by $\delta\phi_{\text{RIN}} = 1/V\sqrt{\text{RIN}}$, where $RIN = \delta P^2/\bar{P}^2$ is the relative intensity noise of source power P, V is the visibility. Therefore the RIN should be reduced down to -130 dB/Hz for resolution of ≤ 1 µrad.

Noise reduction is a hot topic in research and development; and high-quality lasers are now available commercially. Among the various laser schemes, the external cavity frequency-stabilized diode laser (ECDL), the erbium-doped fiber laser incorporated with the phase-shift Bragg grating (DFB-FL), and the diode-pumped Nd:YAG nonplanar ring laser (NPRL) may be the most attractive schemes. Reference [66] describes an DFB-FL with the master oscillator-power amplifier (MOPA) configuration; the RIN is suppressed down to -120 dB/Hz, by stabilizing the 980 nm pump power actively, and the phase noise is suppressed to 1.5 Hz/$\sqrt{\text{Hz}}$ at 1 kHz, by using an MZI as a frequency discriminator and a PZT element to tune the laser fiber. The linewidth of the single-frequency laser was reduced to 1 kHz.

The high-quality source must be carefully packaged to isolate any external vibration and temperature fluctuation. The performance of single-frequency lasers have continuously improved in recent years. Reference [67] gives a detailed review; demonstrating the capability of DFB fiber laser to resolve effective length change of <1 fm/$\sqrt{\text{Hz}}$ at 2 kHz.

The hydrophone technology has progressed greatly during the last three decades. The responsivity of a hydrophone sensor was reported to -127.5 dB relative to 1 rad/µPa, better than those obtained from commercial piezoelectric hydrophones; the resolution reaches below ocean acoustic noise limitation; and large scale systems have been paved on the sea bed [63].

6.3 FIBER FARADAY SENSOR

As introduced in Section 3.4.4, the Faraday effect is widely used in a variety of applications, from optical components to astrophysics research. This section gives just a brief introduction to the sensors based on Faraday rotation.

6.3.1 Faraday Effect in Fiber

Faraday effect is attributed to the motion of electrons in the external electro-magnetic field. According to the elementary theory [68], it obeys the equation of motion under the Lorentzian force:

$$m\frac{d^2r}{dt^2} + \gamma m\frac{dr}{dt} + m\omega_0^2 r = -e\left(E + \frac{dr}{dt} \times B\right), \qquad (6.27)$$

where m is the mass of electron, e is its charge and γ is the damping rate; $m\omega_0^2 r$ stands for the quasi-elastic restoring force, relating with the electron's transitions. For a single frequency optical wave propagating in z-direction, $E = E_0 e^{-j\omega t}$, its solution is expressed as

$$x \pm jy = \frac{e}{m}\frac{(\omega^2 - \omega_0^2 + j\gamma\omega) \mp \omega\omega_c}{(\omega^2 - \omega_0^2 + j\gamma\omega)^2 - \omega^2\omega_c^2}(E_x \pm jE_y), \qquad (6.28a)$$

where $\omega_c = eB_z/m$ is defined as the cyclotron frequency. It indicates that the electrons are moving in a helical manner in the magnetic filed. In case the damping is negligibly small, (6.28a) can be written as

$$x \pm jy = \frac{e/m}{\omega^2 - \omega_0^2 \pm \omega\omega_c}E_\pm \approx \frac{e/m}{\omega(\omega \pm \omega_c)}E_\pm, \qquad (6.28b)$$

where $E_\pm = E_x \pm jE_y$ is the circularly polarized optical waves; the approximation in (6.28b) holds for the case when the optical frequency band is much higher than the electron's eigen resonance frequency. It is shown that the dielectric constant should now be expressed as a tensor; and the refractive indexes for the two circularly polarized waves are deduced to

$$n_\pm = 1 + \frac{Ne^2}{2m\varepsilon_0}\frac{1}{\omega_0^2 - \omega^2 \pm \omega\omega_c}, \qquad (6.29a)$$

where N is the electron number per unit volume. Their difference is

$$n_- - n_+ = \frac{Ne^3}{m^2\varepsilon_0}\frac{\omega B_z}{(\omega_0^2 - \omega^2)^2}. \qquad (6.29b)$$

The polarization rotation is obtained from the index difference

$$\Delta\varphi = \frac{1}{2}k(n_- - n_+)z = V B_z z; \qquad (6.30)$$

and the Verdet constant is expressed as

$$V = \frac{Ne^3}{\varepsilon_0 m^2 c} \frac{\omega^2}{(\omega_0^2 - \omega^2)^2}. \tag{6.31a}$$

Referring to the averaged index $\bar{n} = (n_+ + n_-)/2 \approx n$, it is transformed to

$$V = \frac{e}{mc} \frac{\partial n}{\partial \lambda} \lambda, \tag{6.31b}$$

indicating its relation with the dispersion. In case the magnetic field and/or the material are nonuniform along the optical path, the rotation is written as

$$\Delta \varphi = \frac{e\lambda}{mc} \int_0^z \frac{\partial n}{\partial \lambda}(z) B_z(z) \mathrm{d}z. \tag{6.32}$$

The equations and formulas (6.27–6.32) give just a brief and simple explanation of the Faraday effect. The detailed understanding involves material structures and electron energy band properties; the Verdet constant should be multiplexed by the magneto-optical anomaly factor of the material. More physical phenomena and mechanisms related to Faraday effect have been investigated.

Fiber sensors and devices based on Faraday effect are generally divided into two types: (1) intrinsic and (2) extrinsic sensors. The former utilizes the Faraday effect of fiber itself, whereas the latter is composed of other materials with higher Verdet constant, coupled with fiber pigtails. The Faraday rotations were measured earlier in single mode fibers [69], and in PMFs [70]. The characteristics of Faraday effect were investigated and compared for different materials and for different wavelengths, giving the relation of Faraday rotation with the energy gap [71]. The Verdet constant of fused silica is measured as ~ 3 rad $\cdot$ T$^{-1} \cdot$ m^{-1} (T stands for *Tesla*) at 633 nm band. It is noted that the figure of merit, defined as $M = V/\alpha$, where α is the loss coefficient, is high for low loss silica fibers.

Since the Faraday effect does not affect the strain state in fiber, the Faraday rotation can be simply combined with birefringence and torsion-induced rotation. In case of negligible torsion, the polarization evolution in a PMF with phase retardation of δ can be described [69,72]

$$\boldsymbol{E}(z) = \begin{pmatrix} A & -B \\ B & A^* \end{pmatrix} \boldsymbol{E}(0), \tag{6.33}$$

where $A = \cos(\phi/2) + j \cos \chi \sin(\phi/2)$ and $B = \sin \chi \sin(\phi/2)$ are denoted with $\phi = 2\sqrt{\delta^2/4 + \Delta\varphi^2}$ and $\tan \chi = 2\Delta\varphi/\delta$. It is seen that an elliptically polarized wave results even if the input is a linearly polarized wave; and the polarization rotation, measured by rotation of the ellipsoid axis, is no longer the Faraday rotation $\Delta\varphi$, and is even not proportional to it.

6.3.2 Electric Current Sensor Based on Faraday Rotation

It is attractive to use Faraday effect for the electric current sensor, since current monitoring and metering is of critical importance in the power industry, whereas the traditional transformer becomes cumbersome and costs increasingly more to meet the strict requirement of insulation for high voltages. That is why the optical fiber current sensor is so attractive. The principle of the Faraday current sensor is understandable: the Faraday rotation of the wave in the fiber loop round a power cable is proportional to the line integral of the magnetic field; it is just equal to the current in the wire, according to Ampere theorem. In practical applications, however, several technical problems have to be dealt with and solved.

6.3.2.1 Basic Configurations   Faraday current sensors have been developed for practical field applications. Figure 6.13(a) gives a conceptual diagram to illustrate its operation [73], where the fiber loop may be composed of several circles to increase the rotation angle. The source light is polarized by the polarizer (P); the returned wave is analyzed by a polarization beam splitter (PBS), and received by two detectors. The Faraday rotation is acquired from the two signals P_1 and P_2:

$$\frac{P_1 - P_2}{P_1 + P_2} = C \sin 2\varphi_F, \qquad (6.34)$$

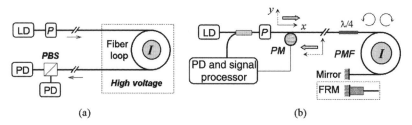

(a)   (b)

Figure 6.13   Faraday sensor: (a) basic structure and (b) reflective sensor.

where C is a constant, dependent on perfectness of the polarizer and PBS. The Faraday rotation is written as $\varphi_F = VB \cdot N2\pi r = NVI$, where N is the circle number, $I = 2\pi r B$ is the current to be measured.

This plain structure has problems of instability and external disturbances. The sensor is then improved greatly for practical applications. Figure 6.13(b) shows schematically a structure well developed [74–76], in which PMF loop is used and a mirror is set at the end of the fiber loop, and the doubled Faraday rotation is measured in the reflected beam, $\theta = 2\varphi_F$. The configuration brings about another merit, that is, reciprocal interferences, such as additional birefringence, can be cancelled. The end mirror can be replaced by a FRM to remove the external disturbances more effectively. A $\lambda/4$ wave plate is inserted to convert the input wave to a circularly polarized wave, which in combination with the polarizer plays the role of isolator; and the polarizer plays also a role of polarization analyzer for the reflected beam. A phase modulator (PM) is inserted to set the quadrature working point and to suppress the noise, similar to the method in fiber optic gyros.

Another configuration is based on SI, as shown in Figure 6.14 [74], in which more reciprocal interferences can be removed. The detected Faraday rotation θ is doubled once again by CW and CCW interference: $\theta = 4\varphi_F$. The depolarized Sagnac loop is also used for current sensing [77]. It is noted that the $\lambda/4$ wave plate used in the structures of Figures 6.13(b) and 6.14(a) is a passive device, since it is located at the high voltage position, together with the sensing fiber coil, and is impossible to adjust electrically or manually. The fixed quarter wavelength

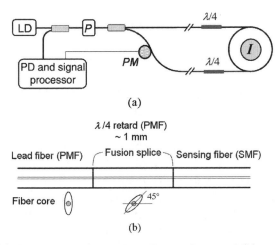

(a)

(b)

Figure 6.14 (a) Faraday sensor based on Sagnag loop and (b) passive fiber phase retard.

waveplate can be fabricated by orthogonally splicing a PMF section [75], as schematically shown in Figure 6.14(b).

Reference [76] summarizes the technologies of the Faraday sensor as one of the polarimetric optical fiber sensors. It is possible to construct a fiber sensor system composed of multiple Faraday sensors to meet practical demand. References [77–79] propose some configurations and discuss their interrogation and other technical problems. The polarization OTDR can also be used to detect distributed current signals [80].

6.3.2.2 *Precision and Stability*   As a current meter and monitor, the Faraday sensor must be highly precise, stable, and have a dynamic range large enough. For example, the nominal primary current is 1,000 A, the dynamic range of 50~1,200 A is required with the measurement precision of 0.5% [73].

Temperature dependence is the main factor of instability. It was measured that the Verdet constant varies with temperature, $V^{-1}(dV/dT) = 0.69 \times 10^{-4} K^{-1}$ for silica at 633 nm; and the thermal expansion of fiber should be taken into account, fortunately its contribution is negligibly small for silica [81]. Materials other than silica were investigated, such as flint glass and BK7 glass [72,81,82]. It is found that the thermal stability of diamagnetic materials is much better than that of paramagnetic materials, though the latter has a higher Verdet constant [73]. The temperature dependence of optical components, such as the coupler and the phase retarder, needs to be dealt with carefully. Mechanical stability is another factor, including influence of vibration and thermal strains; the fiber loop should be packaged robustly [74,75,83].

The Verdet constant is a function of optical frequency (6.31a and 6.31b), and is dependent on material property. It was measured [84,85] that roughly the Verdet constant is proportional to the square of frequency: $\sim 1.28 \, \lambda^{-2} \text{rad}/\text{T} \cdot \text{m}$ (λ in μm). Therefore, shorter wavelength sources are preferable; in practice, the diode laser at 800 nm band is often used.

The instability comes also from the lead fiber, which usually cannot be fixed in position tightly and often suffers vibration and sway, causing the polarization to be disturbed. PMF has proved helpful in reducing the influence [75,83]. A depolarized sensor structure is proposed and studied experimentally [78]. A circular-polarization-maintaining optical fiber is shown to have better immunity against the disturbance of twisting [86].

The external disturbances give serious influence, especially for direct current sensing. In case the sensor is used for alternating currents, the Faraday rotation is modulated by the AC frequency: $\varphi_F = NVI_0 \sin \omega t = \varphi_{F0} \sin \omega t$; and the detected signal is written as

$$i \propto \sin(2\varphi_{F0} \sin \omega t) = 2 \sum_{m=1}^{\infty} J_{2m-1}(2\varphi_{F0}) \sin[(2m-1)\omega t]. \qquad (6.35)$$

The Faraday rotation can then be calculated by a frequency SA, whereas the external disturbances out of the AC frequency are removed greatly. The signal processing also gives out data on the AC frequency and its phase, which are important parameters in the power industry. On the other hand, linear signals in a large dynamic range are obtained by data processing. Schemes of a close loop system with an adjustable PM are developed for the linear signal [75,78].

Faraday effect is also widely used for fiber optic devices, especially the optical isolator [70,72]. Detailed characteristics and analyses can be found in the literature.

6.4 FIBER SENSORS BASED ON SURFACE PLASMON EFFECT

The detection of chemical and biochemical samples is important in various aspects of human life and industrial production. Quite a lot of methods have been developed. Among them optical fiber sensors have attractive features. This section gives a brief introduction to the sensors based on surface plasmon effect.

6.4.1 Surface Plasmon Effect

Electrons in metals and semiconductors can move freely, their behaviors in an electromagnetic field show properties of the plasma. The optical constants of metal are analyzed in [68]. By referring to Equation (6.27), the motion of electrons in cases where there is no external magnetic field, is described as

$$r = \frac{e/m}{\omega(\omega + j\gamma)} E. \qquad (6.36)$$

The dielectric constant is expressed as

$$\varepsilon = 1 + (P_{\text{ion}} - Ner)/E = \varepsilon_L - \frac{Ne^2/m}{\omega(\omega + j\gamma)}, \qquad (6.37)$$

where P_{ion} and ε_L stands for the contributions to electric polarization and dielectric constant from ions and the lattice, respectively. The real and imaginary parts of the dielectric constant are then written as

$$\varepsilon_r = \varepsilon_L - \frac{Ne^2}{m(\omega^2 + \gamma^2)}, \tag{6.38a}$$

$$\varepsilon_i = \frac{Ne^2\gamma}{m\omega(\omega^2 + \gamma^2)}. \tag{6.38b}$$

The expressions are just the well-known Drude formula, considering the damping rate γ is related with the conductivity σ by relation of $\varepsilon_i = \sigma/\omega$. It is shown that ε_r is negative for low frequency and positive for high frequency, and reaches zero at frequency of $\omega_c^2 = Ne^2/m\varepsilon_L - \gamma^2$, which is termed the state of plasma resonance. It is seen that the larger the electron concentration, the higher is the plasma resonance frequency. The resonance frequency of a metal with high conductivity is in the ultraviolet band, much higher than that of semiconductors. At the resonance the imaginary part plays the main role in determining its optical behavior; and the imaginary part of the index reaches its maximum, meaning a loss peak. The refractive index of metals is a complex, usually with $\varepsilon_r < 0$ and $n_i \gg n_r$ in the visible and infrared band. The measured data of the metals are given in reference [68]. Reference [87] gives a formula for the dielectric constant with dependence on wavelength:

$$\varepsilon(\lambda) = 1 - \frac{\lambda^2\lambda_c}{\lambda_p^2(\lambda_c + j\lambda)}. \tag{6.39}$$

The parameters are measured as $\lambda_p = 0.16826$ μm and $\lambda_c = 8.9342$ μm for gold; $\lambda_p = 0.14541$ μm and $\lambda_c = 17.614$ μm for silver. Thus the dielectric constants at 1,550 nm band are calculated as $\varepsilon_{Au} = -80.39 + j14.12$. and $\varepsilon_{Ag} = -111.75 + j9.92$. The dielectric constants measured for other wavelengths are reported: $\varepsilon_{Au} = -12.2 + j1.3$ at 633 nm [88], $\varepsilon_{Ag} = -30.7 + j2.29$ at 830 nm [89].

It is known that in the total internal reflection at the interface between two media the optical wave penetrates the interface, being an evanescent field with amplitude of $\propto \exp(-x/d_{ev})$, where x is perpendicular to the interface. The depth is deduced as

$$d_{ev} = \frac{n_1\lambda}{2\pi\sqrt{\sin^2\theta_i - n_2^2/n_1^2}}, \tag{6.40}$$

where θ_i is the incident angle from medium n_1. When a thin layer of metal is coated on the surface of a medium, or a metal is placed in the evanescent field area, the electrons will be driven by the field, causing an electron density oscillation and a secondary electromagnetic wave, called a surface plasmon wave (SPW). Obviously the plasmon intensity depends on the frequency, and reaches resonance at ω_c. This is the basic physical picture of surface plasmon resonance (SPR).

The *TE* and *TM* incident waves have different characteristics at the interface between the metal and the dielectric medium. The electric field of the *TE* wave is in y-direction, perpendicular to the incident plane; it suffers larger loss in the metal layer. In other words, the *TE* wave is screened from entering the metal, so that $E_y^{(TE)}$ at the interface is minimized with a zero evanescent field. The conclusion can also be derived from the boundary conditions of the electric and magnetic fields. The *TM* wave has components in x- and z-directions; the x-component will penetrate into the thin metal layer. This behavior is similar to the reflections of *P*- and *S*-components of the incident beam at medium surface.

The property of SPW is investigated [87–90] by analyzing the electromagnetic wave at the interface, as shown in Figure 6.15(a), where region of $x < 0$ is the metal; the dielectric constant $\varepsilon_1 = n_1^2$ is a real with a negligible loss in region $x > 0$. The electric fields of a *TM* wave are written as

$$\boldsymbol{E}_1 = (E_{1x}, 0, E_{1z})^T \, \exp[j(k_{1x}x + \beta z)], \qquad (6.41a)$$

$$\boldsymbol{E}_2 = (E_{2x}, 0, E_{2z})^T \, \exp[j(k_{2x}x + \beta z)]. \qquad (6.41b)$$

It is deduced from $\nabla \cdot \vec{E} = 0$ that $k_{1x}E_{1x} + \beta E_{1z} = 0$ and $k_{2x}E_{2x} + \beta E_{2z} = 0$. The boundary conditions require $E_{1z} = E_{2z}$ and $\varepsilon_1 E_{1x} =$

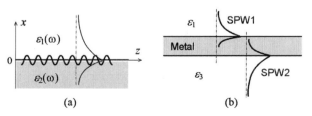

(a) (b)

Figure 6.15 Illustration of SPW.

$\varepsilon_2 E_{2x}$; therefore, $k_{2x}/k_{1x} = \varepsilon_2/\varepsilon_1$. By substituting $k_{1x} = \sqrt{\varepsilon_1 k_0^2 - \beta^2}$ and $k_{2x} = \sqrt{\varepsilon_2 k_0^2 - \beta^2}$, it is deduced that

$$\beta = n_1 k_0 \sqrt{\frac{\varepsilon_2}{\varepsilon_1 + \varepsilon_2}} = n_1 k_0 \sqrt{\frac{\varepsilon_{2r} + j\varepsilon_{2i}}{\varepsilon_1 + \varepsilon_{2r} + j\varepsilon_{2i}}}. \tag{6.42}$$

Its real and imaginary parts are deduced to be

$$\beta_r = \frac{\sqrt{2}}{2} n_1 k_0 \left(\frac{\sqrt{\varepsilon_e^4 + \varepsilon_1^2 \varepsilon_{2i}^2} + \varepsilon_e^2}{(\varepsilon_1 + \varepsilon_{2r})^2 + \varepsilon_{2i}^2} \right)^{1/2}, \tag{6.43a}$$

$$\beta_i = \frac{\sqrt{2}}{2} n_1 k_0 \left(\frac{\sqrt{\varepsilon_e^4 + \varepsilon_1^2 \varepsilon_{2i}^2} - \varepsilon_e^2}{(\varepsilon_1 + \varepsilon_{2r})^2 + \varepsilon_{2i}^2} \right)^{1/2}, \tag{6.43b}$$

where $\varepsilon_e^2 = \varepsilon_{2r}^2 + \varepsilon_{2i}^2 + \varepsilon_1 \varepsilon_{2r}$. The x-components of wave vectors are obtained:

$$k_{1x} = \frac{\varepsilon_1 k_0}{\sqrt{\varepsilon_1 + \varepsilon_2}}, \tag{6.44a}$$

$$k_{2x} = \frac{\varepsilon_2 k_0}{\sqrt{\varepsilon_1 + \varepsilon_2}}. \tag{6.44b}$$

Both of them are complex; for the wave confound at two sides of the interface, the imaginary part of k_{1x} must be positive, whereas that of k_{2x} must be negative. This condition is ensured by the large negative ε_{2r} of the metal. The real and imaginary parts of k_{1x} are deduced to be

$$k_{1x}^{(r)} = \frac{\varepsilon_1 k_0}{\sqrt{2}} \left(\frac{\sqrt{(\varepsilon_1 + \varepsilon_{2r})^2 + \varepsilon_{2i}^2} + (\varepsilon_1 + \varepsilon_{2r})}{(\varepsilon_1 + \varepsilon_{2r})^2 + \varepsilon_{2i}^2} \right)^{1/2}, \tag{6.45a}$$

$$k_{1x}^{(i)} = \frac{\varepsilon_1 k_0}{\sqrt{2}} \left(\frac{\sqrt{(\varepsilon_1 + \varepsilon_{2r})^2 + \varepsilon_{2i}^2} - (\varepsilon_1 + \varepsilon_{2r})}{(\varepsilon_1 + \varepsilon_{2r})^2 + \varepsilon_{2i}^2} \right)^{1/2}. \tag{6.45b}$$

The formulas can be approximated under condition of $|\varepsilon_{2r}| \gg \varepsilon_1 \sim \varepsilon_{2i}$; and the propagation constant in z-direction is obtained as

$$\beta_r \approx \left(1 - \frac{\varepsilon_1}{2\varepsilon_{2r}}\right) n_1 k_0, \qquad (6.46a)$$

$$\beta_i \approx \frac{\varepsilon_1 \varepsilon_{2i}}{2\varepsilon_{2r}^2} n_1 k_0. \qquad (6.46b)$$

It is seen that SPW decays in z-direction, and the attenuation depends on the optical frequency.

For a thin metal layer, SPW will occur on both of its sides, as shown in Figure 6.15(b). The detailed analysis involves a three-layer waveguide with a high loss core [90]. Or phenomenally, ε_2 in (6.42) is regarded as a composite dielectric constant of the metal and medium ε_3. In practical applications, medium ε_3 is usually a fluid; the chemical and biochemical molecules being detected are solved in the fluid and adsorbed partly on the metal surface, contributing to the composite dielectric constant.

6.4.2 Sensors Based on SPW

The SPW is generally excited by the evanescent field in total internal reflection (TIR). Two basic configurations have been developed, as shown in Figure 6.16. Figure 6.16(a) is the Kretschmann configuration, with the evanescent field tunneling through a thin metal layer; and Figure 6.16(b) is the Otto configuration, with the evanescent field extending across a dielectric gap to excite the plasmon on the metal surface [90].

The total internal reflection occurring at one of the surfaces of the glass prism is usually utilized for the SPW sensor. The prism surface is coated with a metal layer, or is pressed by a metal plate. An

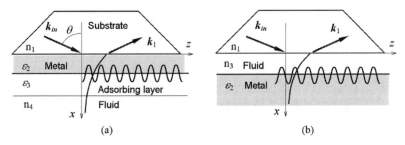

Figure 6.16 Excitations of SPW: (a) Kretschmann and (b) Otto configurations.

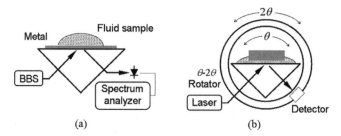

Figure 6.17   Measurements of ATR: (a) varied with wavelength and (b) varied with incident angle.

adsorbing layer may be coated on the metal layer in configuration (a), to enhance the sensitivity and the selectivity of the sensor. The reflected beam suffers certain attenuation caused by the loss of specimen being tested. Characteristics of the configurations can be analyzed as the substrate mode of a three layer waveguide, with a parameter of incident beam wavevector $\beta_r = n_1 k_0 \sin \theta$; or by the theoretical model of multilayer dielectric coating and attenuated total reflection (ATR), as done in [91,92].

The characteristics of ATR are usually interrogated by two methods: measurement of the attenuation varied with the wavelength, and varied with the incident angle. Figures 6.17(a) and (b) show the two methods schematically, where Kretschmann and Otto configurations are exchangeable, that is, both of them can be characterized by the spectrum analysis, or by angle scanning. Reference [91] gives detailed analysis and curve fitting equations. Such apparatus, containing necessary optical components, such as polarizers, and electronics, are already commercialized [93–95].

The development of optical waveguide devices and fiber devices, waveguide and fiber SPW sensors has become a hot topic of R&D, because the evanescent field exists already in waveguide structures, and the fiber sensor is attractive for its small size, low cost, and ease of usage. References [88,96] describe planar waveguide SPW sensors on glass slides. Quite a number of papers show research on optical fibers. Several schemes have been demonstrated. One of them is based on a thinned fiber [92,97–99], as depicted in Figure 6.18(a). The other has D-shaped fibers [89,100,101], as shown in Figure 6.18(b).

Fiber grating SPW sensors has attracted wide interest [101–105]. The long-period gratings have been used to detect the index of the environmental medium; the SPW sensor with metal-coated LPFG shows more functions and improved performance. Several structures are proposed to make SPW sensors with the FBG type, such as thinned and D-shaped

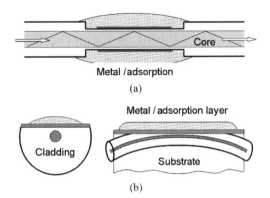

Figure 6.18 SPW sensors with (a) thinned fiber and (b) D-shaped fiber.

structures. A hollow core fiber with metal coating in the inner wall is fabricated as a special SPW sensor [102]; and the tilted FBG is also studied for the SPW sensor applications since quite a lot of cladding modes are excited in the structure [105]. It is noticed that the fiber taper has geometry of thinned tip or a thinned waist, and the core mode is expanded into the cladding region; such features are thought favorable for SPW sensing, and the sensor with a small tip provides convenience in applications [106,107].

Two measurands are of concern in chemical and biochemical sensors: identity of the specimen and the quantity (concentration). It is difficult to fulfill such a task in a single measurement. Reference [93] proposes a self-referencing SPR sensor, which uses a TIR prism with two region coatings: gold layer and gold/Ta_2O_5 double layer, as shown in Figure 6.19(a) and separate SPR peaks are measured in the spectrum. Reference [108] proposes a special prism, shown in Figure 6.19(b), by which SPR spectrum with two different incident angles is obtained in a single measurement.

An issue of especial concern is how to enhance the sensitivity of the sensor. Selection of the metal is important for the SPW sensors. It is seen that the dielectric constant of silver has a larger negative real part

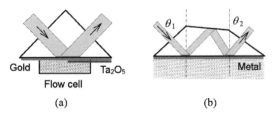

Figure 6.19 Schemes of SPW sensor: (a) double region and (b) two angles.

than gold, which is favorable for excitation of the surface plasmon [97]. However, the chemical stability of gold is better than silver. Therefore a Au/Ag double layer is preferable [99]. A composite layer with multiple dielectric coating and organic cap layer, such as thiol material, is also proposed and demonstrated.

It is also found experimentally that a metal coating with rough surface is beneficial to the SPW excitation than that with smooth surface. Moreover, the surface plasmon will enhance fluorescence emission, which is utilized to develop a new sensor, named the surface plasmon enhanced fluorescence sensor (SPFS) [109].

Similar structures are used in surface enhanced Raman spectroscope (SERS) [110]. In the SERS sensors, nanometer gold balls are deposited onto the surface of the fiber or planar waveguide; and sharp and strong Raman signals are measured. SERS is an attractive topic, but beyond the scope of this book.

PROBLEMS

6.1 How does the phase shift of the Sagnac loop depend on the angular velocity of rotation? How dos it depend on the area of loop? Deduce the formula of the Sagnac phase shift.

6.2 What roles does a PM play in IFOG? How do you set the modulation frequency?

6.3 Why will Rayleigh scattering in the loop fiber induce parasitic signals? How do you reduce the effect?

6.4 Why and how does the Kerr effect induce parasitic signals?

6.5 Why do polarization properties induce noise? Why does the cross-coupling between the two polarization modes play an important role? How do you reduce the effect?

6.6 Why do temperature distribution and its temporal variation in the fiber spool induce nonreciprocity? How do you reduce the effect?

6.7 Why can the fiber optic gyro be used to localize the latitude?

6.8 Describe the structures of IFOG and RFOG. What are their differences and features?

6.9 Describe the most often used structures of acoustic pressure transduction.

6.10 Which interferometers are mostly used in hydrophones? Describe their structures.

6.11 Why should a modulator be inserted in the interferometer of a hydrophone? What is the PGC method? Deduce its basic formulas.

6.12 Why is the sensor array important in practical applications? How do you interrogate the signals from the sensor array?

6.13 Why must a low noise laser source be used in high-sensitivity hydrophones?

6.14 Why can the Faraday effect be used as a current sensor? Deduce the formula of polarization rotation varied with the current.

6.15 Describe the most often used configurations of fiber Faraday sensors. What factors affect the performance of the sensor? How do you improve the precision and stability of the sensor?

6.16 Describe the property of the metal dielectric constant. What is the difference between properties of metal and dielectric medium? What is the SPW? How do you excite the SPW?

6.17 Explain the total internal reflection and attenuated total reflection. Describe the basic structures of SPW sensors. How does the SPW sensor detect index change? How does it detect chemical and biochemical specimens?

REFERENCES

1. Bergh RA, Lefevre HC, Shaw HJ. All-single-mode fiber-optic gyroscope. *Optics Letters* 1981; 6: 198–200.
2. Lefvère HC. *The Fiber-Optic Gyroscope*. London, Boston: Artech House Inc., 1993.
3. Sanders GA, Szafraniec B, Liu RY, Laskoskie C, Strandjord L, Weed G. Fiber optic gyros for space, marine and aviation applications. *Proceeding of SPIE* 1996; 2837: 61–71.
4. Lee B. Review of the present status of optical fiber sensors. *Optical Fiber Technology* 2003; 9: 57–79.
5. Culshaw B. The optical fibre Sagnac interferometer: an overview of its principles and applications. *Measurement Science and Technology* 2006; 17: R1–R16.
6. Leeb WR, Schiffner G, Scheiterer E. Optical fiber gyroscopes: Sagnac or Fizeau effect? *Applied Optics* 1979; 18: 1293–1295.

7. Arditty HJ, Lefevre HC. Sagnac effect in fiber gyroscopes. *Optics Letters* 1981; 6: 401–403.

8. Bergh RA, Lefevre HC, Shaw HJ. An overview of fiber-optic gyroscopes. *Journal of Lightwave Technology* 1984; 2: 91–107.

9. Ruffin PB. Fiber optic gyroscope sensors. Section 8 of *Fiber Optic Sensor*, edited by Yin S, Ruffin PB, Francis TS, Yu FTS. Second Edition. Boca Raton, FL: CRC Press, Taylor & Francis Group, 2008.

10. Bergh RA, Lefevre HC, Shaw HJ. All-single-mode fiber-optic gyroscope with long-term stability. *Optics Letters* 1981; 6: 502–504.

11. Moeller RP, Burns WK, Frigo NJ. Open-loop output and scale factor stability in a fiber-optic gyroscope. *Journal of Lightwave Technology* 1989; 7: 262–269.

12. Trommer GF, Poisel H, Buhler W, Hartl E, Muller R. Passive fiber optic gyroscope. *Applied Optics* 1990; 29: 5360–5365.

13. Kim BY, Shaw HJ. Gated phase-modulation approach to fiber-optic gyroscope with linearized scale factor. *Optics Letters* 1984; 9: 375–377.

14. Pavlath GA. Closed-loop fiber optic gyros. *Proceeding of SPIE* 1996; 2837: 46–60.

15. Cutler CC, Newton SA, Shaw HJ. Limitation of rotation sensing by scattering. *Optics Letters* 1980; 5: 488–490.

16. Bohm K, Russer P, Weidel E, Ulrich R. Low-noise fiber-optic rotation sensing. *Optics Letters* 1981; 6: 64–66.

17. Bohm K, Marten P, Petermann K, Weidel E, Ulrich R. Low-drift fiber gyro using a superluminescent diode. *Electronics Letters* 1981; 17: 352–353.

18. Peng GD, Huang SY, Lin ZQ. Intensity noise characteristics of lasers in fiber-optic gyroscopes. *Optics Letters* 1987; 12: 434–436.

19. Wysocki PF, Digonnet MJF, Kim BY, Shaw HJ. Characteristics of erbium-doped superfluorescent fiber sources for interferometric sensor applications. *Journal of Lightwave Technology* 1994; 12: 550–567.

20. Bergh RA, Culshaw B, Cutler CC, Lefevre HC, Shaw HJ. Source statistics and the Kerr effect in fiber-optic gyroscopes. *Optics Letters* 1982; 7: 563–565.

21. Agrawal GP. *Nonlinear Fiber Optics*. Singapore: Elsevier Science, 2004.

22. Burns WK, Moeller RP, Villarruel CA, Abebe M. Fiber-optic gyroscope with polarization-holding fiber. *Optics Letters* 1983; 8: 540–542.

23. Carrara SLA, Kim BY, Shaw HJ. Bias drift reduction in polarization-maintaining fiber gyroscope. *Optics Letters* 1987; 12: 214–216.

24. Chien PY, Pan CL. Fiber-optic gyroscopes based on polarization scrambling. *Optics Letters* 1991; 16: 189–190.

25. Szafraniec B, Sanders GA. Theory of polarization evolution in interferometric fiber-optic depolarized gyros. *Journal of Lightwave Technology* 1999; 17: 579–590.

26. Shupe DM. Fiber resonator gyroscope: sensitivity and thermal nonreciprocity. *Applied Optics* 1981; 20: 286–289.

27. Ruffin PB, Lofts C, Sung CC, Page JL. Reduction of nonreciprocity noise in wound fiber optic interferometers. *Optical Engineering* 1994; 33: 2675–2679.

28. Lofts CM, Ruffin PB, Parker M, Sung CC. Investigation of the effects of temporal thermal-gradients in fiber optic gyroscope sensing coils. *Optical Engineering* 1995; 34: 2856–2863.

29. Sawyer J, Ruffin PB, Sung CC. Investigation of the effects of temporal thermal gradients in fiber optic gyroscope sensing coils, part 2. *Optical Engineering* 1997; 36: 29–34.

30. Ruffin PB, Baeder J, Sung CC. Study of ultraminiature sensing coils and the performance of a depolarized interferometric fiber optic gyroscope. *Optical Engineering* 2001; 40: 605–611.

31. Kim DH, Kang JU. Sagnac loop interferometer based on polarization maintaining photonic crystal fiber with reduced temperature sensitivity. *Optics Express* 2004; 12: 4490–4495.

32. Bohm K, Petermann K, Weidel E. Sensitivity of a fiber optic gyroscope to environmental magnetic fields. *Optics Letters* 1982; 7: 180–182.

33. Hotate K, Tabe K. Drift of an optical fiber gyroscope caused by the Faraday effect: influence of the earth's magnetic field. *Applied Optics* 1986; 25: 1086–1092.

34. Ojeda L, Chung H, Borenstein J. Precision-calibration of fiber-optic gyroscopes for mobile robot navigation. *Proceedings of the IEEE International Conference on Robotics and Automation* 2000: 2064–2069, San Francisco. vol 4.

35. Zheng J. Birefringent fibre frequency-modulated continuous-wave Sagnac gyroscope. *Electronics Letters* 2004; 40: 1520–1522.

36. Zarinetchi F, Smith SP, Ezekiel S. Stimulated Brillouin fiber-optic laser gyroscope. *Optics Letters* 1991; 16: 229–231.

37. Tanaka Y, Yamasaki S, Hotate K. Brillouin fiber-optic gyro with directional sensitivity. *IEEE Photonics Technology Letters* 1996; 8: 1367–1369.

38. Cole JH, Johnson RL, Bhuta PG. Fiber-optic detection of sound. *Journal of Acoustical Society America* 1977; 62: 1136–1138.

39. Bucaro JA, Dardy HD, Carome EF. Fiber-optic hydrophone. *Journal of Acoustical Society America* 1977; 62: 1302–1304.

40. Shajenko P, Flatley JP, Moffett MB. Fiberoptic hydrophone sensitivity. *Journal of Acoustical Society America* 1978; 64: 1286–1288.

41. Stanton TK, Pridham RG, Mccollough WV, Sanguinetti MP. Fiber-optic hydrophone noise-equivalent pressure. *Journal of Acoustical Society America* 1979; 66: 1893–1894.
42. Spillman WB, Gravel RL. Moving fiber-optic hydrophone. *Optics Letters* 1980; 5: 30–31.
43. Rashleigh SC. Acoustic sensing with a single coiled monomode fiber. *Optics Letters* 1980; 5: 392–394.
44. Giallorenzi TG, Bucaro JA, Dandridge A, Sigel GH, Cole JH, Rashleigh SC, Priest RG. Optical fiber sensor technology. *IEEE Journal of Quantum Electronics* 1982; 18: 626–665.
45. Garrett SL, Brown DA, Beaton BL, Wetterskog K, Serocki J. A general-purpose fiber-optic hydrophone made of castable epoxy. *Proceeding of SPIE* 1990; 1367: 13–29.
46. Dandridge A, Cogdell GB. Fiber optic sensors for navy applications. *IEEE LCS* 1991; 2: 81–89.
47. Nash P. Review of interferometric optical fibre hydrophone technology. *IEE Proceedings-Radar Sonar and Navigation* 1996; 143: 204–209.
48. Wang Z, Hu Y, Meng Z, Ni M. Fiber-optic hydrophone using a cylindrical Helmholtz resonator as a mechanical anti-aliasing filter. *Optics Letters* 2008; 33: 37–39.
49. Beverini N, Falciai R, Maccioni E, Morganti M, Sorrentino F, Trono C. Developing fiber lasers with Bragg reflectors as deep sea hydrophones. *Annals of Geophysics* 2006; 49: 1157–1165.
50. Bagnoli PE, Beverini N, Falciai R, Maccioni E, Morganti M, Sorrentino F, Stefani F, Trono C. Development of an erbium-doped fibre laser as a deep-sea hydrophone. *Journal of Optics A-Pure and Applied Optics* 2006; 8: S535–S539.
51. Hill DJ, Nash PJ, Jackson DA, Webb DJ, O'Neill SF, Bennion I, Zhang L. A fiber laser hydrophone array. *Proceedings SPIE* 1999; 3860: 55–66.
52. Hill DJ, Hodder B, De Freitas J, Thomas SD, Hickey L. DFB fibre-laser sensor developments. *Proceedings SPIE* 2005; 5855: 904–907.
53. Peng GD, Chu PL. Optical fiber hydrophone systems, Section 9 of *Fiber Optic Sensor*, edited by Yin S, Ruffin PB, Yu FTS. Second Edition. Boca Raton, FL: CRC Press, Taylor & Francis Group, 2008.
54. Digonnet MJF, Vakoc BJ, Hodgson CW, Kino GS. Acoustic fiber sensor arrays. *Proceedings SPIE* 2004; 5502: 39–50.
55. Koo KP, Tveten AB, Dandridge A. Passive stabilization scheme for fiber interferometers using (3 × 3) fiber directional couplers. *Applied Physics Letters* 1982; 41: 616–618.
56. Fritsch K, Adamovsky G. Simple circuit for feedback stabilization of a single-mode optical fiber interferometer. *Review of Scientific Instruments* 1981; 52: 996–1000.

57. Dandridge A, Tveten AB, Giallorenzi TG. Homodyne demodulation scheme for fiber optic sensors using phase generated carrier. *IEEE* Journal of *Quantum Electronics* 1982; 18: 1647–1653.

58. Kersey AD, Dandridge A, Tveten AB. Time-division multiplexing of interferometric fiber sensors using passive phase-generated carrier interrogation. *Optics Letters* 1987; 12: 775–777.

59. Dandridge A, Tveten AB, Kersey AD, Yurek AM. Multiplexing of interferometric sensors using phase carrier techniques. *Journal of Lightwave Technology* 1987; 5: 947–952.

60. Lim TK, Zhou Y, Lin Y, Yip YM, Lam YL. Fiber optic acoustic hydrophone with double Mach-Zehnder interferometers for optical path length compensation. *Optics Communications* 1999; 159: 301–308.

61. Nash P, Cranch G, Cheng LK, de Bruijn D, Crowe I. A 32 element TDM optical hydrophone array. *Proceedings SPIE* 1998; 3483: 238–242.

62. Cranch GA, Nash PJ. Large-scale multiplexing of interferometric fiber-optic sensors using TDM and DWDM. *Journal of Lightwave Technology* 2001; 19: 687–699.

63. Cranch GA, Nash PJ, Kirkendall CK. Large-scale remotely interrogated arrays of fiber-optic interferometric sensors for underwater acoustic applications. *IEEE Sensors Journal* 2003; 3: 19–30.

64. Cranch GA, Kirkendall CK, Daley K, Motley S, Bautista A, Salzano J, Nash PJ, Latchem J, Crickmore R. Large-scale remotely pumped and interrogated fiber-optic interferometric sensor array. *IEEE Photonics Technology Letters* 2003; 15: 1579–1581.

65. Brooks JL, Moslehi B, Kim BY, Shaw HJ. Time-domain addressing of remote fiber-optic interferometric sensor arrays. *Journal of Lightwave Technology* 1987; 5: 1014–1023.

66. Cranch GA. Frequency noise reduction in erbium-doped fiber distributed-feedback lasers by electronic feedback. *Optics Letters* 2002; 27: 1114–1116.

67. Cranch GA, Flockhart GMH, Kirkendall CK. Distributed feedback fiber laser strain sensors. *IEEE Sensors Journal* 2008; 8: 1161–1172.

68. Born M, Wolf E. *Principles of Optics*. Seventh Edition. Cambridge: Cambridge University Press, 1999.

69. Smith AM. Polarization and magneto-optic properties of single-mode optical fiber. *Applied Optics* 1978; 17: 52–56.

70. Stolen RH, Turner EH. Faraday rotation in highly birefringent optical fibers. *Applied Optics* 1980; 19: 842–845.

71. Munin E, Roversi JA, Villaverde AB. Faraday effect and energy gap in optical-materials. *Journal of Physics D-Applied Physics* 1992; 25: 1635–1639.

72. Shiraishi K, Sugaya S, Kawakami S. Fiber Faraday rotator. *Applied Optics* 1984; 23: 1103–1106.

73. Papp A, Harms H. Magneto-optical current transformer. 1: principles. *Applied Optics* 1980; 19: 3729–3734.

74. Bohnert K, Gabus P, Nehring J, Brandle H. Temperature and vibration insensitive fiber-optic current sensor. *Journal of Lightwave Technology* 2002; 20: 267–276.

75. Bohnert K, Gabus P, Kostovic J, Brandle H. Optical fiber sensors for the electric power industry. *Optics and Lasers in Engineering* 2005; 43: 511–526.

76. Michie C. Polarimetric optical fiber sensors. Section 3 of *Fiber Optic Sensor*, edited by Yin S, Ruffin PB, Francis TS, Yu FTS. Second Edition. CRC Press, Taylor & Francis Group, 2008.

77. Lin H, Huang SC. Fiber-optics multiplexed interferometric current sensors. *Sensors and Actuators A* 2005; 121: 333–338.

78. Takahashi M, Sasaki K, Ohno A, Hirata Y, Terai K. Sagnac interferometer-type fibre-optic current sensor using single-mode fibre down leads. *Measurement Science and Technology* 2004; 15: 1637–1641.

79. Guan ZG, He S. Coherence multiplexing system based on asymmetric mach-zehnder interferometers for Faraday sensors. *IEEE Photonics Technology Letters* 2007; 19: 1907–1909.

80. Kim BY, Park D, Choi SS. Use of polarization-optical time domain reflectometry for observation of the Faraday effect in single-mode fibers. *IEEE Journal of Quantum Electronics* 1982; 18: 455–456.

81. Williams PA, Rose AH, Day GW, Milner TE, Deeter MN. Temperature-dependence of the Verdet constant in several diamagnetic glasses. *Applied Optics* 1991; 30: 1176–1178.

82. Kurosawa K, Yoshida S, Sakamato K, Masuda I, Yamashita T. An optical fiber-type current sensor utilizing the Faraday effect of the flint glass fiber. *Proceeding of SPIE Tenth Optical Fiber Sensors Conference* 1994; 24–27, Glasgow, vol. 2360.

83. Short SX, Tantaswadi P, deCarvalho RT. An experimental study of acoustic vibration effects in optical fiber current sensors. *IEEE Transactions on Power Delivery* 1996; 11: 1702–1706.

84. Rose AH, Etzel SM, Wang CM. Verdet constant dispersion in annealed optical fiber current sensors. *Journal of Lightwave Technology* 1997; 15: 803–807.

85. Cruz JL, Andres MV, Hernandez MA. Faraday effect in standard optical fibers: dispersion of the effective Verdet constant. *Applied Optics* 1996; 35: 922–927.

86. Huang H. Practical circular-polarization-maintaining optical fiber. *Applied Optics* 1997; 36: 6968–6975.

87. Homola J. On the sensitivity of surface plasmon resonance sensors with spectral interrogation. *Sensors and Actuators B* 1997; 41: 207–211.

88. Homola J, Ctyroky J, Skalsky M, Hradilova J, Kolarova P. A surface plasmon resonance based integrated optical sensor. *Sensors and Actuators B* 1997; 39: 286–290.

89. Homola J. Optical fiber sensor based on surface plasmon excitation. *Sensors and Actuators B* 1995; 29: 401–405.

90. Yeatman EM. Resolution and sensitivity in surface plasmon microscopy and sensing. *Biosensors and Bioelectronics* 1996; 11: 635–649.

91. Kurihara K, Nakamura K, Suzuki K. Asymmetric SPR sensor response curve-fitting equation for the accurate determination of SPR resonance angle. *Sensors and Actuators B* 2002; 86: 49–57.

92. Lin WB, Lacroixa M, Chovelon JM, Jaffrezic-Renault N, Gagnaire H. Development of a fiber-optic sensor based on surface plasmon resonance on silver film for monitoring aqueous media. *Sensors and Actuators B* 2001; 75: 203–209.

93. Boozer C, Yu Q, Chen S, Lee CY, Homola J, Yee SS, Jiang S. Surface functionalization for self-referencing surface plasmon resonance (SPR) biosensors by multi-step self-assembly. *Sensors and Actuators B* 2003; 90: 22–30.

94. Naimushin AN, Soelberg SD, Bartholomew DU, Elkind JL, Furlong CE. A portable surface plasmon resonance (SPR) sensor system with temperature regulation. *Sensors and Actuators B* 2003; 96: 253–260.

95. Johnston KS, Booksh KS, Chinowsky TM, Yee SS. Performance comparison between high and low resolution spectrophotometers used in a white light surface plasmon resonance sensor. *Sensors and Actuators B* 1999; 54: 80–88.

96. Karlsen SP, Johnston KS, Yee SS, Jung CC. First-order surface plasmon resonance sensor system based on a planar light pipe. *Sensors and Actuators B* 1996; 32: 137–141.

97. Iga M, Seki A, Watanabe K. Hetero-core structured fiber optic surface plasmon resonance sensor with silver film. *Sensors and Actuators B* 2004; 101: 368–372.

98. Peng W, Banerji S, Kim YC, Booksh KS. Investigation of dual-channel fiber-optic surface plasmon resonance sensing for biological applications. *Optics Letters* 2005; 30: 2988–2990.

99. Dostalek J, Vaisocherova H, Homola J. Multichannel surface plasmon resonance biosensor with wavelength division multiplexing. *Sensors and Actuators B* 2005; 108: 758–764.

100. Slavik R, Homola J, Ctyroky J. Single-mode optical fiber surface plasmon resonance sensor. *Sensors and Actuators B* 1999; 54: 74–79.

101. Tripathi SM, Kumar A, Marin E, Meunier JP. Side-polished optical fiber grating-based refractive index sensors utilizing the pure surface plasmon polariton. *Journal of Lightwave Technology* 2008; 26: 1980–1985.

102. Nenova G, Kashyap R. Modeling of plasmon-polariton refractive-index hollow core fiber sensors assisted by a fiber Bragg grating. *Journal of Lightwave Technology* 2006; 24: 3789–3796.

103. He YJ, Lo YL, Huang JF. Optical-fiber surface-plasmon-resonance sensor employing long-period fiber gratings in multiplexing. *Journal of Optical Society America B* 2006; 23: 801–811.

104. Nemova G, Kashyap R. Fiber-Bragg-grating-assisted surface plasmon-polariton sensor. *Optics Letters* 2006; 31: 2118–2120.

105. Shevchenko YY, Albert J. Plasmon resonances in gold-coated tilted fiber Bragg gratings. *Optics Letters* 2007; 32: 211–213.

106. Grunwald B, Holst G. Fibre optic refractive index microsensor based on white-light SPR excitation. *Sensors and Actuators A* 2004; 113: 174–180.

107. Monzon-Hernandez D, Villatoro J, Talavera D, Luna-Moreno D. Optical-fiber surface-plasmon resonance sensor with multiple resonance peaks. *Applied Optics* 2004; 43: 1216–1220.

108. Gupta BD, Sharma AK. Sensitivity evaluation of a multi-layered surface plasmon resonance-based fiber optic sensor: a theoretical study. *Sensors and Actuators B* 2005; 107: 40–46.

109. Attridge JW, Daniels PB, Deacon JK, Robinson GA, Davidson GP. Sensitivity enhancement of optical immunosensors by the use of a surface-plasmon resonance fluoroimmunoassay. *Biosensors and Bioelectronics* 1991; 6: 201–214.

110. Haynes CL, McFarland AD, Van Duyne RP. Surface-enhanced Raman spectroscopy. *Analytical Chemistry* 2005; 77: 338a–346a.

CHAPTER 7

EXTRINSIC FIBER FABRY–PEROT INTERFEROMETER SENSOR

Chapter 7 is engaged mainly in the extrinsic fiber Fabry–Perot interferometer (EFFPI) sensor, which is usually composed of a fiber end facet and a mirror. Some intrinsic FFP sensors, in which the Fabry–Perot cavity is built inside the fiber without other components incorporated, are described also. The EFFP sensors are used widely in displacement sensing, pressure sensing, acoustic and ultrasonic sensing, pickup of atomic force microscope (AFM) signals, and other applications. The structures and principles of EFFPI sensors are introduced in Section 7.1. The theoretical analysis on the Gaussian beam Fabry–Perot interferometer is presented in Section 7.2. The basic characteristics and performances of EFFPI sensors are discussed in Section 7.3. The last section introduces applications of EFFPI sensors and discusses the technical issues involved.

7.1 BASIC PRINCIPLES AND STRUCTURES OF EXTRINSIC FIBER F-P SENSORS

The extrinsic fiber FPI sensors were demonstrated earlier in the 1980s [1–3], showing their features and usefulness. The EFFPI sensors have

Fundamentals of Optical Fiber Sensors, First Edition.
Zujie Fang, Ken K. Chin, Ronghui Qu, and Haiwen Cai.
© 2012 John Wiley & Sons, Inc. Published 2012 by John Wiley & Sons, Inc.

since been studied and developed greatly. The typical sensor structures are introduced as follows.

7.1.1 Structures of EFFP Devices

The EFFP device is typically composed of a fiber with a cleaved or polished end facet and a nonfiber-optical component with a reflective surface, such as a diaphragm fixed on a hollow support, and a cantilever with a planar surface. When a probe optical beam inputs the fiber, two backward beams are obtained: one is the beam reflected partly from the polished end facet; the other is the reflected beam from the external mirror and coupled back into the fiber. The two-beam interference gives signals dependent on the mirror's position. The light wave is reflected back and force multiplied in the cavity composed of the fiber facet and the mirror, leading to an effect of F-P interference.

Figure 7.1(a) shows schematically a typical structure of a diaphragm fiber optic sensor (DFOS) [4–12]. The diaphragm is deformed under external static pressure or acoustic pressure; the spacing between the fiber facet and the diaphragm changes, and the related interference signal is generated from the backward beam. The diaphragm is usually fabricated by microelectronics technology, or micro electromechanical system (MEMS) technology, as described in references [13,14]. The sensor size can be reduced to near or equal the fiber diameter, when a short section of multimode fiber (MMF) is used as the mirror as shown in Figure 7.1(b) [15,16]. The reflective mirror can be fabricated just on the fiber end facet as shown in Figure 7.1(c) [17–19]. Figure 7.1(d) shows a structure composed of two thin pieces of pure silica core fiber and with a short section of MMF with its core etched off as the cavity of FPI

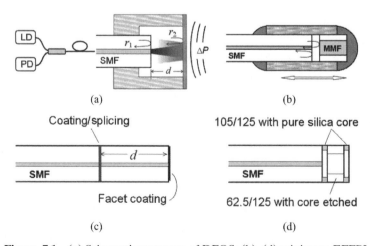

Figure 7.1 (a) Schematic structure of DFOS; (b)–(d) miniature EFFPI.

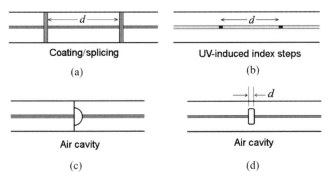

Figure 7.2 In-fiber FPI: (a) by coating and splicing; (b) UV-induced index steps; (c) by chemical etching; and (d) by ion milling or femtosecond laser processing.

[20,21]. A micro cavity can be made on the tip of the fiber by chemical etching and discharge fusing [22].

Both the reflection and transmission of the FPI can be utilized as sensors. The F-P cavity can be made inside the fiber with input and output fiber pigtails. Four examples are shown in Figure 7.2: (a) two reflective dielectric coatings made on the fiber end facets and concatenated by splicing or packaging [23,24]; (b) UV-induced index steps [25] for low-finesse F-P; and FBG pair for high-finesse F-P, as discussed in Section 4.3.1; (c) a micro air cavity fabricated by chemical etching [26–28]; and (d) by ion milling or femtosecond laser processing [29]. If F-P cavity mirrors are with lower reflectivity, multiple FPIs can be connected in series to serve as a quasi-distributed sensor system, as discussed in Section 5.5.5. For different applications, certain structures of transduction are needed to transfer the parameters to be sensed to the EFFPI sensor probe. Reference [30] gives a summarized review on the EFFPI sensors.

An important application of the EFFPI sensor is for signal pickup for the AFM, where the cantilever, coated with gold or other reflective coating, is taken as a mirror to compose the F-P cavity, the gap between the fiber facet and the mirror varies correspondingly during the cantilever scanning along the sample [3,31–35]. Figure 7.3 gives a schematic diagram of the fiber optic pickup used in AFM.

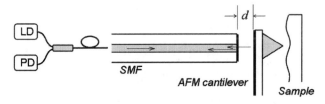

Figure 7.3 Schematic structure of an EFFPI used in AFM.

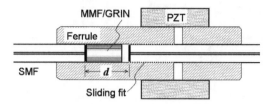

Figure 7.4 Structure of a tunable fiber F-P filter.

As one of the best optical filters, with features of narrow linewidth and tunability, the FFPI is widely used in optical communications and sensor technology. Figure 7.4 depicts a typical structure of tunable fiber F-P filter. Two cleaved fibers with high reflective coated ends are put in a ferrule to form an F-P cavity. One of them can be translated by piezo transducer (PZT), causing the cavity length to change. A short section of MMF (gradient-index fiber in the figure) is fixed inside the ferrule to reduce the coupling loss of beam divergence.

7.1.2 Basic Characteristics of a Fabry–Perot Interferometer

It is seen that two effects occur and act in the devices with structures shown in the figures above: interference between waves reflected from the fiber end facet and the mirror, and the coupling loss dependent on displacement of the mirror. The common physical principle is the theory of FPI, which is presented as follows.

The basic theory of FPI can be found in many textbooks [36,37]. A multiple-beam interference model of a FPI composed of two parallel planes is shown in Figure 7.5, where the two interfaces are coated with

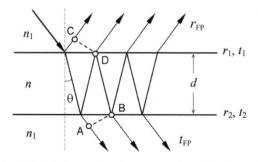

Figure 7.5 Multiple-beam interference model of F-P interferometer.

field reflectivity of $r_1 = \sqrt{R_1}$ and $r_2 = \sqrt{R_2}$. The field transmission and reflection are deduced as

$$t_{FP} = \frac{t_1 t_2 e^{j\Delta\phi/2}}{1 - r_1 r_2 e^{j\Delta\phi}}, \tag{7.1a}$$

$$r_{FP} = \frac{-r_1 + r_2 e^{j\Delta\phi}}{1 - r_1 r_2 e^{j\Delta\phi}}, \tag{7.1b}$$

where $\Delta\phi = 4\pi nd \cos\theta/\lambda$ is the phase difference between every two adjacent beams, as marked by points A and B, or C and D in the figure.

The intensity transmission and reflection are written as

$$T_{FP} = \frac{T_1 T_2}{1 + R_1 R_2 - 2\sqrt{R_1 R_2}\cos\Delta\phi}, \tag{7.2a}$$

$$R_{FP} = \frac{R_1 + R_2 - 2\sqrt{R_1 R_2}\cos\Delta\phi}{1 + R_1 R_2 - 2\sqrt{R_1 R_2}\cos\Delta\phi}, \tag{7.2b}$$

where $T_{1,2} = |t_{1,2}|^2 = 1 - R_{1,2}$. The latter equality holds when losses at the interfaces are negligibly small, resulting in $T_{FP} + R_{FP} = 1$. For conventional FPI, the interfaces are coated to $R_1 = R_2 = R \simeq 1$. Taking the attenuation A at the interfaces into account, we have $R + T + A = 1$. Then (7.2a) is rewritten as

$$T_{FP} = \frac{T_{max}}{1 + F^2 \sin^2(\Delta\phi/2)}, \tag{7.3}$$

with $F^2 = 4R/(1 - R)^2$ and $T_{max} = T^2/(1 - R)^2 = T^2/(T + A)^2$. It is shown by (7.2a), (7.2b) and (7.3) that a series of resonances occurs at $\Delta\phi = 2m\pi$, that is,

$$\lambda_m = 2nd \cos\theta/m, \tag{7.4}$$

termed Airy peaks. The spectral structure of a light wave can be measured by FPI based on the resonance characteristics; it is thus called a F-P etalon. From (7.4) the free spectral region (FSR) of F-P cavity is obtained:

$$\Delta\lambda = \lambda^2/(2nd\cos\theta). \tag{7.5}$$

From (7.3) the linewidth of resonance peaks [full width of half maximum (FWHM)] is deduced to be

$$\delta\lambda = \frac{\Delta\lambda}{\pi}\sin^{-1}\frac{1}{F} \approx \frac{\Delta\lambda}{\pi}\frac{1-R}{2\sqrt{R}}. \tag{7.6}$$

The finesse of the F-P cavity is then obtained as

$$F = \frac{\Delta\lambda}{\delta\lambda} = \frac{\pi\sqrt{R}}{1-R}. \tag{7.7}$$

It means that high-finesse requires high reflectivity near unity as much as possible; and the quality of high reflective coating must be assured to have a negligible loss with $A \approx 0$. Near the resonant peaks, (7.3) can be rewritten as $T_{FP} \approx T_{max}/[1 + (\lambda - \lambda_m)^2/\delta\lambda^2]$, showing a Lorentzian line shape. Figure 7.6 shows spectra of a high-finesse FPI with $R_1 = R_2 = 0.95$, and low-finesse FP cavities with $R_1 = 0.035$, $R_2 = 0.95$, and $R_1 = R_2 = 0.035$.

When the rear interface is a total reflective mirror, $R_2 = 1$, we obtain a Gires–Tournois Interferometer (GTI) with $T_{GT} = 0$ and $R_{GT} = 1$. The reflected wave experiences a phase shift:

$$r_{GT} = \frac{e^{j\Delta\phi} - r_1}{1 - r_1 e^{j\Delta\phi}} = \exp j\phi_{GT}, \tag{7.8a}$$

$$\phi_{GT} = \tan^{-1}\frac{(1 - r_1^2)\sin\Delta\phi}{(1 + r_1^2)\cos\Delta\phi - 2r_1}. \tag{7.8b}$$

It is an all-pass filter, as discussed in Section 3.3.

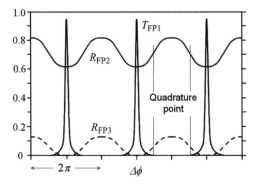

Figure 7.6 Spectra of FPI: FP1, $R_1 = R_2 = 0.95$; FP2, $R_1 = 0.035$, $R_2 = 0.95$; and FP3, $R_1 = R_2 = 0.035$.

If the medium inside the cavity is with loss or gain, its refractive index becomes a complex, and so does the phase factor: $\Delta\phi = \Delta\phi_r + j\Delta\phi_i$, where $\Delta\phi_i = \alpha d \cos\theta$ with loss α. In such cases, the reflectance and transmission of an F-P cavity with $R_1 = R_2 = R$ are expressed as

$$R_{\text{FP}} = \frac{R[(1 - e^{-\Delta\phi_i})^2 + 4e^{-\Delta\phi_i}\sin^2(\Delta\phi_r/2)]}{(1 - Re^{-\Delta\phi_i})^2 + 4Re^{-\Delta\phi_i}\sin^2(\Delta\phi_r/2)}, \qquad (7.9a)$$

$$T_{\text{FP}} = \frac{T^2 e^{-\Delta\phi_i}}{(1 - Re^{-\Delta\phi_i})^2 + 4Re^{-\Delta\phi_i}\sin^2(\Delta\phi_r/2)}. \qquad (7.9b)$$

It is shown that the finesse decreases with loss increasing and increases with gain increasing. The modulation depth of the fringes (the visibility) is obtained to be

$$M = \frac{T_{\max} - T_{\min}}{T_{\max} + T_{\min}} = \frac{2Re^{-\Delta\phi_i}}{1 + R^2 e^{-2\Delta\phi_i}}. \qquad (7.10)$$

7.2 THEORY OF A GAUSSIAN BEAM FABRY–PEROT INTERFEROMETER

7.2.1 Basic Model and Theoretical Analysis

Two fundamental differences between an EFFPI and the plane-wave F-P etalon are noticeable: (1) the beam propagating in the cavity is generally a divergent beam, with the widely accepted Gaussian beam approximation and (2) the diaphragm or the mirror has more freedoms of motion than that described in Figure 7.5. The mirror may move in parallel and/or deflective directions; the deflection may be around two perpendicular axes. It is, therefore, necessary to derive a Gaussian beam FPI theory. References [13,38–44] analyze the characteristics of Gaussian beam coupling in the EFFPI and related devices in detail.

As discussed in Section 2.3, the near field of an SMF has a Gaussian distribution approximately, and the output beam can be described as a Gaussian beam:

$$E = E_0 \frac{w_0}{w(z)} \exp\frac{-r^2}{w^2(z)} \exp j\left[kz + \frac{kr^2}{2R(z)} + \Gamma(z)\right], \qquad (7.11)$$

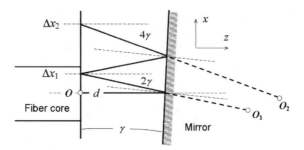

Figure 7.7 Multiple reflected beams between two un-parallel planes.

where $w(z) = (w_0^2 + 4z^2/k^2w_0^2)^{1/2}$ is the beam width, and $R(z) = \sqrt{z^2 + z_0^2}$ is the curvature radius of wave front in far field; $\Gamma(z) = \tan^{-1}(2z/kw_0^2) = \tan(z/z_0)$ is Guoy phase, with Rayleigh range z_0 denoted. Its divergent angle is expressed as

$$\tan\theta = \lim_{z \to \infty} \frac{w(z)}{z} = \frac{2}{kw_0} = \frac{w_0}{z_0}. \tag{7.12}$$

It is obvious that the power coupled back into the fiber will be attenuated by the beam divergence. For the multiple reflections in the F-P cavity, it is necessary to investigate the composite attenuation.

Generally, the normal of the mirror may not coincide with the fiber axis, and its direction may change in motion. Figure 7.7 shows a geometric relation of the beam axes, where the mirror is tilted with angle γ to the fiber axis in x–z-plane. It is shown that the incident angle θ_m to the fiber facet is multiplied by the times of reflection m. The incident point displacement Δx_m and the accumulated propagation distance from the fiber end L_m are expressed approximately for small γ as

$$\theta_m = 2m\gamma, \tag{7.13a}$$

$$\Delta x_m \approx 2m^2\gamma d, \tag{7.13b}$$

$$L_m \approx 2md. \tag{7.13c}$$

The coupling coefficient of the reflected beam into the SMF is proportional to the overlap integral of the Gaussian beam and the fundamental mode of SMF, expressed as

$$T_m \propto \frac{\iint E_m^* \cdot E_{01} dS}{\left[\iint |E_m|^2 \, dS \cdot \iint |E_{01}|^2 \, dS \right]^{1/2}}, \tag{7.14}$$

where E_m is the mth reflected beam coupled into the fiber facet, E_{01} is the LP_{01} mode, approximately expressed as $E_{01} \propto (\sqrt{2}/w_0)$ $\exp(-r^2/w_0^2)$ as in (2.114) and (2.116).

Since the coupling decays fast with the reflection times for a divergent beam, let us consider the first reflected beam, which is regarded as a Gaussian bean launched from point O_1 in Figure 7.7; and a coordinate conversion has to be operated as follows:

$$x_1 = \Delta x_1 + x \cos \theta_1 + z \sin \theta_1, \tag{7.15a}$$

$$y_1 = y, \tag{7.15b}$$

$$z_1 = L_1 + z \cos \theta_1 - x \sin \theta_1. \tag{7.15c}$$

The module of the radial vector and beam width are then written as

$$
\begin{aligned}
r_1 &= [(\Delta x_1 + x \cos \theta_1 + z \sin \theta_1)^2 + y^2]^{1/2} \\
&\approx [(x \cos 2\gamma + 2\gamma d + 2d \sin 2\gamma)^2 + y^2]^{1/2},
\end{aligned} \tag{7.16}
$$

$$
\begin{aligned}
w(z) &= [w_0^2 + 4(L_1 + z \cos \theta_1 - x \sin \theta_1)^2/k^2 w_0^2]^{1/2} \\
&\approx [w_0^2 + 16d^2(1 + \cos 2\gamma - \gamma \sin 2\gamma)^2/k^2 w_0^2]^{1/2} \\
&\approx (w_0^2 + 64d^2/k^2 w_0^2)^{1/2}.
\end{aligned} \tag{7.17}
$$

The approximated equalities in the expressions are for the positions at the fiber facet. For the power coupling, the phase factors in Gaussian beam expression (7.11) can be omitted. It is noted that the coupling coefficient has polarization dependence, that is, for x-polarization an additional factor of $\cos^2 \theta_m$ should be multiplied. By using (7.16) and (7.17), the first order coupling coefficient is obtained as [13,45]

$$
\begin{aligned}
T_1 &\propto \frac{2w_0 w_1}{[(w_1^2 + w_0^2)(w_1^2 + w_0^2 \cos^2 2\gamma)]^{1/2}} \exp\left[-\frac{4d^2(\gamma + \sin 2\gamma)^2}{w_1^2 + w_0^2 \cos^2 2\gamma}\right] \\
&\approx \frac{2w_0 w_1}{w_1^2 + w_0^2} \exp\left(-\frac{36\gamma^2 d^2}{w_1^2 + w_0^2}\right) \approx \frac{kw_0^2}{4d} \exp\left(-\frac{9k^2 w_0^2 \gamma^2}{16}\right).
\end{aligned} \tag{7.18}
$$

The approximations hold for $\gamma \simeq 0$ and for $d \gg w_0$, respectively. The higher order reflections can similarly be deduced. For the sum of multiple reflections, the composite reflectance is

$$r_{\mathrm{FP}} = -r_1 + r_2 t_1^2 T_1 e^{j\Delta\phi_1} + r_2 t_1^2 T_2 r_1 r_2 e^{j\Delta\phi_2} + r_2 t_1^2 T_3 r_1^2 r_2^2 e^{j\Delta\phi_3} + \cdots,$$

$$(7.19)$$

where $t_1^2 = 1 - R_1$ is the Fresnel transmission of the interface. The optical path difference $\Delta L = L_{m+1} - L_m$ actually increases with m, since the tilt angle increases. For the Gaussian beam, the Guoy phase should be added to phase factor: $\phi_m = nkL_m \cos\theta_m + \Gamma_m$. Therefore, the phase shift between the successive reflected waves is not the same for the different reflection times; and the summation cannot be reduced to an analytical expression. Digital calculations show that the effect of the tilting is not negligible for $\gamma > 0.5^0$.

When both the mirror and the fiber facet have low reflectivity, the transmission of EFFPI can also be used as a sensor signal, which is just the case of sensors shown in Figure 7.2. Due to the beam divergence optical loss takes place every round trip. By introducing coupling coefficients η_1 and η_2 at the fiber facet and at the mirror, the transmission of F-P cavity for the normal incident case is expressed as

$$
\begin{aligned}
t_{\mathrm{FP}} &= t_1 e^{j\Delta\phi} \eta_2 t_2 + t_1 e^{j\Delta\phi} \eta_2 r_2 e^{j\Delta t} \eta_1 r_1 e^{j\Delta\phi} \eta_2 t_2 + \cdots \\
&= \frac{t_1 \eta_2 t_2 e^{j\Delta\phi}}{1 - \eta_1 \eta_2 r_1 r_2 e^{j2\Delta\phi}} = \frac{\eta t^2 e^{j\Delta\phi}}{1 - \eta^2 r^2 e^{j2\Delta\phi}},
\end{aligned}
$$

$$(7.20)$$

where the last expression is for the symmetric FFPI sensor.

7.2.2 Approximation as a Fizeau Interferometer

Even if there is no tilting, the beam divergence results in large attenuations with reflection times. In such cases, the interferometer should be regarded as a low-finesse FPI, or as a Fizeau interferometer [36,45,46], which takes the interference of just the first two terms of (7.19) into account

$$r_{Fiz} = -r_1 + r_2 t_1^2 T_1 e^{j\Delta\phi_1};$$

$$(7.21)$$

$$R_{Fiz} = R_1 + R_2(1 - R_1)^2 T_1^2 - 2\sqrt{R_1 R_2}(1 - R_1)T_1 \cos\Delta\phi.$$

$$(7.22)$$

It is seen that the interference fringe is a sinusoidal curve. The phase shift is written as

$$\Delta\phi = nkd(\cos\gamma + \cos 2\gamma) + \tan^{-1}(2d/z_0) \approx 2kd, \qquad (7.23)$$

where the last approximated expression is for a negligible tilt and for an air cavity; and the Guoy phase variation is neglected also for $d \gg z_0$. The free spectral range of the Fizeau interferometer is the same as the F-P: $\Delta\lambda = \lambda^2/2d$. It is obvious that the deformation of the diaphragm, or the displacement of the mirror, changes the cavity length d, and consequently the detected optical power. For a sensitive EFFPI sensor, a higher visibility of fringe is preferred, which depends on the reflectivity of the two cavity facets R_1 and R_2. It is deduced from (7.22) that the visibility reaches the maximum at $R_1 = R_2(1 - R_1)^2 T_1^2$, where T_1 is a function of cavity length d. For a cleaved fiber end facet, the Fresnel reflectivity is about 3.5%, deduced from $R_1 = [(n - 1)/(n + 1)]^2$; therefore, a reflective coating on the fiber facet is needed. For fixed R_1 and R_2, the visibility varies with d, which is designed and adjusted in sensor fabrication for a higher sensitivity.

Figure 7.8 gives a simulated interference signal varying with the cavity length, based on (7.22) for the fixed reflectivity of $R_1 = 0.04$ and $R_2 = 0.95$, and $\lambda = 1550$ nm; and the coupling coefficient of $T_1 \approx z_0/2d$ is taken for the case without tilting [45]. It is seen that the interference maxima decay with the length monotonically, whereas the minima reaches zero at a certain position (367 μm in the simulation); a maximum visibility is obtained at the position. Experimentation verifies such characteristics.

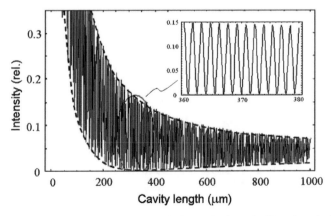

Figure 7.8 Interference signal varying with the cavity length. (Reprinted with permission from reference [45].)

7.3   BASIC CHARACTERISTICS AND PERFORMANCES OF EFFPI SENSORS

The EFFPI sensors are mostly used for pressure measurement, acoustic detection, and ultrasonic detection. The sensor can also be used to measure the index of the surrounding medium. These practical applications require high-performance sensors, such as those with high sensitivity, broad frequency response, good stability, and a larger dynamic range.

7.3.1   Sensitivity of an EFFPI Sensor

The sensitivity is determined by two main factors: one is the ratio of the displacement of the mirror over the pressure to be detected; the other is the conversion of the displacement to the interference phase.

The EFFPI sensor structure is generally composed of the fiber tip and an elastic diaphragm, which is usually made of a thin elastic plate with its edge fixed on a support, such as a cylindrical base, as shown in Figure 7.1(a). The deformation and movement of the diaphragm is governed by the elasticity of a thin plate [8,12]. Figure 7.9 gives a schematic diagram for the analysis.

The basic equation for z-direction movement $\zeta(r, \varphi)$ under a static pressure P, which is the difference between the external and internal pressure, is expressed as

$$D\nabla^4\zeta = P, \tag{7.24}$$

where $D = Yh^3/12(1 - \nu^2)$ is the *flexural rigidity* of the plate and h is its thickness. For a cylindrical symmetric structure, the differential operation is

$$\nabla^2 = \frac{\partial^2}{\partial r^2} + \frac{1}{r}\frac{\partial}{\partial r} + \frac{1}{r^2}\frac{\partial^2}{\partial\varphi^2}.$$

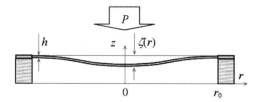

Figure 7.9   Schematic diagram of an elastic plate supported by a cylinder.

Under a spatially uniform force, the diaphragm deformation is a function of r only. The operation is then deduced as

$$\nabla^4 = \frac{d^4}{dr^4} + \frac{2}{r}\frac{d^3}{dr^3} - \frac{1}{r^2}\frac{d^2}{dr^2} + \frac{1}{r^3}\frac{d}{dr}.$$

The equation is solved under boundary conditions of $\zeta = 0$ and $d\zeta/dr = 0$ at $r = r_0$:

$$\zeta = \zeta_0 \left(1 - \frac{r^2}{r_0^2}\right)^2, \tag{7.25a}$$

where ζ_0 is the displacement at the center of diaphragm:

$$\zeta_0 = \frac{3(1 - v^2)r_0^4}{16Y h^3} P. \tag{7.25b}$$

The above formulas hold for smaller deformations. For a larger displacement, expressions (7.25b) have to be revised to take the change of the plate thickness into account; a precise solution is given as [14]

$$\zeta_0 + \frac{(1 + v)(7 - v)}{16}\frac{\zeta_0^3}{h^2} = \frac{3(1 - v^2)r_0^4}{16Y h^3} P. \tag{7.26}$$

It is seen that the sensitivity of the sensor depends strongly on the thickness and diameter of the diaphragm; and a thinner diaphragm is beneficial for a high-sensitivity EFFPI sensor, which is limited mainly by the processing and material strength.

The flexural rigidity of the plate is determined by Young's modulus and Poisson's ratio of the material. For the most often used materials, silicon and silica, the data are listed here for reference: $Y_{Si} = 13 \times 10^{10}$ N/m^2, $v_{Si} = 0.28$; $Y_{silica} = 7.3 \times 10^{10}$ N/m^2, and $v_{silica} = 0.17$. If an equivalent modulus is defined as $Y_e = 16Y/3(1 - v^2)$, the parameter for the two materials are calculated as: $Y_{Si} = 75 \times 10^{10}$ N/m^2 and $Y_{silica} = 40 \times 10^{10}$ N/m^2.

The second factor of sensitivity involves phase variation with the displacement:

$$dI \propto \frac{\partial \cos \Delta\phi}{\partial d}\zeta \propto k\zeta \sin \Delta\phi_0. \tag{7.27}$$

It is indicated if the interference amplitude reaches its maximum or minimum, that is $\cos \Delta\phi_0 = \pm 1$, the sensitivity to the phase change is toward zero. The original phase should be biased at the so-called quadrature point: $\Delta\phi_0 = (m + 1/2)\pi$ to acquire the maximum sensitivity. It can be realized by two methods: one is by setting the original cavity length d_0; the other is by adjusting the wavelength of the source. Reference [47] proposed a scheme where two fiber tips are packaged into one sensor head with distances to the mirror of $\pi/2$ phase difference, so that by detecting the two signals simultaneously the quadrature is guaranteed.

On the other hand, however, some of the applications require a lower sensitivity to the pressure, for example, the sensors used for measurement of deep level liquid pressure and combustion pressure [5,6], where a thick and strong plate is used as the cavity mirror to avoid damage under high pressure.

7.3.2 Linear Range and Dynamic Range of Measurement

It is indicated by (7.27) that even if the sensor works at the quadrature, the linearity of measurement is limited in a range of fraction of $\pi/2$. The signal will be distorted if the induced phase is beyond the range. For the exact analysis, the signal for a single-frequency acoustic vibration $\zeta = \zeta_0 \cos \Omega t$ is expanded by Fourier series:

$$
\begin{aligned}
I \propto &\left[J_0(k\zeta_0) + 2 \sum_{n=1}^{\infty} (-1)^n J_{2n}(k\zeta_0) \cos 2n\Omega t \right] \cos \Delta\phi_0 \\
&- 2 \left[\sum_{n=1}^{\infty} (-1)^{n+1} J_{2n-1}(k\zeta_0) \cos(2n-1)\Omega t \right] \sin \Delta\phi_0 \qquad (7.28) \\
&\to 2 \sum_{n=1}^{\infty} (-1)^n J_{2n-1}(k\zeta_0) \cos(2n-1)\Omega t,
\end{aligned}
$$

where the last expression is for the case at quadrature. This means that the detected frequency spectrum may be much different from the original acoustic one.

A large dynamic range of measurement is required for many applications. For a large displacement, the interference signal may cover several periods. It is therefore necessary to record the fringe number. The precision of fringe counting is supposed to be the half wavelength, far below the usual requirement. Schemes of enlarging the dynamic range with acceptable precision have been proposed and demonstrated.

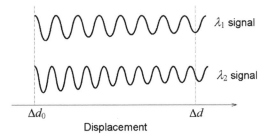

Figure 7.10   Measurement of larger displacement by duel wavelength.

Reference [48] used two laser diodes with different wavelengths and detected separately. As the cavity length changes in a large range, two fringe numbers are counted for the two different wavelengths, respectively. The problem lies in evaluating the remainder from the amplitudes of the two wavelength signals. The algorithm is similar to the vernier, as shown schematically in Figure 7.10. Two wavelengths generate two sets of interference fringes, with two different free spectral ranges, $\Delta\lambda_{1,2} = \lambda_{1,2}^2/2d$. The reminded phase difference can be evaluated by a combined equation of the two signals. The FSR is a function of the cavity length: $\partial(\Delta\lambda)/\partial d = -\lambda^2/d^2$. By measuring the variation of FSR with d, a larger displacement can be calculated, so long as the device is stable enough and the source power is high enough.

   In the detection of a large-amplitude vibration, if the phase shift covers several periods, the temporal waveform will be distorted dramatically as shown schematically in Figure 7.11(a). An experimentally recorded waveform is shown in Figure 7.11(b) [49–51]. Based on the fundamental relation of (7.28), the large amplitude displacement and

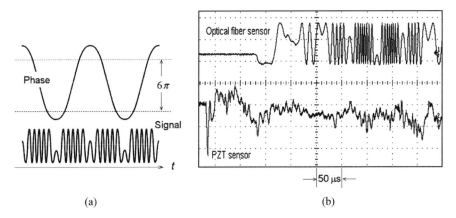

Figure 7.11   (a) Interference signal for larger amplitude vibration and (b) experimentally measured waveform. (Reprinted with permission from reference [51]. © 2006 IEEE.)

high fidelity acoustic wave can be retrieved by a high frequency spectrum analyzer, so long as the band of detection is broad enough [52]. The lower trace of Figure 7.11(b) is the signal of a commercial PZT sensor measured at the same time [51], indicating that the sensitivity of DFOS is much higher than the traditional sonar.

7.3.3 Interrogation and Stability

The interrogations of EFFPI sensors include two types basically. One of them is by using a broadband source combined with a spectrum analyzer, such as a fiber Mach–Zehnder interferometer, a Michelson interferometer, scanning F-P cavity, and a grating spectrometer [8,53–55]. This method is also suitable for transmitted signals; especially for serially concatenated low-finesse FFPI networks, by using broadband superluminescence LD (SLD) to localize the sensor, as discussed in Section 5.5.

The other interrogation scheme is by using a narrow linewidth laser source, suitable for the low-finesse FFPI. Figure 7.12 shows a typical configuration, where a distributed feedback LD (DFB-LD) is used as the source.

In the deduction of the formulas in Section 7.2, a single-frequency light source is assumed. Obviously the linewidth of the source is a factor to reduce the visibility. It is obtained from (7.23) that a linewidth $\delta\lambda$ corresponds to an uncertainty of $\delta d = (d_0/\lambda)\delta\lambda$. The length d_0 in practical sensors is typically several tens to few hundreds micrometer; and the most used light source is at 1.5 μm band. The spectrum of the source intensity is typically described by a Gaussian function as [46,54] $S(k) = (\sqrt{\pi}\Delta k)^{-1}\exp[-(k-\bar{k})^2/\Delta k^2]$, where Δk is the half-width of the wave vector at the $1/e$ intensity maximum. The signal detected by a PD will be reduced to

$$\langle I\rangle \propto \int S(k)\cos(2kd)\mathrm{d}k = \cos 2\bar{k}d \cdot \exp(-d^2\Delta k^2). \qquad (7.29)$$

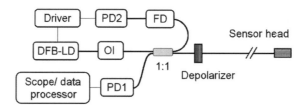

Figure 7.12 Typical configuration of EFFPI sensor.

indicating that the source linewidth affects directly the visibility of the interference signal.

The stability of the laser source is critical for the sensor's performance. An optical isolator (OI) is used to avoid the reflected beam from the downward fiber and components to disturb the state of the laser. As analyzed in Section 7.2, the sensor should work at the qaudrature point for the highest and linear sensitivity. The quadrature depends directly on the stability of the lasing wavelength. As well known, the wavelength of LD is determined by its pump current and the temperature. Therefore, the laser driver must provide a stable current and a function of temperature control and stabilization. The spectral characteristics of the source may also change with aging; careful technical measures are needed. For this purpose, PD2 is used to receive the laser output, as a feedback signal to control the laser power and the quadrature; and an optical frequency discriminator (FD), such as a tunable filter, may be inserted to provide feedback signals to the laser driver.

It is shown experimentally that although a single-frequency source is necessary to interrogate the signal with a required resolution, the linewidth should not be too narrow. Serious coherent noise will occur when the coherence length is too large because any small scattering in the fiber loop will be enhanced by coherence. In addition, any external disturbance applied on the lead fiber will also be detected with a high coherent source, which is a spurious signal to disturb the signal detected from the sensor head. Figure 7.13 shows a detected signal when the lead fiber is being stretched, when a narrow line DFB is used; the waveform is believed a demonstration of the strain wave along the fiber. These spurious signals should be removed, giving good selectivity to the signal being detected by the sensor head. Technical methods are adopted to reduce the coherence to a certain degree, such as to

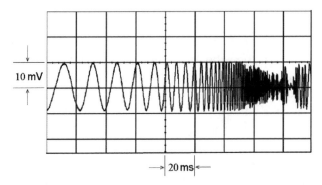

Figure 7.13 Detected signal of the lead fiber being stretched.

intentionally modulate the LD at frequency higher than the signal band. It is also helpful to introduce a little external reflection to the DFB laser, which will cause its line width to broaden in some degree. There is thus a trade-off between the resolution and the noise reduction. Another factor of interference comes from the polarization effect. A depolarizer can be inserted in the fiber to suppress the polarization noise, as depicted in Figure 7.12.

The instability comes also from the sensor head itself. One of the causes is thermal stress, since the thermal expansion coefficients (TEC) of the materials used in the structure are not the same. For example, silicon diaphragm is often adopted, whose TEC is 4.2×10^{-6} K^{-1}, whereas that of silica is 0.56×10^{-6} K^{-1}. The large difference causes the cavity length to vary with temperature seriously, and even the deformation of the diaphragm, resulting in the sensor deviating from the quadrature, and its sensitivity to fade and fluctuate. To avoid such an effect sensors with silica diaphragm and silica support were proposed, showing low temperature sensitivity [56]. For the pressure sensor with a sealed cavity, temperature change also induces internal pressure change, as predicted by Boyle's law. Such issues must be taken into consideration for practical applications. The thickness of the diaphragm cannot be too thin to reduce the influence of thermal expansion.

To avoid the interference noise completely, a sensor based on coupling loss is developed. Formula (7.18) indicates that the coupling efficiency changes also with the cavity length variation. If a broadband light source is used, term $\cos 2\Delta\phi$ in (7.22) may be averaged to zero. However, the detected signal I responses still to d variation. Approximately, we have

$$\frac{\partial I}{\partial d} \propto \frac{-2z_0^2}{d^3} = \frac{-2T_1^2}{d}. \tag{7.30}$$

In practice two fiber tips are used, one for input beam and the other for picking up the reflected beam. To enhance sensitivity, slanted beams are used, as shown in Figure 7.14. It is deduced from geometric relations that a displacement of the diaphragm Δd corresponds to a transverse displacement of $x = \Delta d \sin \theta$ as shown in the figure. The coupling loss thus varies correspondingly. It is shown that a larger angle is beneficial to higher sensitivity, and a shorter working distance d is needed for higher coupling coefficients between the two fiber ends. In the system, a light source with a broad spectrum, such as a light emission diode (LED), is usually used. The sensor has the merits of higher stability, simple and low-cost, though with lower sensitivity.

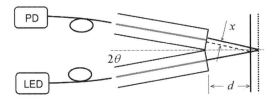

Figure 7.14 Schematic structure of coupling loss sensor.

7.3.4 Frequency Response

For the acoustic detection, the EFFPI sensor must have a broad frequency response band, especially when used as a microphone, which has to transmit the sound signal with high fidelity.

The frequency response should also match the sensitivity of the human ear, which is shown in Figure 7.15 [57,58], where the sound intensity is relative to the reference sound intensity $I_{ref} = 10^{-12}$ W/m^2, corresponding to acoustic pressure of 2×10^{-5} Pa.

On the other hand, the EFFPI sensor is also used for ultrasonic detection, such as the partial discharge detection in the power industry. A high frequency response is needed for such applications. It is therefore necessary to analyze the dynamic behavior of the diaphragm. Several factors should be taken into consideration: (1) the dynamic properties of the diaphragm; (2) the air pressure inside the cavity; (3) the air passage, and (4) other damping factors. Quite a number of references were engaged in the investigation of the frequency response of EFFPI [8,11,59,60].

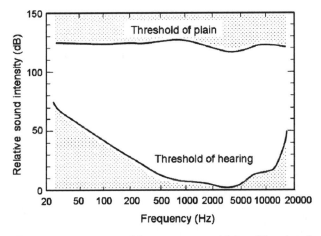

Figure 7.15 Frequency spectrum of human ear sensitivity. (Reprinted with permission from reference [57].)

For a temporally varied external pressure, e.g. a sound, the dynamic equation of the diaphragm is written as

$$\rho h \frac{\partial^2 \zeta}{\partial t^2} = P(t) - D\nabla^4 \zeta, \tag{7.31}$$

where ρ is the density of the diaphragm material. Assuming $P(t) = p_0 \exp(j\omega t)$ for a single-frequency vibration, and a trial solution of $\zeta(t) = \bar{\zeta}(r, \varphi) \exp(j\omega t)$, the equation is transformed to

$$\nabla^4 \bar{\zeta} - k_S^4 \bar{\zeta} = p_0/D, \tag{7.32}$$

with $k_S^4 = \rho h \omega^2 / D$. The lowest order solution under the same boundary condition as that in static case is a cylindrically symmetric function, expressed as

$$\bar{\zeta}(r) = \frac{p_0}{\rho h \omega^2} \left[\frac{I_1(k_S r_0)J_0(k_S r) + J_1(k_S r_0)I_0(k_S r)}{I_1(k_S r_0)J_0(k_S r_0) + J_1(k_S r_0)I_0(k_S r_0)} - 1 \right], \tag{7.33a}$$

where J_0, J_1, I_0, and I_1 are the Bessel functions for the real and imaginary arguments. The vibration amplitude at the center is expressed as

$$\bar{\zeta}(0) = \frac{p_0}{\rho h \omega^2} \left[\frac{I_1(k_S r_0) + J_1(k_S r_0)}{I_1(k_S r_0)J_0(k_S r_0) + J_1(k_S r_0)I_0(k_S r_0)} - 1 \right]. \tag{7.33b}$$

It is seen that the vibration amplitude at the diaphragm center is proportional to the external pressure; but the coefficient is a function of vibration frequency. It is noticed that resonances occur under the condition of

$$I_1(k_S r_0)J_0(k_S r_0) + J_1(k_S r_0)I_0(k_S r_0) = 0. \tag{7.34}$$

The resonant frequencies are deduced as

$$\omega_n = \frac{h\mu_n^2}{2r_0^2} \sqrt{\frac{Y}{3\rho(1 - v^2)}}, \tag{7.35}$$

where $\mu_n = k_S r_0$ are the roots of equation (7.34): $\mu_n \approx 3.20, 6.30, 9.44, \cdots, \sim m\pi$, given by the numerical calculation. The dynamic characteristics of the DFOS for vibration and sound can thus be analyzed by the above formulas.

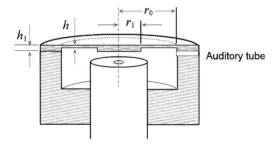

Figure 7.16 EFFPI structure with an embossed diaphragm and an auditory tube.

The shape of the diaphragm may take forms different from the circle, such as a square and a rectangle. Equations (7.24) and (7.31) are still valid, but with different boundary conditions. Generally, a numerical method is needed to solve them. A diaphragm with non-uniform thickness will modify its frequency response. Figure 7.16 shows a sensor structure by using an embossed diaphragm with the thickness of the central part thicker than the edge part, $h_1 > h$ [12,13,60]. The embossed part increases the mass of the diaphragm, whereas the elastic force is affected just a little. For a cylindrical geometry, the displacement is deduced theoretically as

$$\zeta(r_1) = \frac{P}{64D}[r_0^4 - r_1^4 - 4r_0^2 r_1^2 \ln(r_0/r_1)], \qquad (7.36)$$

where r_1 is the radius of the embossed part. The deformation of central embossment is negligibly small, so that its thickness does not appear in the formula. It is seen that its response to the pressure is reduced in some degree, whereas the resonant frequency is decreased also. Reference [60] gives simulations on the width's ratio of a rectangular embossment and on h_1/h, showing changes of frequency response.

It is noticed that the pressure in the above formulas is actually the difference between the external pressure and the pressure inside the cavity. The volume of the cavity V_C will decrease under a positive external pressure; thus the internal pressure increases according to Boyle's law: $p_{in} V_C = N k_B T$, where N is the number of molecules inside the cavity, and k_B is Boltzmann's constant. Roughly the volume change is proportional to the diaphragm displacement, written as $\delta V_C = S \zeta_0$, where S is an equivalent area of transverse cross section; and $\delta p_{in} = -p_{in} S \zeta_0 / V_C$. It is seen that a larger cavity volume is helpful to mitigate the effect. The pressure difference is thus $P = p_{ext} - p_{in} = p_{ext} + q \zeta_0$ with $q = p_{in} S / V_C$, resulting in a reduced sensitivity.

Moreover, the effect changes the frequency response, just as we hear the sound when the air pressure changes suddenly. Such an effect is diminished by the auditory tube between the human ear and oral cavity. It is therefore needed to open a small passage in the sensor structure between the F-P cavity and the outside. Figure 7.16 shows a structure with such an auditory tube [7,13]. Due to the auditory tube, the air will flow through the tube, meaning N is no longer a constant, and the internal pressure is rewritten as $\delta p_{in} = [\delta(Nk_BT) - p_{in}S\zeta_0]/V_C$. Phenomenally the variation of the molecular number is written as $k_BT\,\Delta N \propto b(\omega)e^{j\omega\tau}p_{ext}$ by speculating that the flow is a function of the sound frequency and with a phase retard related to the acoustic vibration. Parameters b and τ are determined by the structure of auditory tube. The mechanism involves an analysis based on hydrodynamics.

By denoting the elastic stress of the diaphragm as a recovering force $\eta\zeta(0)$ phenomenally with coefficient of $\eta = \rho h\omega^2/[(I_1 + J_1)/(I_1 J_0 + J_1 I_0) - 1]$, the equation of diaphragm motion at the center point is now modified to

$$\rho h \frac{\partial^2 \zeta_0}{\partial t^2} = p_{ext}(1 - be^{j\omega\tau}) - g\frac{\partial \zeta_0}{\partial t} - (\eta - q)\zeta_0, \qquad (7.37)$$

where an additional damping factor g is introduced for other dissipation mechanisms, which exist inevitably in practical sensor structures. Obviously the auditory tube should not be opened too big; or the pressure difference will vanish, corresponding to $be^{j\omega\tau} \to 1$ and $\zeta_0 \to 0$. The frequency response is then described by the solution of (7.37):

$$\zeta_0 = \frac{p_{ext}(1 - be^{j\omega\tau})}{\eta - q - \rho h\omega^2 + j\omega g} = \frac{(p_{ext}/\rho h)(1 - be^{j\omega\tau})e^{j\vartheta}}{[(\omega_D^2 - \omega_V^2 - \omega^2)^2 + \omega^2\omega_g^2]^{1/2}}, \qquad (7.38)$$

where $\omega_D = \sqrt{\eta/\rho h}$ is the diaphragm-related frequency, $\omega_V = \sqrt{q/\rho h}$ is the cavity volume-related frequency, $\omega_g = g/\rho h$ is the other damping mechanism-related frequency; and $\vartheta = \tan^{-1}[g\omega/(\rho h\omega^2 + q - \eta)]$ is an additional phase retard. ω_D is actually just the resonant frequency ω_n expressed in (7.35). If the frequency relation of factor $be^{j\omega\tau}$ in the numerator of (7.38) is omitted for simplicity, the resonant frequency is obtained:

$$\omega_R = \sqrt{\omega_n^2 - \omega_V^2 - \omega_g^2/2}. \qquad (7.39)$$

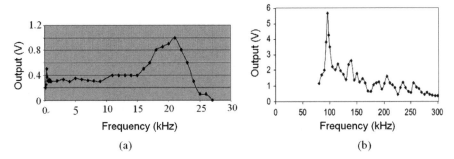

Figure 7.17 Frequency responses for (a) ultrasonic band and (b) acoustic band. (Reprinted with permission from reference [60].)

The detailed frequency response can be analyzed by solving equation (7.37) numerically. Figure 7.17 shows the experimentally measured frequency response for the acoustic band [61] and for the ultrasonic band [60] of the respective sensors.

There are other performances concerned in practical applications, such as long term reliability and robustness. The EFFPI sensor and its applications are being developed to meet practical requirements.

7.4 APPLICATIONS OF THE EFFPI SENSOR AND RELATED TECHNIQUES

The EFFPI sensors have been widely used in various applications. There are a number of technical issues related to the application requirements and field conditions.

7.4.1 Localization of the Sound Source

In the power industry, the partial discharge (PD) in transformers and other related apparatus is one of the factors that could lead to power failure and installation damage. PD usually emits ultrasonic signals in quite a short duration. The diaphragm fiber optical sensor (DFOS) is promising in detecting the ultrasonic and in localizing the position of partial discharge [9,10,49–52]. For the localization, several sensor probes are set at different positions, constituting a network, as shown in Figure 7.18. The ultrasonic signal will reach the probe at different moments based on time of flight (TOF). It is noted that the TOF is different from that in the air, since the transformer is usually soaked in the electrically isolating oil. Measuring the TOF of the different sensors

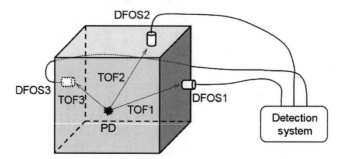

Figure 7.18   Localization of the partial discharge by a DFOS network.

the position of PD can be localized by the triangle algorithm. The reflection of the ultrasonic at the wall of the transformer container may bring about interference, more detailed analysis is needed.

The localization function is useful also in battlefield, geological prospecting and similar applications. The various applications of DFOS show good commercial prospects [61,62].

7.4.2   Applications in an Atomic Force Microscope

As described in [3,31–35], the EFFPI sensor is useful in signal pickup of AFM. Special features are required of the sensor, such as high resolution, low noise, easy and quick change of sample to be observed and cantilever probe, large dynamic range for convenient collimating and adjusting. The environment of the sensor is often at low temperature and high vacuum; and sometimes in liquid for bio samples. Therefore a high precision mechanical structure, usually with piezoelectric driven multi-dimensional translators for the sample stage alignment and for the probe cantilever alignment is critical; and carefully designed electronics, and effective data processing are necessary.

The characteristics of the laser source is a key issue. The laser should meet the following requirements: wavelength stabilized, low spontaneous emission, single longitudinal mode with high side-mode suppression, proper linewidth to avoid the coherence noise (low-phase noise). An OI with isolation more than 30dB is inserted in the laser output port to avoid the optical feedback noise (OFN). The fiber sections in the system should be fusion spliced to reduce additive reflections; and bend-insensitive fiber is helpful in mitigating external disturbances.

The radio frequency modulation of the laser (e.g., 200 MHz) is useful to broaden its linewidth in some degree with a proper coherence to reduce the optical interference noise (OIN). The cantilever can be modulated by the PZT, which makes the interference signal have a single-frequency modulation bias, and used in lock-in amplification. The modulation can also enable a function of monitoring and stabilizing the quadrature working point, enhancing the Q value and the sensitivity of the AFM, i.e. the so-called frequency modulated AFM (FM-AFM).

The power of the laser beam must be high enough to reduce the shot noise:

$$\frac{\langle (\Delta i)^2 \rangle}{\bar{i}^2} = \frac{2e\Delta f}{R_{PD}P}, \tag{7.40}$$

where e is the charge of an electron, Δf is the band width of the detection system, R_{PD} is the response of the photodiode, and P is the power of the laser beam with $\bar{i} = R_{PD}P$. However the optical power should not be too large, because the motion of the cantilever will be affected by a high intensity laser beam by photo pressure and by photo thermal effect. The latter is mainly due to the heating of one side of the cantilever, and the thermal expansion induced bending. Reference [33] investigates the effect of photoinduced forces both theoretically and experimentally. The motion of cantilever is described by the equation:

$$m\ddot{\zeta} + \frac{m\omega_0}{Q}\dot{\zeta} + c_\zeta \zeta = c_\zeta z_0 \cos \omega_m t + F_{ph}(d), \tag{7.41}$$

where $F_{ph}(d)$ is the photoinduced force, as a function of cavity length d, c_ζ is the coefficient of elastic recovering, $z_0 \cos \omega_m t$ is the cantilever modulation. It is found that a resonant frequency exists for the cantilever vibration, and the quality factor (Q) of the cantilever changes with the cavity length, and it depends also on the coating of the cantilever.

High performances of AFM incorporated with the EFFPI have been reported, based on several advanced techniques and improvements. Resolution of 40 fm/$\sqrt{Hz}$ above 20 Hz is achieved, and the shot noise limitation of 20 fm/$\sqrt{Hz}$ at 1 kHz is approached for an optical power of 10 μW [34]; the lowest noise density of 2 fm/$\sqrt{Hz}$ is achieved [35].

7.4.3 More Application Examples

Magnetic field sensor. The basic function of EFFPI is to detect displacement; besides the stress-induced deformation and sound-induced vibration, any other factor-induced displacement can be detected in principle. Reference [63] demonstrates a magnetic field sensor by using EFFPI to measure the magnetostrictive effect of a selected material.

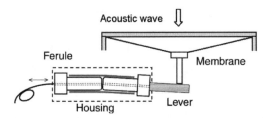

Figure 7.19 Sensor head using acousto-induced fiber deflection.

Refractive index measurement. The optical length of F-P cavity should be multiplied by the index. Therefore the interference fringe depends on the index of the medium in the cavity. Reference [64] built an FPI with inlet and outlet for the fluid whose index is to be measured. By another way, reference [29] utilized the dependence of the reflectivity on the index of the surrounding medium: $r_2 = (n_{silica} - n_{medium})/(n_{silica} + n_{medium})$. The experiment results are noticeable: when the index of the surrounding medium is higher than that of silica fiber, r_2 changes its sign, and the interference fringes acquires a phase shift of π.

Localization and monitoring of vehicles. Reference [65] shows applications of EFFPI sensor in the airport to locate aircrafts landing and taking-off, and even to distinguish them via their characteristic noise spectra. The demonstrating EFFPI sensor uses a horn to receive more acoustic energy, as shown in Figure 7.19. The F-P cavity is composed of two cleaved fiber facets, one is the lead fiber, the other is a sensing fiber section, which is pushed in transverse direction by the acoustic vibration, leading to the deflection of the fiber section, and consequently the deflection of its end facet. Thus, the sound signals are converted to the output of the F-P interferometer.

Characterization of MEMS device. It is of interest to understand the dynamic properties of the fabricated MEMS devices, to see if it behaves as designed. The EFFPI can serve that purpose. Figure 7.20

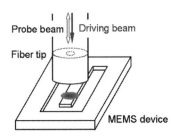

Figure 7.20 Active sensor for characterizing dynamic properties of a MEMS cantilever.

shows schematically an example of an experiment setup for measuring the vibration characteristics of a MEMS micro cantilever, which is driven by a high peak power pulsed laser beam, based mainly on the photothermal effect. The basic principle is similar to the analysis of the photoinduced force of AFM cantilever, but with a pulsed laser beam to drive the cantilever intentionally.

There are many other types of extrinsic fiber sensors incorporated with functional optical materials and components. Their applications are being developed. Interested readers can find more in the literature.

PROBLEMS

7.1 Deduce the basic formula of a F-P interferometer. Why is the phase factor related to $d \cos \theta$, not to $d / \cos \theta$?

7.2 Deduce the expression of FSR of FPI. Can you design an F-P cavity with resonant frequency and FSR independently? Can you tune the resonance wavelength with a constant FSR?

7.3 Deduce the expression of finesse. How is the finesse affected by the loss of reflectors and the loss of cavity?

7.4 Deduce the line shape near a resonance wavelength. Under what conditions is it approximated to a Lorentzian form?

7.5 Deduce the Gaussian beam expression from a Maxwell equation (Helmholtz equation). Under what approximation is it obtained? Discuss the effect of beam divergence on the EFFPI, for working distances much more and much less than Rayleigh length, respectively.

7.6 What is the quadrature in the EFFPI sensor? Why it is a key factor to sensor sensitivity? How do you set up or adjust the interferometer to reach the quadrature?

7.7 At the quadrature, what parameters determine the linear dynamic range? For a large pressure, how does the interference signal behave? How do you retrieve the pressure from the interference waveform?

7.8 Why is the EFFPI usually regarded as a Fizeau interferometer? How does the reflectance of the two cavity mirrors affect the extinction ratio of interference fringes?

7.9 Why is a single-mode laser required for a DFOS? If a Gaussian spectrum is assumed, how does the linewidth of the source affect the sensitivity?

7.10 If a DFOS sensor is composed of a circle diaphragm, deduce its displacement varied with the pressure, that is, formulas (7.25a and 7.25b). If a silica diaphragm has a thickness of 10 μm and diameter of 2 mm, estimate how much it is displaced under 20 μPa pressure.

7.11 What parameters and physical conditions determine the frequency response of the EFFPI? How do you design the sensor head for ultrasonic detection and for an acoustic sensor that fits human hearing?

7.12 Referring to Figure 7.18 for PD detection, how many sensors are needed to localize the PD position? If the container is a cubic box, how do you arrange the sensors?

7.13 Apart from pressure and acoustic sensing, what other signals can the EFFPI sensor measure and detect? Discuss their principles and characteristics.

7.14 What noise sources do you think the most possible in applications of EFFPI sensors? What other problems will occur in practical applications? Propose your solutions.

REFERENCES

1. Petuchowski SJ, Giallorenzi TG, Sheem SK. A sensitive fiber-optic Fabry-Perot interferometer. *IEEE Journal of Quantum Electronics*, 1981; 17: 2168–2170.

2. Lee CE, Taylor HF. Interferometric optical fiber sensors using internal mirror. *Electronics Letters* 1988; 24: 193–194.

3. Rugar D, Mamin HJ, Guethnera P. Improved fiber-optic interferometer for atomic force microscopy. *Applied Physics Letters* 1989; 55: 2588–2590.

4. Lequime M, Lecot C, Jouve P, Pouleau J. Fiber optic pressure and temperature sensor for down-hole applications. *Proceeding SPIE* 1991; 1511: 244–249.

5. Rouhet J, Graindorge P, Laloux B, Girault M, Martin P, Lefèvre H, Desforges FX. Applications of fiber optic sensors to cryogenic spacecraft engines. *Proceedings SPIE* 1997; 3000: 29–36.

6. Wlodarczyk MT. Fiber-optic combustion pressure sensor for automotive engine controls. *Proceedings SPIE* 1997; 3000: 51–59.

7. Wang WJ, Lin RM, Sun TT, Guo DG, Ren Y. Performance-enhanced Fabry–Perot microcavity structure with a novel non-planar diaphragm. *Microelectronic Engineering* 2003; 70: 102–108.

8. Yu M, Balachandran B. Acoustic measurements using a fiber optic sensor system. *Journal of Intelligent Material Systems and Structures* 2003; 14: 409–414.

9. Wang X, Li B, Xiao Z, Lee SH, Roman H, Russo OL, Chin KK, Farmer KR. An ultra-sensitive optical MEMS sensor for partial discharge detection. *Journal of Micromechanics and Microengineering* 2005; 15: 521–527.

10. Yu B, Kim DW, Deng J, Xiao H, Wang A. Fiber Fabry–Perot sensors for detection of partial discharges in power transformers. *Applied Optics* 2003; 42: 3241.

11. Xu J, Wang X, Cooper KL, Wang A. Miniature all-silica fiber optic pressure and acoustic sensors. *Optics Letters* 2005; 30: 3269–3271.

12. Lü T, Yang S. Extrinsic Fabry–Perot cavity optical fiber liquid-level sensor. *Applied Optics* 2007; 46: 3682–3687.

13. Chin KK, Sun Y, Feng G, Georgiou GE, Guo K, Niver E, Roman H, Noe K. Fabry–Perot diaphragm fiber-optic sensor. *Applied Optics* 2007; 46: 7614–7619.

14. Ge Y, Wang M, Chen X, Rong H. An optical MEMS pressure sensor based on a phase demodulation method. *Sensors and Actuators A*, 2008; 143: 224–229.

15. Tran TA, Miller WV, Murphy KA, Vengsarkar AM, Claus RO. Stabilized extrinsic fiber-optic for surface acoustic wave Fizeau sensor detection. *Journal of Lightwave Technology* 1992; 10: 1499–1506.

16. Arya V, de Vries M, Murphy KA, Wang A, Claus RO. Exact analysis of the extrinsic Fabry-Perot interferometric optical fiber sensor using Kirchhoff's diffraction formalism. *Optical Fiber Technology* 1995; 1: 380–384.

17. Leilabady PA, Corke M. All-fiber-optic remote sensing of temperature employing interferometric techniques. *Optics Letters* 1987; 12: 772–774.

18. Inci MN, Kidd SR, Barton JS, Jones JDC. Fabrication of single-mode fiber optic Fabry-Perot interferometers using fusion spliced titanium dioxide optical coatings. *Measurement Science and Technology* 1992; 3: 678–684.

19. Dorighi JF, Krishnaswamy S, Achenbach JD. Stabilization of an embedded fiber optic Fabry-Perot sensor for ultrasound detection. *IEEE Transaction on Ultrasonics, Ferroelectrics, and Frequency Control* 1995; 42: 820–824.

20. Zhu Y, Wang A. Miniature fiber-optic pressure sensor. *IEEE Photonics Technology Letters* 2005; 17: 447–449.

21. Wang X, Xu J, Zhu Y, Cooper KL, Wang A. All-fused-silica miniature optical fiber tip pressure sensor. *Optics Letters* 2006; 31: 885–887.

22. Ma J, Ju J, Jin L, Jin W, Wang D. Fiber-tip micro-cavity for temperature and transverse load sensing. *Optics Express* 2011; 19: 12418–12426.

23. Lee CE, Gibler WN, Atkins RA, Taylor HF. In-line fiber Fabry-Perot interferometer with high-reflectance internal mirrors. *Journal* of *Ligtwave Technology* 1992; 10: 1376–1379.

24. Zhao Y, Ansari F. Quasi-distributed fiber-optic strain sensor: principle and experiment. *Applied Optics* 2001; 40: 3176–3181.

25. Shen F, Peng W, Cooper K, Pickrell G, Wang A. UV-induced intrinsic Fabry-Perot interferometric fiber sensors. *Proceedings SPIE* 2004; 5590: 47–56.

26. Chen X, Shen F, Wang Z, Huang Z, Wang A. Micro-air-gap based intrinsic Fabry–Perot interferometric fiber-optic sensor. *Applied Optics* 2006; 45: 7760–7766.

27. Cibula E, Donlagic D. In-line short cavity Fabry-Perot strain sensor for quasi distributed measurement utilizing standard OTDR. *Optics Express* 2007; 15: 8719–8730.

28. Donlagic D, Cibula E. All-fiber high-sensitivity pressure sensor with SiO_2 diaphragm. *Optics Letters* 2005; 30: 2071–2073.

29. Ran ZL, Rao YJ, Liu WJ, Liao X, Chiang KS. Laser-micromachined Fabry-Perot optical fiber tip sensor for high-resolution temperature independent measurement of refractive index. *Optics Express* 2008; 16: 2252–2263.

30. Taylor HF. Fiber optic sensors based upon the Fabry–Perot interferometer. Section 2 of *Fiber Optic Sensor*, edited by Yin S, Ruffin PB, Yu FTS. Second Edition. CRC Press, Taylor & Francis Group, Boca Raton, London, New York, 2008.

31. Suehira N, Tomiyoshi Y, Sugawara Y, Morita S. Low-temperature non-contact atomic-force microscope with quick sample and cantilever exchange mechanism. *Review of Scientific Instruments* 2001; 72: 2971–2976.

32. Oral A, Grimble RA, Özer HÖ, Pethica JB. High-sensitivity noncontact atomic force microscope/scanning tunneling microscope (nc AFM/STM) operating at subangstrom oscillation amplitudes for atomic resolution imaging and force spectroscopy. *Review of Scientific Instruments* 2003; 74: 3656–3663.

33. Hölscher H, Milde P, Zerweck U, Eng LM, Hoffmann R. The effective quality factor at low temperatures in dynamic force microscopes with Fabry–Pérot interferometer detection. *Applied Physics Letters* 2009; 94: 223514-1–223514-3.

34. Smith DT, Pratt JR, Howard LP. A fiber-optic interferometer with subpicometer resolution for dc and low-frequency displacement measurement. *Review of Scientific Instruments* 2009; 80: 035105-1–035105-8.

35. Rasool HI, Wilkinson PR, Stieg AZ, Gimzewski JK. A low noise all-fiber interferometer for high resolution frequency modulated atomic force microscopy imaging in liquids. *Review of Scientific Instruments* 2010; 81: 023703-1–023703-10.

36. Born M, Wolf E. *Principles of Optics*. Seventh Edition. Cambridge University Press, Cambridge, 1999.

37. Saleh BEA, Teich MC. *Fundamentals of Photonics*. John Wiley & Sons, Hoboken New Jersey, 2007.

38. Nemoto S, Makimoto T. Analysis of splice loss in single-mode fibers using a Gaussian field approximation. *Optical and Quantum Electronics* 1979; 11: 447–457.

39. Arya V, de Vries MJ, Athreya M, Wang A, Claus RO. Analysis of the effect of imperfect fiber endfaces on the performance of extrinsic Fabry-Perot interferometric optical fiber sensors. *Optical Engineering* 1996; 35: 2262–2265.

40. Han M, Wang A. Exact analysis of low-finesse multimode fiber extrinsic Fabry–Perot interferometers. *Applied Optics* 2004; 43: 4659–4666.

41. St-Amant Y, Gariépy D, Rancourt D. Intrinsic properties of the optical coupling between axisymmetric Gaussian beams. *Applied Optics* 2004; 43: 5691–5704.

42. Guo D, Lin R, Wang W. Gaussian-optics-based optical modeling and characterization of a Fabry–Perot microcavity for sensing applications. *Journal of Optical Society America A* 2005; 22: 1577–1588.

43. Han M, Wang A. Mode power distribution effect in white-light multimode fiber extrinsic Fabry–Perot interferometric sensor systems. *Optics Letters* 2006; 31: 1202–1204.

44. Chin KK. Interference of fiber-coupled Gaussian beam multiply reflected between two planar interfaces. *IEEE Photonics Technology Letters* 2007; 19: 1643–1645.

45. Chen J, Chen D, Geng J, Li J, Cai H, Fang Z. Stabilization of optical Fabry–Perot sensor by active feedback control of diode laser. *Sensors and Actuators A* 2008; 148: 376–380.

46. Yu B, Wang A, Pickrell GR. Analysis of fiber Fabry–Perot interferometric sensors using low-coherence light sources. *Journal of Lightwave Technology* 2006; 24: 1758–1767.

47. Murphy KA, Gunther MF, Vengsarkar AM, Claus RO. Quadrature phase-shifted, extrinsic Fabry-Perot optical fiber sensors. *Optics Letters* 1991; 16: 273–275.

48. Wright OB. Stabilized dual-wavelength fiber-optic interferometer for vibration measurement. *Optics Letters* 1991; 16: 56–58.

49. Macià-Sanahuja C, Lamela H, García-Souto JA. Fiber optic interferometric sensor for acoustic detection of partial discharges. *Journal of Optical Technology* 2007; 74: 122–126.

50. Gangopadhyay TK, Chakravorti S, Chatterjee S, Bhattacharya K. Time-frequency analysis of multiple fringe and nonsinusoidal signals obtained from a fiber-optic vibration sensor using an extrinsic Fabry–Pérot interferometer. *Journal of Lightwave Technology* 2006; 24: 2122–2131.

51. Wang X, Li B, Roman HT, Russo OL, Chin K, Farmer KR. Acousto-optical PD detection for transformers. *IEEE Transaction on Power Delivery* 2006; 21: 1068–1073.

52. Wang X, Li B, Liu Z, Roman HT, Russo OL, Chin KK, Farmer KR. Analysis of partial discharge signal using the Hilbert-Huang transform. *IEEE Transactions on Power Delivery* 2006; 21: 1063–1067.

53. Bhatia V, Murphy KA, Claus RO, Jones ME, Grace JL, Tran TA, Greene JA. Optical fiber based absolute extrinsic Fabry–Perot interferometric sensing system. *Measurement Sciences and Technology* 1996; 7: 58–61.

54. Qi B, Pickrell GR, Xu JC, Zhang P, Duan Y, Peng W, Huang Z, Huo W, Xiao H, May RG, Wang A. Novel data processing techniques for dispersive white light interferometer. *Optical Engineering* 2003; 42: 3165–3171.

55. Depiereux F, Lehmann P, Pfeifer T, Schmitt R. Fiber-optical sensor with miniaturized probe head and nanometer accuracy based on spatially modulated low-coherence interferogram analysis. *Applied Optics* 2007; 46: 3425–3431.

56. Xu J, Pickrell G, Wang X, Peng W, Cooper K, Wang A. A novel temperature-insensitive optical fiber pressure sensor for harsh environments. *IEEE Photonics Technology Letters* 2005; 17: 870–872.

57. Du G, Zhu Z, Gong X. *The Fundamentals of Acoustics*. Press of Nanjing University, Nanjing, 2001. [In Chinese]

58. Greywall DS. Micromachined optical-interference microphone. *Sensors and Actuators* 1999; 75: 257–268.

59. Shen F, Xu J, Wang A. Measurement of the frequency response of a diaphragm-based pressure sensor by use of a pulsed excimer laser. *Optics Letters* 2005; 30: 1935–1937.

60. Suna Y, Feng F, Georgiou G, Niver E, Noe K, Chin K. Center embossed diaphragm design guidelines and Fabry–Perot diaphragm fiber optic sensor. *Microelectronics Journal* 2008; 39: 711–716.

61. Chin KK, Feng GH, Pedro I, Roman H. Aligned embossed diaphragm based fiber optic sensor. US patent 7,697,797, issued April 13, 2010.

62. Chin KK, Feng GH, Roman H. MEMS fiber optic microphone. US patent 7,561,277, issued July 14, 2009.

63. Oh KD, Wang A, Claus RO. Fiber-optic extrinsic Fabry–Perot dc magnetic field sensor. *Optics Letters* 2004; 29: 2115–2117.

64. Liu X, Cao Z, Shen Q, Huang S. Optical sensor based on Fabry–Perot resonance modes. *Applied Optics* 2003; 36: 7137–7140.

65. Füurstenau N, Horack H, Schmidt W. Extrinsic Fabry–Perot interferometer fiber-optic microphone. *IEEE Transactions on Instrumentation and Measurement* 1998; 47: 138–142.

APPENDICES

APPENDIX 1: MATHEMATICAL FORMULAS

A1.1 Bessel Equations and Bessel Functions

The Bessel equation has two linearly independent solutions: the first and second kinds Bessel functions, $J_\nu(x)$ and $N_\nu(x)$:

$$x^2 f'' + x f' + (x^2 - \nu^2) f = 0, \qquad (A1.1)$$

$$J_\nu(x) = \sum_{k=0}^{\infty} \frac{(-1)^k}{k! \Gamma(\nu + k + 1)} \left(\frac{x}{2}\right)^{\nu + 2k}, \qquad (A1.2)$$

where $\Gamma(x)$ is gamma function; for an integer argument, $\Gamma(m) = (m - 1)!$.

$$N_\nu(x) = \frac{J_\nu(x) \cos \nu\pi - J_{-\nu}(x)}{\sin \nu\pi}. \qquad (A1.3)$$

Fundamentals of Optical Fiber Sensors, First Edition.
Zujie Fang, Ken K. Chin, Ronghui Qu, and Haiwen Cai.
© 2012 John Wiley & Sons, Inc. Published 2012 by John Wiley & Sons, Inc.

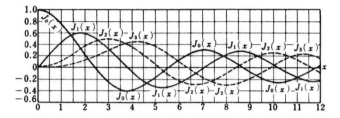

Figure A1.1   Bessel functions.

$N_\nu(x)$ is also called Neumann function. Their linear combinations are Hankel functions:

$$H_\nu^{(1)}(x) = J_\nu(x) + jN_\nu(x), \qquad (A1.4a)$$

$$H_\nu^{(1)}(x) = J_\nu(x) - jN_\nu(x). \qquad (A1.4b)$$

The modified Bessel equation is its imaginary argument counterpart:

$$x^2 f'' + xf' - (x^2 + \nu^2)f = 0. \qquad (A1.5)$$

It has also two linearly independent solutions: $I_\nu(x)$ and $K_\nu(x)$:

$$I_\nu(x) = (-j)^\nu J_\nu(jx) = \sum_{k=0}^{\infty} \frac{1}{k!\Gamma(\nu + k + 1)} \left(\frac{x}{2}\right)^{\nu+2k}, \qquad (A1.6)$$

$$K_\nu(x) = \frac{\pi}{2} \frac{I_{-\nu}(x) - I_\nu(x)}{\sin \nu\pi}. \qquad (A1.7)$$

Figures A1.1 and A1.2 are the first several order Bessel functions and the modified Bessel functions.

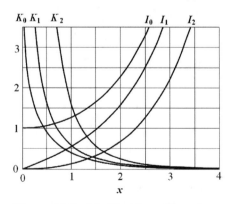

Figure A1.2   Modified Bessel functions.

A1.1.1 Recurrence Relations (for a Non-Negative Integer m)

$$J_{-m}(x) = (-1)^m J_m(x), \tag{A1.8}$$

$$J_m(x) = \frac{x}{2m}[J_{m+1}(x) + J_{m-1}(x)], \tag{A1.9}$$

$$J_m'(x) = \frac{m}{x}J_m(x) - J_{m+1}(x) = -\frac{m}{x}J_m(x) + J_{m-1}(x)$$
$$= \frac{1}{2}[J_{m-1}(x) - J_{m+1}(x)], \tag{A1.10}$$

$$K_{-m}(x) = K_m(x), \tag{A1.11}$$

$$K_m(x) = \frac{x}{2m}[K_{m+1}(x) - K_{m-1}(x)], \tag{A1.12}$$

$$K_m'(x) = \frac{m}{x}K_m(x) - K_{m+1}(x) = -\left[\frac{m}{x}K_m(x) + K_{m-1}(x)\right]$$
$$= \frac{-1}{2}[K_{m-1}(x) + K_{m+1}(x)]; \tag{A1.13}$$

$$I_{-m}(x) = I_m(x), \tag{A1.14}$$

$$I_m(x) = \frac{x}{2m}[I_{m-1}(x) - I_{m+1}(x)], \tag{A1.15}$$

$$I_m'(x) = \frac{m}{x}I_m(x) + I_{m+1}(x) = -\frac{m}{x}I_m(x) + I_{m-1}(x)$$
$$= \frac{1}{2}[I_{m-1}(x) + I_{m+1}(x)]; \tag{A1.16}$$

$$J_0'(x) = -J_1(x) \quad K_0'(x) = -K_1(x) \quad I_0'(x) = I_1(x). \tag{A1.17}$$

A1.1.2 Limiting Forms for Small Argument (x → 0)

$$J_m(x) \approx \frac{1}{m!} \left(\frac{x}{2}\right)^m ; \; I_m(x) \approx \frac{1}{m!} \left(\frac{x}{2}\right)^m ;$$

$$K_m(x) \approx \frac{(m-1)!}{2} \left(\frac{2}{x}\right)^m \cdots (m \geq 1); \qquad \text{(A1.18)}$$

$$N_0(x) \approx \frac{2}{\pi} \ln x, \quad N_m(x) \approx -\frac{(m-1)!}{\pi} \left(\frac{2}{x}\right)^m \cdots (m \geq 1); \quad \text{(A1.19)}$$

$$J_0(x) \approx 1 - \frac{1}{4}x^2 + \frac{1}{64}x^4, \;\; J_1(x) \approx \frac{1}{2}x - \frac{1}{16}x^3; \;\; J_2(x) \approx \frac{1}{24}x^2 - \frac{1}{384}x^4;$$

$$\text{(A1.20)}$$

$$I_0(x) \approx 1 + \frac{1}{4}x^2 + \frac{1}{64}x^4; \;\; I_1(x) \approx \frac{1}{2}x + \frac{1}{16}x^3; \;\; I_2(x) \approx \frac{1}{24}x^2 + \frac{1}{384}x^4;$$

$$\text{(A1.21)}$$

$$K_0(x) \approx \ln \frac{2}{x} - \gamma, \text{ (with Eular Constant } \gamma \approx 0.5772). \qquad \text{(A1.22)}$$

A1.1.3 Asymptotic Expressions for Large Argument (x → ∞)

$$J_v(x) \approx \sqrt{\frac{2}{\pi x}} \cos \left(x - \frac{v\pi}{2} - \frac{\pi}{4}\right); \qquad \text{(A1.23)}$$

$$N_v(x) \approx \sqrt{\frac{2}{\pi x}} \sin \left(x - \frac{v\pi}{2} - \frac{\pi}{4}\right); \qquad \text{(A1.24)}$$

$$I_v(x) \approx \frac{1}{\sqrt{2\pi x}} \left(1 - \frac{4v^2 - 1}{8x}\right) e^x; \qquad \text{(A1.25)}$$

$$K_\nu(x) \approx \sqrt{\frac{\pi}{2x}} \left(1 + \frac{4\nu^2 - 1}{8x}\right) e^{-x};$$
(A1.26)

$$\frac{K_{\nu\pm1}(x)}{K_\nu(x)} \approx 1 + \frac{1 \pm 2\nu}{2x}.$$
(A1.27)

A1.1.4 Integral Expression and Orthogonality Relation

$$J_m(x) = \frac{1}{\pi} \int_0^\pi \cos(x \sin t - mt)\mathrm{d}t = \frac{1}{2\pi} \int_{-\pi}^\pi e^{j(mt - x \sin t)}\mathrm{d}t;$$
(A1.28)

$$\int_0^1 J_m(a_i t)J_m(a_j t)t\,\mathrm{d}t = \frac{1}{2}J_{m+1}^2(a_i)\delta_{ij}. \quad [m > -1; a_{i,j}: \text{roots of } J_m(x) = 0].$$
(A1.29)

$$\int J_m(pt)J_m(qt)t\,\mathrm{d}t = \frac{t}{p^2 - q^2}[pJ_{m+1}(pt)J_m(qt) - qJ_{m+1}(qt)J_m(pt)];$$
(A1.30)

$$\int J_m^2(pt)t\,\mathrm{d}t = \frac{t^2}{2}\left[J_m^2(pt) - J_{m+1}(pt)J_{m-1}(pt)\right];$$
(A1.31)

$$\int K_m(pt)K_m(qt)t\,\mathrm{d}t = \frac{t}{p^2 - q^2}[qK_m(pt)K_{m+1}(qt) - pK_m(qt)K_{m+1}(pt)];$$
(A1.32)

$$\int K_m^2(pt)\mathrm{d}t = \frac{t^2}{2}\left[K_m^2(pt) - K_{m+1}(pt)K_{m-1}(pt)\right];$$
(A1.33)

$$\int x J_0(x)\mathrm{d}x = x J_1(x), \quad \int x^{-1}J_1^2(x)\mathrm{d}x = \frac{-1}{2}[J_0^2(x) + J_1^2(x)];$$
(A1.34)

$$\int x J_0^2(x)\mathrm{d}x = \frac{x^2}{2}\left[J_0^2(x) + J_1^2(x)\right], \quad \int x^2 J_0(x)J_1(x)\mathrm{d}x = \frac{x^2}{2}J_1^2(x);$$

$$(A1.35)$$

$$\int x K_0^2(x)\mathrm{d}x = \frac{x^2}{2}\left[K_0^2(x) - K_1^2(x)\right], \quad \int x^2 K_0(x)K_1(x)\mathrm{d}x = \frac{-x^2}{2}K_1^2(x).$$

$$(A1.36)$$

A1.1.5   Bessel Series of Trigonometric Functions

$$\cos(a \sin x) = J_0(a) + 2 \sum_{m=1}^{\infty} J_{2m}(a) \cos 2mx, \qquad (A1.37a)$$

$$\sin(a \sin x) = 2 \sum_{m=1}^{\infty} J_{2m-1}(a) \sin(2m - 1)x, \qquad (A1.37b)$$

$$\cos(a \cos x) = J_0(a) + 2 \sum_{m=1}^{\infty} (-1)^m J_{2m}(a) \cos 2mx, \qquad (A1.37c)$$

$$\sin(a \cos x) = 2 \sum_{m=1}^{\infty} (-1)^{m+1} J_{2m-1}(a) \cos(2m - 1)x. \qquad (A1.37d)$$

A1.2   Runge–Kutta Method

The two-variable first-order differential equation is expressed as

$$\begin{cases} \dfrac{\mathrm{d}y}{\mathrm{d}x} = f(x, y, z) \\ \dfrac{\mathrm{d}z}{\mathrm{d}x} = g(x, y, z). \end{cases} \qquad (A1.38)$$

It is solved by Runge–Kutta method, with solution in forms of

$$\begin{cases} y_{n+1} = y_n + (k_1 + 2k_2 + 2k_3 + k_4)/6 \\ z_{n+1} = z_n + (l_1 + 2l_2 + 2l_3 + l_4)/6, \end{cases} \qquad (A1.39)$$

where $k_1 = hf(x_n, y_n, z_n)$, $\quad k_2 = hf(x_n + h/2, y_n + k_1/2, z_n + l_1/2)$,

$\quad k_3 = hf(x_n + h/2, y_n + k_2/2, z_n + l_2/2)$,

$\quad k_4 = hf(x_n + h, y_n + k_3, z_n + l_3)$;

$\quad l_1 = hg(x_n, y_n, z_n)$, $\quad l_2 = hg(x_n + h/2, y_n + k_1/2, z_n + l_1/2)$,

$\quad l_3 = hg(x_n + h/2, y_n + k_2/2, z_n + l_2/2)$,

$\quad l_4 = hg(x_n + h, y_n + k_3, z_n + l_3)$;

with step of $h = x_{i+1} - x_i$.

A1.3 The First-Order Linear Differential Equation

The solution of equation $dy/dx = P(x)y + Q(x)$ is written as

$$y = \exp\left(\int P\,dx\right)\left[C + \int Q \exp\left(-\int P\,dx\right)dx\right]. \qquad (A1.40)$$

A1.4 Riccati Equation

The nonlinear differential equation

$$\frac{dy}{dx} = P(x) + Q(x)y + R(x)y^2 \qquad (A1.41)$$

is called a Riccati equation. If $y_1(x)$ is a known particular solution of (A1.42), its general solution is rewritten as $y = y_1 + u$; then (A1.42) is transformed to an equation of u:

$$\frac{du}{dx} - (Q + 2y_1 R)u = Ru^2, \qquad (A1.42)$$

which is a Bernoulli equation with $n = 2$. By introducing conversion of $w = u^{-1}$, the equation can be reduced to the linear equation

$$\frac{dw}{dx} + [Q(x) + 2y_1(x)R]w = -R(x). \qquad (A1.43)$$

A1.5 Airy Equation and Airy Functions

The Airy equation is expressed as

$$\frac{d^2 F}{dx^2} - xF = 0. \tag{A1.44}$$

Its two linearly independent solutions are

$$\text{Ai}(x) = \frac{1}{\pi} \int_0^\infty \cos(t^3/3 + xt)dt, \tag{A1.45a}$$

$$\text{Bi}(x) = \frac{1}{\pi} \int_0^\infty [\exp(-t^3/3 + xt) + \sin(t^3/3 + xt)]dt. \tag{A1.45b}$$

They can be rewritten as

$$\text{Ai}(x) = \frac{1}{\pi}\sqrt{\frac{x}{3}} K_{1/3}\left(\frac{2}{3}x^{3/2}\right), \tag{A1.46a}$$

$$\text{Bi}(x) = \sqrt{\frac{x}{3}} \left[I_{1/3}\left(\frac{2}{3}x^{3/2}\right) + I_{-1/3}\left(\frac{2}{3}x^{3/2}\right) \right], \tag{A1.46b}$$

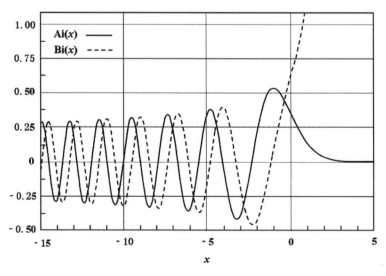

Figure A1.3 Airy functions.

$$\mathrm{Ai}(-x) = \frac{1}{3}\sqrt{x}\left[J_{1/3}\left(\frac{2}{3}x^{3/2}\right) + J_{-1/3}\left(\frac{2}{3}x^{3/2}\right)\right], \qquad (A1.47a)$$

$$\mathrm{Bi}(-x) = \sqrt{\frac{x}{3}}\left[J_{-1/3}\left(\frac{2}{3}x^{3/2}\right) - J_{1/3}\left(\frac{2}{3}x^{3/2}\right)\right]. \qquad (A1.47b)$$

Their asymptotic expressions as $x \to \infty$ are written as

$$\mathrm{Ai}(x) \sim \frac{\exp[-(2/3)x^{3/2}]}{2\sqrt{\pi}x^{1/4}}, \quad \mathrm{Bi}(x) \sim \frac{\exp[(2/3)x^{3/2}]}{\sqrt{\pi}x^{1/4}}; \qquad (A1.48a)$$

$$\mathrm{Ai}(-x) \sim \frac{\sin[(2/3)x^{3/2} + \pi/4]}{\sqrt{\pi}x^{1/4}}, \quad \mathrm{Bi}(-x) \sim \frac{\cos[(2/3)x^{3/2} + \pi/4]}{\sqrt{\pi}x^{1/4}}. \qquad (A1.48b)$$

APPENDIX 2: FUNDAMENTALS OF ELASTICITY

A2.1   Strain, Stress, and Hooke's Law

A2.1.1   Definition of Strain and Stress   By denoting the position vector as $\boldsymbol{r}$ and displacement vector as $\boldsymbol{u} = \boldsymbol{r}' - \boldsymbol{r}$, the strain tensor is defined as

$$e_{ij} = \frac{1}{2}\left(\frac{\partial u_i}{\partial x_j} + \frac{\partial u_j}{\partial x_i}\right) \quad (i, j = x, y, z). \qquad (A2.1)$$

The strain tensor has six independent components: three normal strains e_{xx}, e_{yy}, e_{zz} and three shear strains e_{yz}, e_{zx}, e_{xy} in the Cartesian coordinate. The volume relative change is expressed as

$$\Delta V/V = e_{xx} + e_{yy} + e_{zz}. \qquad (A2.2)$$

From the definitions, the strains satisfy differential relations of

$$\frac{\partial^2 e_{xx}}{\partial y^2} + \frac{\partial^2 e_{yy}}{\partial x^2} = \frac{1}{2}\frac{\partial^2 e_{xy}}{\partial x \partial y}, \quad \frac{\partial^2 e_{zz}}{\partial y^2} + \frac{\partial^2 e_{yy}}{\partial z^2} = \frac{1}{2}\frac{\partial^2 e_{yz}}{\partial y \partial z},$$

$$\frac{\partial^2 e_{xx}}{\partial z^2} + \frac{\partial^2 e_{zz}}{\partial x^2} = \frac{1}{2}\frac{\partial^2 e_{xz}}{\partial x \partial z},$$

$$\frac{\partial}{\partial x}\left(-\frac{\partial e_{yz}}{\partial x}+\frac{\partial e_{xz}}{\partial y}+\frac{\partial e_{xy}}{\partial z}\right)=\frac{\partial^2 e_{xx}}{\partial y \partial z},$$

$$\frac{\partial}{\partial y}\left(\frac{\partial e_{yz}}{\partial x}-\frac{\partial e_{xz}}{\partial y}+\frac{\partial e_{xy}}{\partial z}\right)=\frac{\partial^2 e_{yy}}{\partial x \partial z},$$

$$\frac{\partial}{\partial z}\left(\frac{\partial e_{yz}}{\partial x}+\frac{\partial e_{xz}}{\partial y}-\frac{\partial e_{xy}}{\partial z}\right)=\frac{\partial^2 e_{zz}}{\partial x \partial y}. \tag{A2.3}$$

The stress is a tensor describing the force per unit area on the interface; it has three normal components $\sigma_{xx}, \sigma_{yy}, \sigma_{zz}$ and three shear components $\sigma_{yz}, \sigma_{zx}, \sigma_{xy}$. Denoting the force per unit volume as f, the relation between stresses and f is expressed as

$$f_i = \frac{\partial \sigma_{ix}}{\partial x}+\frac{\partial \sigma_{iy}}{\partial y}+\frac{\partial \sigma_{iz}}{\partial z} \quad (i=x, y, z). \tag{A2.4}$$

The force acting on a volume is written as

$$F_i = \int f_i dV = \oint (\sigma_{ix} ds_x + \sigma_{iy} ds_y + \sigma_{iz} ds_z) \, (i=x, y, z), \tag{A2.5}$$

where ds is the surface element vector. The momentum acting on a volume is written as

$$M_{ij} = \int (f_i x_j - f_j x_i)dV = \sum_{l=x,y,z} \oint (\sigma_{il} x_j - \sigma_{jl} x_i)ds_l$$

$$(i, j, l = x, y, z; \quad x_{i,j,l} = x, y, z). \tag{A2.6}$$

In equilibrium, the internal stresses in a volume must balance; taking the gravity into account, the equations of equilibrium are

$$\frac{\partial \sigma_{ix}}{\partial x}+\frac{\partial \sigma_{iy}}{\partial y}+\frac{\partial \sigma_{iz}}{\partial z}+\rho g_i = 0 \quad (i=x, y, z), \tag{A2.7}$$

where g is the gravitational acceleration vector and ρ is the medium density. At the boundary of a volume, the internal stresses must balance the external forces in equilibrium. In nonequilibrium, the equation of motion is

$$\frac{\partial \sigma_{ix}}{\partial x}+\frac{\partial \sigma_{iy}}{\partial y}+\frac{\partial \sigma_{iz}}{\partial z}+f_i^{[b]}=\rho a_i \quad (i=x, y, z), \tag{A2.8}$$

where a is the acceleration vector of the volume element, $f_i^{[b]}$ stands for body forces per unit volume, including the gravity. NB: strain e_{ii} and stress σ_{ii} are occasionally simplified as e_i and σ_i. The shear strain is defined as $e_{ij} = \partial u_i/\partial x_j + \partial u_j/\partial x_i$ for $i \neq j$ in some books. Correspondingly related formulas have to be modified.

A2.1.2 Hooke's Law

In the range of elastic deformation, a linear dependence exists between the strains and the stresses, expressed as

$$e_i = \sum_{j=1}^{6} h_{ij}\sigma_j \quad (i, j = xx, yy, zz, yz, zx, xy). \tag{A2.9}$$

The 6×6 coefficients are not independent. By the symmetry, $h_{ij} = h_{ji}$. For isotropic media, $h_{ij} = h_{11} + (h_{12} - h_{11})(1 - \delta_{ij}) \cdots (i, j = 1, 2, 3)$, $h_{ij} = h_{ji} = h_{44}\delta_{ij} \cdots (i, j = 4, 5, 6)$, and $h_{44} = (h_{11} - h_{12})/2$. Equation (A2.9) is then rewritten as

$$\begin{pmatrix} e_{xx} \\ e_{yy} \\ e_{zz} \\ e_{yz} \\ e_{zx} \\ e_{xy} \end{pmatrix} = \frac{1}{Y} \begin{pmatrix} 1 & -\nu & -\nu & 0 & 0 & 0 \\ -\nu & 1 & -\nu & 0 & 0 & 0 \\ -\nu & -\nu & 1 & 0 & 0 & 0 \\ 0 & 0 & 0 & 1+\nu & 0 & 0 \\ 0 & 0 & 0 & 0 & 1+\nu & 0 \\ 0 & 0 & 0 & 0 & 0 & 1+\nu \end{pmatrix} \begin{pmatrix} \sigma_{xx} \\ \sigma_{yy} \\ \sigma_{zz} \\ \sigma_{yz} \\ \sigma_{zx} \\ \sigma_{xy} \end{pmatrix}, \tag{A2.10}$$

with Young's modulus Y (the modulus of extension) and Poisson's ratio ν (the ratio of the transverse compression to the longitudinal extension). ν is in the range of $0 < \nu < 0.5$.

The stresses can be written as functions of strains by inversion of (A2.10), expressed as

$$\begin{pmatrix} \sigma_{xx} \\ \sigma_{yy} \\ \sigma_{zz} \\ \sigma_{yz} \\ \sigma_{zx} \\ \sigma_{xy} \end{pmatrix} = \begin{pmatrix} 2\mu+\lambda & \lambda & \lambda & 0 & 0 & 0 \\ \lambda & 2\mu+\lambda & \lambda & 0 & 0 & 0 \\ \lambda & \lambda & 2\mu+\lambda & 0 & 0 & 0 \\ 0 & 0 & 0 & \mu & 0 & 0 \\ 0 & 0 & 0 & 0 & \mu & 0 \\ 0 & 0 & 0 & 0 & 0 & \mu \end{pmatrix} \begin{pmatrix} e_{xx} \\ e_{yy} \\ e_{zz} \\ e_{yz} \\ e_{zx} \\ e_{xy} \end{pmatrix},$$

where $\lambda = Y\nu/(1+\nu)(1-2\nu)$ and $\mu = Y/2(1+\nu)$ are termed Lamé coefficients. The bulk modulus $K = Y/3(1-2\nu)$ is the ratio of pressure to volume change.

Equation (A2.10) can be expanded as

$$e_{xx} = [\sigma_{xx} - \nu(\sigma_{yy} + \sigma_{zz})]/Y, \tag{A2.10a}$$

$$e_{yy} = [\sigma_{yy} - \nu(\sigma_{zz} + \sigma_{xx})]/Y, \tag{A2.10b}$$

$$e_{zz} = [\sigma_{zz} - \nu(\sigma_{xx} + \sigma_{yy})]/Y, \tag{A2.10c}$$

$$e_{ij} = (1+\nu)\sigma_{ij}/Y \quad (i, j = x, y, z; \quad i \neq j). \tag{A2.10d}$$

By substituting them into (A2.8) and using (A2.1), the equilibrium equation is expressed as

$$\rho a = f^{[b]} + \frac{Y}{2(1+\nu)}\left[\nabla^2 u + \frac{1}{1-2\nu}\nabla(\nabla \cdot u)\right], \tag{A2.11a}$$

or

$$\rho a = f^{[b]} + \frac{Y}{2(1+\nu)}\left[\frac{2(1-\nu)}{1-2\nu}\nabla(\nabla \cdot u) - \nabla \times \nabla \times u\right]. \tag{A2.11b}$$

In the static case, with the gravity and body forces neglected, the equation is rewritten as

$$(1-2\nu)\nabla^2 u + \nabla(\nabla \cdot u) = 0, \quad \text{or } \nabla(\nabla \cdot u) = \frac{1-2\nu}{2(1-\nu)}\nabla \times \nabla \times u, \tag{A2.12}$$

where $\nabla \cdot u$ is just the relative volume change (A2.2). Its divergence is deduced as $2(1-\nu)\nabla^2(\nabla \cdot u) = 0$. By taking the Laplacian, it is obtained that

$$\Delta^2 u = 0, \tag{A2.13}$$

that is, the displacement vector satisfies the biharmonic equation in equilibrium.

A2.2   Conversions Between Coordinates

Conversions between the cylindrical polar coordinate and Cartesian coordinate are

$$x = r\cos\varphi, \quad y = r\sin\varphi, \quad r = \sqrt{x^2 + y^2}, \quad \varphi = \arctan(y/x).$$

The unit vectors are expressed as: $\hat{x} = \hat{r} \cos \varphi - \hat{\varphi} \sin \varphi$, $\hat{y} = \hat{r} \sin \varphi + \hat{\varphi} \cos \varphi$; and $\hat{r} = \hat{x} \cos \varphi + \hat{y} \sin \varphi$, $\hat{\varphi} = -\hat{x} \sin \varphi + \hat{y} \cos \varphi$. Their derivative operations are

$$\frac{\partial}{\partial x} = \cos \varphi \frac{\partial}{\partial r} - \frac{\sin \varphi}{r} \frac{\partial}{\partial \varphi}, \qquad \frac{\partial}{\partial y} = \sin \varphi \frac{\partial}{\partial r} + \frac{\cos \varphi}{r} \frac{\partial}{\partial \varphi}; \qquad (A2.14a)$$

$$\frac{\partial}{\partial r} = \cos \varphi \frac{\partial}{\partial x} + \sin \varphi \frac{\partial}{\partial y}, \qquad \frac{1}{r} \frac{\partial}{\partial \varphi} = -\sin \varphi \frac{\partial}{\partial x} + \cos \varphi \frac{\partial}{\partial y}. \qquad (A2.14b)$$

The strains are written as

$$e_{rr} = \frac{\partial u_r}{\partial r}, \qquad e_{r\varphi} = \frac{1}{2} \left(\frac{\partial u_\varphi}{\partial r} - \frac{u_\varphi}{r} + \frac{1}{r} \frac{\partial u_r}{\partial \varphi} \right), \qquad e_{rz} = \frac{1}{2} \left(\frac{\partial u_r}{\partial z} + \frac{\partial u_z}{\partial r} \right),$$

$$e_{\varphi\varphi} = \frac{1}{r} \frac{\partial u_\varphi}{\partial \varphi} + \frac{u_r}{r}, \qquad e_{\varphi z} = \frac{1}{2} \left(\frac{\partial u_z}{\partial \varphi} + \frac{\partial u_\varphi}{\partial z} \right), \qquad e_{zz} = \frac{\partial u_z}{\partial z}. \qquad (A2.15)$$

The conversions between the two coordinates are

$$e_{xx} = e_{rr} \cos^2 \varphi + e_{\varphi\varphi} \sin^2 \varphi - e_{r\varphi} \sin 2\varphi, \qquad (A2.16a)$$

$$e_{yy} = e_{rr} \sin^2 \varphi + e_{\varphi\varphi} \cos^2 \varphi + e_{r\varphi} \sin 2\varphi, \qquad (A2.16b)$$

$$e_{xy} = e_{r\varphi} \cos 2\varphi + (e_{rr} - e_{\varphi\varphi}) \sin \varphi \cos \varphi, \qquad (A2.16c)$$

$$e_{rr} = e_{xx} \cos^2 \varphi + e_{yy} \sin^2 \varphi + e_{xy} \sin 2\varphi, \qquad (A2.17a)$$

$$e_{\varphi\varphi} = e_{xx} \sin^2 \varphi + e_{yy} \cos^2 \varphi - e_{xy} \sin 2\varphi, \qquad (A2.17b)$$

$$e_{r\varphi} = e_{xy} \cos 2\varphi - (e_{xx} - e_{yy}) \sin \varphi \cos \varphi. \qquad (A2.17c)$$

It is derived that the relative transverse area change is $e_{xx} + e_{yy} = e_{rr} + e_{\varphi\varphi}$. From the related vector operations:

$$\nabla \cdot \boldsymbol{u} = \frac{1}{r} \frac{\partial (r u_r)}{\partial r} + \frac{1}{r} \frac{\partial u_\varphi}{\partial \varphi} + \frac{\partial u_z}{\partial z}, \qquad (A2.18)$$

$$\nabla^2 = \frac{1}{r} \frac{\partial}{\partial r} \left(r \frac{\partial}{\partial r} \right) + \frac{1}{r^2} \frac{\partial^2}{\partial \varphi^2} + \frac{\partial^2}{\partial z^2}, \qquad (A2.19)$$

the motion equations are written as

$$\frac{\partial \sigma_{rr}}{\partial r} + \frac{1}{r}\frac{\partial \sigma_{r\varphi}}{\partial \varphi} + \frac{\partial \sigma_{rz}}{\partial z} + \frac{1}{r}(\sigma_{rr} - \sigma_{\varphi\varphi}) + f_r^{[b]} = \rho a_r, \qquad (A2.20)$$

$$\frac{\partial \sigma_{r\varphi}}{\partial r} + \frac{1}{r}\frac{\partial \sigma_{\varphi\varphi}}{\partial \varphi} + \frac{\partial \sigma_{z\varphi}}{\partial z} + \frac{2}{r}\sigma_{r\varphi} + f_\varphi^{[b]} = \rho a_\varphi, \qquad (A2.20b)$$

$$\frac{\partial \sigma_{rz}}{\partial r} + \frac{1}{r}\frac{\partial \sigma_{z\varphi}}{\partial \varphi} + \frac{\partial \sigma_{zz}}{\partial z} + \frac{1}{r}\sigma_{rz} + f_z^{[b]} = \rho a_z. \qquad (A2.20c)$$

Hooke's law in the cylindrical polar coordinate for isotropic media is the same as (A2.10) with corresponding subscripts:

$$e_i = \sum_{j=1}^{6} h_{ij}\sigma_j \quad (i, j = rr, \varphi\varphi, zz, \varphi z, zr, r\varphi). \qquad (A2.21)$$

From the conversion between two Cartesian coordinates with angle of θ in $(x - y)$ plane:

$$\begin{array}{ll} x_1 = x\cos\varphi - y\sin\varphi \\ y_1 = x\sin\varphi + y\cos\varphi \end{array}, \text{ and } \begin{array}{ll} x = x_1\cos\varphi + y_1\sin\varphi \\ y = -x_1\sin\varphi + y_1\cos\varphi \end{array}, \qquad (A2.22)$$

the strain conversions are expressed as

$$e_{xx1} = e_{xx}\cos^2\theta + e_{yy}\sin^2\theta - e_{xy}\sin\theta\cos\theta, \qquad (A2.23a)$$

$$e_{yy1} = e_{xx}\sin^2\theta + e_{yy}\cos^2\theta + e_{xy}\sin\theta\cos\theta, \qquad (A2.23b)$$

$$e_{xy1} = e_{xy}\cos 2\theta + (e_{xx} - e_{yy})\sin\theta\cos\theta. \qquad (A2.23c)$$

A2.3 Plane Deformation

When it is unnecessary for one of three-dimensional strains or stresses to be taken into consideration, the three-dimensional problem is degraded to a two-dimensional one.

In case 1, with no deformation in z-direction, that is, $u_z = 0$, for example, the medium is bounded by rigid bodies, it is derived that $e_{zz} = e_{zx} = e_{zy} = 0$ and $\sigma_{zx} = \sigma_{zy} = 0$. The strains to be solved are

e_{xx}, e_{yy}, e_{xy}; the stresses to be solved are σ_{xx}, σ_{yy} with $\sigma_{zz} = \nu(\sigma_{xx} + \sigma_{yy})$. Hooke's law is expressed as

$$e_{xx} = (1 + \nu)[(1 - \nu)\sigma_{xx} - \nu\sigma_{yy}]/Y, \qquad (A2.24a)$$

$$e_{yy} = (1 + \nu)[(1 - \nu)\sigma_{yy} - \nu\sigma_{xx}]/Y, \qquad (A2.24b)$$

$$e_{xy} = (1 + \nu)\sigma_{xy}/Y; \qquad (A2.24c)$$

and

$$e_{xx} + e_{yy} = (1 + \nu)(1 - 2\nu)(\sigma_{xx} + \sigma_{yy})/Y. \qquad (A2.24d)$$

The equilibrium equations without body forces are

$$\frac{\partial \sigma_{xx}}{\partial x} + \frac{\partial \sigma_{xy}}{\partial y} = 0, \quad \frac{\partial \sigma_{xy}}{\partial x} + \frac{\partial \sigma_{yy}}{\partial y} = 0. \qquad (A2.25)$$

Their general solutions take forms of

$$\sigma_{xx} = \partial^2 \chi/\partial y^2, \quad \sigma_{yy} = \partial^2 \chi/\partial x^2, \quad \sigma_{xy} = -\partial^2 \chi/\partial x \partial y, \qquad (A2.26)$$

where χ is an arbitrary function of x and y, satisfying the related boundary conditions, called the stress function. It is seen that

$$\nabla^2 \chi = \sigma_{xx} + \sigma_{yy} = Y(e_{xx} + e_{yy})/(1 + \nu)(1 - 2\nu). \qquad (A2.27)$$

Therefore, the stress function satisfies the biharmonic equation: $\Delta^2 \chi = 0$. The longitudinal stress is then expressed as $\sigma_{zz} = \nu \nabla^2 \chi$.

In case 2, with no stress in z-direction, that is, $\sigma_{zz} = \sigma_{zx} = \sigma_{zy} = 0$, for example, no external force in z-direction exists, it is derived that

$$e_{xx} = (\sigma_{xx} - \nu\sigma_{yy})/Y, \qquad (A2.28a)$$

$$e_{yy} = (\sigma_{yy} - \nu\sigma_{xx})/Y, \qquad (A2.28b)$$

$$e_{zz} = -\nu(\sigma_{xx} + \sigma_{yy})/Y, \qquad (A2.28c)$$

$$e_{xy} = (1 + \nu)\sigma_{xy}/Yd; \qquad (A2.28d)$$

and

$$e_{xx} + e_{yy} = (1 - \nu)(\sigma_{xx} + \sigma_{yy})/Y. \qquad (A2.28e)$$

It is seen that the same equilibrium equations as those of case 1 hold; and a stress function with the same properties exists.

The plane deformation in the polar coordinate is described by the following formulas. Three strain components are considered:

$$e_{rr} = \frac{\partial u_r}{\partial r}, \quad e_{\varphi\varphi} = \frac{1}{r}\frac{\partial u_\varphi}{\partial \varphi} + \frac{u_r}{r}, \quad e_{r\varphi} = \frac{1}{2}\left(\frac{\partial u_\varphi}{\partial r} - \frac{u_\varphi}{r} + \frac{1}{r}\frac{\partial u_r}{\partial \varphi}\right).$$

$$(A2.29)$$

The equilibrium equations for the isotropic medium without body forces are expressed as

$$\frac{\partial \sigma_{rr}}{\partial r} + \frac{1}{r}\frac{\partial \sigma_{r\varphi}}{\partial \varphi} + \frac{1}{r}(\sigma_{rr} - \sigma_{\varphi\varphi}) = 0, \qquad (A2.30a)$$

$$\frac{\partial \sigma_{r\varphi}}{\partial r} + \frac{1}{r}\frac{\partial \sigma_{\varphi\varphi}}{\partial \varphi} + \frac{2}{r}\sigma_{r\varphi} = 0. \qquad (A2.30b)$$

Hooke's law is expressed as

$$e_{rr} = \frac{1}{Y}(\sigma_{rr} - \nu\sigma_{\varphi\varphi}), \quad e_{\varphi\varphi} = \frac{1}{Y}(\sigma_{\varphi\varphi} - \nu\sigma_{rr}), \quad e_{r\varphi} = \frac{1}{Y}(1 + \nu)\sigma_{r\varphi}.$$

$$(A2.31)$$

The biharmonic equation for the stress function is written as

$$\left(\frac{\partial^2}{\partial r^2} + \frac{1}{r}\frac{\partial}{\partial r} + \frac{1}{r^2}\frac{\partial^2}{\partial \varphi^2}\right)^2 \chi = 0$$

and the relations between the stresses and the stress function are written as

$$\sigma_{rr} = \frac{1}{r}\frac{\partial \chi}{\partial r} + \frac{1}{r^2}\frac{\partial \chi^2}{\partial \varphi^2}, \quad \sigma_{\varphi\varphi} = \frac{\partial^2 \chi}{\partial r^2}, \quad \sigma_{r\varphi} = -\frac{\partial}{\partial r}\left(\frac{1}{r}\frac{\partial \chi}{\partial \varphi}\right). \qquad (A2.32)$$

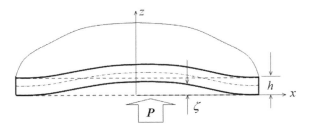

Figure A2.1   Diagram of a bent plate.

A2.4 Equilibrium of Plates and Rods

A2.4.1 The Equilibrium Equation for a Thin Plate   A plate is regarded as a thin plate, if its thickness is much thinner than its transverse size, typically several tenths. Its deformation is regarded as a problem of small deflection, if the displacement occurs only in the direction perpendicular to the plane before deformation, and the strain in the direction is negligibly small, that is, $e_{zz} \approx 0$, and $\sigma_{zz} \approx 0$. Figure A2.1 shows a section of plate. It is noted that the part on the convex side is stretched, whereas the part on the concave side is compressed; and a neutral surface exists at the middle, on which there is no extension, nor compression.

The equilibrium equation for the thin plate is deduced as

$$D\Delta^2\zeta = P, \qquad (A2.33)$$

where $D = Yh^3/[12(1 - v^2)]$ is called the flexural rigidity of the plate, $P(x, y)$ is the pressure difference between the two sides. In dynamic states, the motion equation is expressed as

$$\rho\frac{\partial^2\zeta}{\partial t^2} = P - D\Delta^2\zeta. \qquad (A2.34)$$

A2.4.2 Bending of a Rod   For a rod bent in x–z-plane, its strain can be regarded as a plane deformation if the deflection is small enough so that the deformation in y-direction is neglected, that is, with $e_{yy} = e_{yz} = e_{yx} = 0$. The bending induced stretching and compression cause stains in z-direction:

$$e_{zz} = x/R = x\xi_z'', \qquad (A2.35)$$

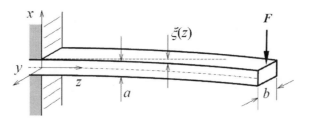

Figure A2.2   Diag ram of a bent cantilever with one end clamped.

where x is measured from the neutral surface, R is the curvature radius of bending, $R = (1 + \xi_z'^2)^{3/2}/\xi_z'' \approx 1/\xi_z''$ with $\xi(z)$ denoted as the displacement in x-direction. The stress is written as $\sigma_{zz} = Yx/R$.

Considering a rectangular rod, as shown in Figure A2.2, the moment of the force on the cross section of an infinitesimal element between z and $z + dz$ is written as

$$M_y = \frac{Y}{R} \int x^2 dx dy = \frac{Ya^3b}{12R} = YI\xi_z'', \qquad (A2.36)$$

where $I = ba^3/12$ is the moment of inertia about the y-axis. In equilibrium, $M_y(z) - M_y(z + dz) = T_x dz$, where T_x is the shear force of the adjacent element applied on the element, i.e., $T_x = -\partial M_y/\partial z$. The sum of the shear forces at the left side and right side of the element accelerates it as $T_x(z) - T_x(z + dz) = (\rho dz)\xi_t''$. Thus a motion equation is deduced:

$$\rho\frac{\partial^2\xi}{\partial t^2} = -\frac{\partial T_x}{\partial z} = \frac{\partial^2 M_y}{\partial z^2} = YI\frac{\partial^4\xi}{\partial z^4}. \qquad (A2.37)$$

For a circular cross section with radius r, the moment of inertia is $I = \pi r^4/4$. If the rod is not uniform in z-direction, that is, the moment of inertia is a function of z, the motion equation is expressed as

$$\rho\frac{\partial^2\xi}{\partial t^2} = Y\left(I\frac{\partial^4\xi}{\partial z^4} + 2\frac{\partial I}{\partial z}\frac{\partial^3\xi}{\partial z^3} + \frac{\partial^2 I}{\partial z^2}\frac{\partial^2\xi}{\partial z^2}\right) + f_x(z, t), \qquad (A2.38)$$

where $f_x(z, t)$ denotes the external distributed force, including gravity.

When a force F is applied at the free end $(z = L)$ of a uniform rod, the displacement along z-direction is then solved as

$$\xi = \frac{F}{2YI}(L - z/3)z^2. \tag{A2.39}$$

In dynamic cases, its motion for a single frequency is described as

$$\xi = \xi_0[(\cos \kappa z - \cosh \kappa z) + b(\sin \kappa z - \sinh \kappa z)] \exp(i\omega t), \tag{A2.40}$$

where the wave vector $\kappa^2 = \omega\sqrt{\rho/YI}$ obeys the eigen equation of

$$1 + \cos \kappa L \cosh \kappa L = 0. \tag{A2.41}$$

The eigenvalues are solved as $\kappa L = \mu_i = 1.875, 4.694, \cdots$; and the eigen frequencies are written as $\omega_i = (\mu_i/L)^2 \sqrt{YI/\rho}$.

A2.4.3 Torsion of Rods

Let us inspect a twisted thin rod with its transverse dimension R much smaller than its length, as shown in Figure A2.3. If the twisting rate $\tau = d\phi/dz$ is small enough to meet condition of $\tau R \ll 1$, that is, the relative displacement of adjoining transverse sections is small, the torsional deformation is described by only shear strains, with $e_{xx} = e_{yy} = e_{zz} = 0$ and $e_{xy} = 0$; the last expression means that no deformation of its cross section occurs. With relations of $x = r \cos \varphi$ and $y = r \sin \varphi$ shear strains in z direction are deduced to be $e_{zx} = -\tau y$ and $e_{zy} = \tau x$; the shear stresses $\sigma_{xz} = -Y\tau y/(1 + \nu)$ and $\sigma_{yz} = Y\tau x/(1 + \nu)$ are then obtained.

Consider a solid rod with circular cross section as a simplified case. The moment of twisting force is $M = \int r \times F = \hat{z} \int (x\sigma_{yz} - y\sigma_{xz})dxdy = \hat{z}Y\tau\pi R^4/[2(1 + \nu)]$. The torsional rigidity is defined as

$$C = \frac{M}{\tau} = \frac{\pi Y R^4}{2(1 + \nu)}. \tag{A2.42}$$

For a circular tube, $C = \pi Y (R_2^4 - R_1^4)/[2(1 + \nu)]$.

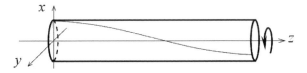

Figure A2.3 Diagram of a twisted rod.

A2.5 Photoelastic Effect

The dielectric constant of a solid medium is generally a 3×3 tensor with six independent components. It will be affected by the strains, expressed as

$$\Delta \left(\frac{1}{\varepsilon} \right) = \sum_{j=1}^{6} p_{ij} e_j \quad (i, j = xx, yy, zz, yz, zx, xy), \qquad (A2.43)$$

where p_{ij} are strain-optic coefficients. The coefficients possess symmetric properties, similar to Hooke's law. For isotropic media the effect is expressed as

$$\Delta \left(\frac{1}{\varepsilon} \right) = \frac{-1}{\varepsilon^2} \begin{pmatrix} \Delta \varepsilon_x \\ \Delta \varepsilon_y \\ \Delta \varepsilon_z \\ \Delta \varepsilon_{yz} \\ \Delta \varepsilon_{zx} \\ \Delta \varepsilon_{xy} \end{pmatrix} = \begin{pmatrix} p_{11} & p_{12} & p_{12} & 0 & 0 & 0 \\ p_{12} & p_{11} & p_{12} & 0 & 0 & 0 \\ p_{12} & p_{12} & p_{11} & 0 & 0 & 0 \\ 0 & 0 & 0 & p_{44} & 0 & 0 \\ 0 & 0 & 0 & 0 & p_{44} & 0 \\ 0 & 0 & 0 & 0 & 0 & p_{44} \end{pmatrix} \begin{pmatrix} e_{xx} \\ e_{yy} \\ e_{zz} \\ e_{yz} \\ e_{zx} \\ e_{xy} \end{pmatrix}$$

$$(A2.44)$$

with relation of $p_{44} = (p_{11} - p_{12})/2$. It means that only two coefficients are independent. It can be rewritten as the refractive index change, for example, with shear strains neglected:

$$\begin{pmatrix} \Delta n_x \\ \Delta n_y \\ \Delta n_z \end{pmatrix} = \frac{-n^3}{2} \begin{pmatrix} p_{11} & p_{12} & p_{12} \\ p_{12} & p_{11} & p_{12} \\ p_{12} & p_{12} & p_{11} \end{pmatrix} \begin{pmatrix} e_{xx} \\ e_{yy} \\ e_{zz} \end{pmatrix}. \qquad (A2.45)$$

APPENDIX 3: FUNDAMENTALS OF POLARIZATION OPTICS

A3.1 Polarized Light and Jones Vector

A single-frequency planar optical wave propagating in z-direction is expressed generally as

$$E = \begin{pmatrix} E_x \\ E_y \end{pmatrix} = E \begin{pmatrix} \cos \alpha \\ \sin \alpha e^{j\varphi} \end{pmatrix} e^{j(kz - \omega t)}, \qquad (A3.1)$$

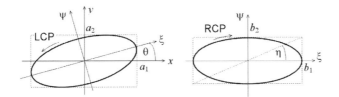

Figure A3.1 Traces of electric field of an elliptically polarized light.

where $E = |E|$, $\cos\alpha = |E_x|/E$, $\sin\alpha = |E_y|/E$, and $\tan\alpha = |E_y|/|E_x|$. The Jones vector

$$J = \begin{pmatrix} \cos\alpha \\ e^{j\varphi}\sin\alpha \end{pmatrix} \qquad (A3.2)$$

is used to describe any polarized light wave, including linear polarized light wave with $\varphi = 0$ and the polarization direction α, and circularly polarized light waves with $\varphi = \pm\pi/2$ and $\alpha = \pi/4$, where the sign is positive for right circular polarization (RCP) and negative for left circular polarization (LCP). NB: RCP is defined as the rotation of E vector in CW way when viewed towards $-z$ direction; and LCP is in CCW way. Their Jones vectors are written as $J_+ = (1 + j)^T/\sqrt{2}$ and $J_- = (1 - j)^T/\sqrt{2}$.

Figure A3.1 shows the electric field trace in the transverse plane of an elliptically polarized light. From the relation between the two Cartesian coordinates: $x{\sim}y$ and $\xi{\sim}\psi$, conversions of polarization parameters are deduced as

$$\tan 2\theta = \pm\tan 2\alpha \cos\varphi, \qquad (A3.3)$$

$$\sin 2\eta = \pm\sin 2\alpha \sin\varphi, \qquad (A3.4)$$

where positive and negative signs correspond to right and left elliptically polarized waves. Jones vector is then converted to $J_1 = (\cos\eta \; j\sin\eta)^T$ in $\xi{\sim}\psi$ coordinate. The ellipticity is expressed as $e = \sqrt{1 - b_2^2/b_1^2} = \sqrt{1 - \tan^2\eta}$.

A3.2 Stokes Vector and Poincaré Sphere

The Jones vector is a good description of a fully polarized single frequency lightwave. In general cases, the Stokes vector is needed, which has four components, expressed as

$$S_0 = \langle E_x E_x^* + E_y E_y^* \rangle, \tag{A3.5a}$$

$$S_1 = \langle E_x E_x^* - E_y E_y^* \rangle, \tag{A3.5b}$$

$$S_2 = \langle E_x E_y^* + E_y E_x^* \rangle, \tag{A3.5c}$$

$$S_3 = j\langle E_x E_y^* - E_y E_x^* \rangle, \tag{A3.5d}$$

where $\langle \cdots \rangle$ stands for the statistic average to take the degree of coherence into consideration. For an elliptically polarized wave, described by Jones vector $E = (|E_x|\ |E_y|e^{j\varphi})^T$, Stokes components are simplified as $S_0 = |E_x|^2 + |E_y|^2 = I$, $S_1 = |E_x|^2 - |E_y|^2 = I\cos 2\alpha$, $S_2 = 2|E_x||E_y|\cos\varphi = I\sin 2\alpha\cos\varphi$, and $S_3 = 2|E_x||E_y|\sin\varphi = I\sin 2\alpha\sin\varphi$, satisfying relation of $S_1^2 + S_2^2 + S_3^2 = S_0^2$. It is shown that S_0 is the intensity of the optical beam; S_1 stands for the wave linearly polarized in x-direction; $-S_1$ stands for the wave linearly polarized in y-direction; $\pm S_2$ stand for the wave linearly polarized in $\pm 45°$-directions; $\pm S_3$ stand for right and left circularly polarized waves, respectively.

For a partly polarized wave, $S_1^2 + S_2^2 + S_3^2 < S_0^2$; and for a natural light beam without polarization, $S_1^2 + S_2^2 + S_3^2 = 0$. The degree of polarization is defined as

$$DOP = \sqrt{S_1^2 + S_2^2 + S_3^2}/S_0. \tag{A3.6}$$

Poincaré sphere is a sphere in a space with three components of the Stokes vector as its axes, as shown in Figure A3.2. Points in the sphere

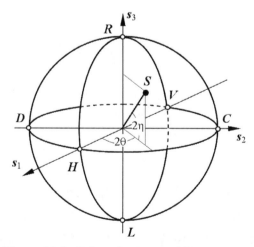

Figure A3.2 Poincaré sphere and Stokes vector.

correspond to the different polarization states: the points on the equator (circle $HCVD$) stands for a linearly polarized wave; the north and south poles (point R and L) stand for right and left circularly polarized waves, respectively; the north and south halves stand for the right and left elliptically polarized waves, respectively. Points inside the sphere stand for partly polarized waves with $DOP < 1$; the center is for the nonpolarized wave. It is seen for from (A3.3) and (A3.4) that $S_2/S_1 = \tan\theta$, and $S_3/\sqrt{S_1^2 + S_2^2} = \tan 2\eta$, as depicted in the Poincaré sphere.

A3.3 Optics of Anisotropic Media

For the anisotropic media, especially the electro-optic crystals, the dielectric constant is a tensor, expressed as

$$
\begin{pmatrix} D_x \\ D_y \\ D_z \end{pmatrix} = \begin{pmatrix} \varepsilon_{xx} & \varepsilon_{xy} & \varepsilon_{xz} \\ \varepsilon_{yx} & \varepsilon_{yy} & \varepsilon_{yz} \\ \varepsilon_{zx} & \varepsilon_{zy} & \varepsilon_{zz} \end{pmatrix} \begin{pmatrix} E_x \\ E_y \\ E_z \end{pmatrix} \rightarrow \begin{pmatrix} \varepsilon_{xx} & 0 & 0 \\ 0 & \varepsilon_{yy} & 0 \\ 0 & 0 & \varepsilon_{zz} \end{pmatrix} \begin{pmatrix} E_x \\ E_y \\ E_z \end{pmatrix}. \quad \text{(A3.7)}
$$

The latter form is the expression in the principal axis coordinate. In general, the direction of vector D is no longer in the direction of electric vector E. Therefore, the direction of the optical wave vector does not coincide with the direction of energy flow, that is, the Poynting vector, since the former is $k \propto D \times H$, different from the latter, $S \propto E \times H$. On the interface of the anisotropic medium the wave vector refracts according to the condition of phase matching, whereas the optical beam propagates in different directions, which, moreover, depends on its polarization, resulting in the double refraction (birefringence).

Two kinds of anisotropic crystals are found: uniaxial crystals with $\varepsilon_{ii} = \varepsilon_{jj} \neq \varepsilon_{ll}$, and biaxial crystals with ε_{ii} different from each other. To understand the properties of the anisotropic media, the index ellipsoid (Optical Indicatrix) is used as a representation of electric impermeability (the inverse tensor of the dielectric constant). In the principal axis coordinate, it is expressed as

$$
\frac{x_1^2}{n_1^2} + \frac{x_2^2}{n_2^2} + \frac{x_3^2}{n_3^2} = 1. \quad \text{(A3.8)}
$$

For the uniaxial crystal, $n_1 = n_2 = n_o, n_3 = n_e$, where the subscripts o and e stand for the ordinary wave and extraordinary wave. They have

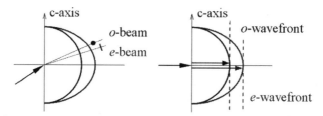

Figure A3.3 Optical paths of *o*-wave and *e*-wave in a birefringent crystal.

perpendicular polarizations. For the ordinary wave, D is parallel to E; whereas for the extraordinary wave, they are not parallel.

Based on the birefringence, the medium is used to make a polarizer, a phase retarder, and other optical components. Figure A3.3 shows the optical paths of *o*-wave and *e*-wave in a phase retarder, which is also called a wave plate. The phase difference between *o*-wave and *e*-wave in normal incidence, passing through the crystal with thickness d, is $\delta = (n_e - n_o)k_0 d$. The most widely used wave plates are the half wavelength plate with $\delta = (2m + 1)\pi$, and the quarter wavelength plate with $\delta = (2m + 1)\pi/2$. It is obvious that the phase retard depends on the wavelength. Therefore, the wave plate has a working linewidth; the smaller the integer m, that is, the thinner the thickness d, the large the linewidth.

A3.4 Jones Matrix and Mueller Matrix

When a beam passes through a device, Jones vector of the output is expressed as $J_1 = TJ_0$. Matrix T characterizes the device's action. If the device is lossless, its Jones matrix must be unitary: $T^\dagger T = I$, that is,

$$T_{11}^* T_{11} + T_{21}^* T_{21} = T_{12}^* T_{12} + T_{22}^* T_{22} = 1,$$
$$T_{12}^* T_{11} + T_{22}^* T_{21} = T_{11}^* T_{12} + T_{21}^* T_{22} = 0. \tag{A3.9}$$

It is deduced from the relation that $|T_{22}| = |T_{11}|$, and $|T_{12}| = |T_{21}| = \sqrt{1 - |T_{11}|^2}$.

The matrix of a wave plate is expressed in the principal axis coordinate as

$$T_\delta = \begin{pmatrix} 1 & 0 \\ 0 & e^{j\delta} \end{pmatrix}, \tag{A3.10}$$

where the fast axis is set as the x-axis. The Jones matrix in the coordinate that is rotated θ to the principal axis of the waveplate is

$$
\begin{aligned}
T_\delta(\theta) &= \begin{pmatrix} \cos\theta & -\sin\theta \\ \sin\theta & \cos\theta \end{pmatrix} \begin{pmatrix} 1 & 0 \\ 0 & e^{j\delta} \end{pmatrix} \begin{pmatrix} \cos\theta & \sin\theta \\ -\sin\theta & \cos\theta \end{pmatrix} \\
&= \begin{pmatrix} \cos^2\theta + e^{j\delta}\sin^2\theta & (1 - e^{j\delta})\sin\theta\cos\theta \\ (1 - e^{j\delta})\sin\theta\cos\theta & \sin^2\theta + e^{j\delta}\cos^2\theta \end{pmatrix}.
\end{aligned}
\tag{A3.11}
$$

For a half wavelength plate, $\delta = \pi$,

$$
T_{\lambda/2} = \begin{pmatrix} \cos 2\theta & \sin 2\theta \\ \sin 2\theta & -\cos 2\theta \end{pmatrix}.
\tag{A3.12}
$$

When a linearly polarized wave $E_0 = [1, 0]^T$ inputs, the output $J_1 = [\cos 2\theta, \sin 2\theta]^T$ is obtained, meaning that the output beam remains a linearly polarized wave, but its polarization is rotated 2θ. In case of $\theta = \pi$, the polarization direction coincides with that of the input.

For a quarter wavelength plate, $\delta = \pi/2$,

$$
T_{\lambda/4} = \begin{pmatrix} \cos^2\theta + j\sin^2\theta & (1 - j)\sin\theta\cos\theta \\ (1 - j)\sin\theta\cos\theta & \sin^2\theta + j\cos^2\theta \end{pmatrix}.
\tag{A3.13}
$$

When the principal axis is $45°$-off the input wave, which is polarized linearly on x-axis (or y-axis), we have

$$
T_{\lambda/4}\left(\frac{\pi}{4}\right) = \frac{1}{2}\begin{pmatrix} 1 + j & 1 - j \\ 1 - j & 1 + j \end{pmatrix};
\tag{A3.14}
$$

the output will be a circularly polarized wave.

A polarization controller, which converts the input wave to an output with desired polarizations, can thus be composed by concatenating quarter waveplates and half waves with their relative orientation rotatable.

The Jones matrix of a linear polarization analyzer (polarizer) is expressed as

$$
\begin{aligned}
T_A &= \begin{pmatrix} \cos\theta & -\sin\theta \\ \sin\theta & \cos\theta \end{pmatrix} \begin{pmatrix} t_\parallel & 0 \\ 0 & t_\perp \end{pmatrix} \begin{pmatrix} \cos\theta & \sin\theta \\ -\sin\theta & \cos\theta \end{pmatrix} \\
&= \begin{pmatrix} t_\parallel\cos^2\theta + t_\perp\sin^2\theta & (t_\parallel - t_\perp)\cos\theta\sin\theta \\ (t_\parallel - t_\perp)\cos\theta\sin\theta & t_\parallel\sin^2\theta + t_\perp\cos^2\theta \end{pmatrix}.
\end{aligned}
\tag{A3.15a}
$$

where $t_\parallel$ and $t_\perp$ are the transmissions of incident beams with polarizations parallel and perpendicular to the axis of analyzer, respectively. For an ideal analyzer with $t_\parallel = 1$ and $t_\perp = 0$,

$$T_A = \begin{pmatrix} \cos^2\theta & \cos\theta\sin\theta \\ \cos\theta\sin\theta & \sin^2\theta \end{pmatrix}. \tag{A3.15b}$$

When a wave is denoted by (A3.2) inputs, the output Jones vector is obtained:

$$J = (\cos\theta\cos\alpha + e^{j\varphi}\sin\theta\sin\alpha)\begin{pmatrix} \cos\theta \\ \sin\theta \end{pmatrix}. \tag{A3.16}$$

Its power is expressed as

$$P = (\cos\theta\cos\alpha + \sin\theta\sin\alpha)^2 - \sin 2\alpha\sin 2\theta\sin^2\frac{\varphi}{2}. \tag{A3.17}$$

In case partly polarized waves are involved, the Stokes vector must be used; and the device is characterized by the Mueller matrix: $S_1 = MS_0$.

1. Mueller matrix for coordinate rotation:
   Corresponding to $J_1 = \begin{pmatrix} \cos\theta & \sin\theta \\ -\sin\theta & \cos\theta \end{pmatrix} J_0$, the Mueller matrix is

$$\begin{pmatrix} S_{11} \\ S_{12} \\ S_{13} \\ S_{14} \end{pmatrix} = \begin{pmatrix} 1 & 0 & 0 & 0 \\ 0 & \cos 2\theta & \sin 2\theta & 0 \\ 0 & -\sin 2\theta & \cos 2\theta & 0 \\ 0 & 0 & 0 & 1 \end{pmatrix}\begin{pmatrix} S_{01} \\ S_{02} \\ S_{03} \\ S_{04} \end{pmatrix}. \tag{A3.18}$$

2. Mueller matrix of a polarization analyzer:

$$M_A = \frac{1}{2}\begin{pmatrix} 1 & \cos 2\theta & \sin 2\theta & 0 \\ \cos 2\theta & \cos^2 2\theta & \sin 2\theta\cos 2\theta & 0 \\ \sin 2\theta & \sin 2\theta\cos 2\theta & \sin^2 2\theta & 0 \\ 0 & 0 & 0 & 0 \end{pmatrix}. \tag{A3.19}$$

3. Mueller matrix of a quarter wave plate:

$$M_{\lambda/4} = \begin{pmatrix} 1 & 0 & 0 & 0 \\ 0 & \cos^2 2\theta & \sin 2\theta \cos 2\theta & -\sin 2\theta \\ 0 & \sin 2\theta \cos 2\theta & \sin^2 2\theta & \cos 2\theta \\ 0 & \sin 2\theta & -\cos 2\theta & 0 \end{pmatrix}. \quad \text{(A3.20)}$$

4. Mueller matrix of a half wave plate:

$$M_{\lambda/2} = \begin{pmatrix} 1 & 0 & 0 & 0 \\ 0 & \cos 4\theta & \sin 4\theta & 0 \\ 0 & \sin 4\theta & -\cos 4\theta & 0 \\ 0 & 0 & 0 & -1 \end{pmatrix}. \quad \text{(A3.21)}$$

A3.5   Measurement of Jones Vector and Stokes Vector

The polarization characteristics of an optical wave are measured by an instrument composed of a polarizer, an analyzer, and wave plates, as shown in Figure A3.4, where DUT is a device under test.

A3.5.1   Measurement of Jones Vector   The input optical wave is described by $E_0 = [H, K e^{j\varphi}]^T$. First, powers are measured by using an analyzer in three orientations: $\theta = 0, \pi/2,$ and $\pi/4$. The results are $I_{01} = H^2, I_{02} = K^2,$ and $I_{03} = \frac{1}{2}(H^2 + 2HK \cos \varphi + K^2)$. Thus, the parameters of Jones vector are obtained: $H = \sqrt{I_{01}}, K = \sqrt{I_{02}},$ and $\cos \varphi = (2I_{03} - I_{01} - I_{02})/2\sqrt{I_{01}I_{02}}$.

To determine the sign of φ, powers are measured again with a quarter wave plate inserted. The wave behind the wave plate is $E_1 = [H, jK e^{j\varphi}]^T$ under condition that the axis of wave plate is aligned coincident with the axis of the analyzer. The measured powers are obtained to be $I_{11} = H^2 = I_{01}, I_{12} = K^2 = I_{02},$ and $I_{13} = \frac{1}{2}(H^2 - 2HK \sin \varphi + K^2)$. It is obtained that $\sin \varphi = (I_{11} + I_{12} - 2I_{13})/2\sqrt{I_{11}I_{12}}$; or $\tan \varphi = (I_0 - 2I_{13})/(2I_{03} - I_0)$ with $I_0 = I_{01} + I_{02} = I_{11} + I_{12}$.

To reduce the errors of alignments, more measurements are made with $\theta = 180°, 270°, \cdots$, and the measured data are averaged.

A3.5.2   Measurement of Stokes Vector   First, powers of the input wave polarized in x- and y-directions with a polarization analyzer are

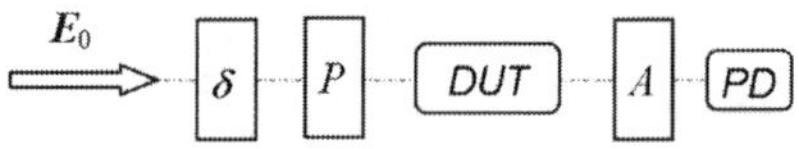

Figure A3.4   Schematic diagram of polarization measurement.

measured; that is I_1 for $\boldsymbol{E}_1 = \boldsymbol{P}_x \boldsymbol{E}_0$ and I_2 for $\boldsymbol{E}_2 = \boldsymbol{P}_y \boldsymbol{E}_0$. It is then obtained that $S_0 = I_1 + I_2$ and $S_1 = I_1 - I_2$.

Second, powers of the input wave polarized in $\pm 45°$ directions are measured; that is, I_3 for $\boldsymbol{E}_3 = \boldsymbol{P}_{45} \boldsymbol{E}_0$ and I_4 for $\boldsymbol{E}_4 = \boldsymbol{P}_{-45} \boldsymbol{E}_0$. From Mueller matrix (A3.20), $I_3 = \frac{1}{2}(I + S_2)$ and $I_4 = \frac{1}{2}(I - S_2)$. Thus, $S_2 = I_3 - I_4$ is obtained.

Third, let the input pass through a quarter wave plate with $\theta = 0$. It is transformed to $\boldsymbol{E}_{01} = \boldsymbol{W}_{\lambda/4} \boldsymbol{E}_0 = [S_0, S_1, S_3, -S_2]^T$. Powers I_5 for $\boldsymbol{E}_5 = \boldsymbol{P}_{45} \boldsymbol{E}_{01}$ and I_6 for $\boldsymbol{E}_6 = \boldsymbol{P}_{-45} \boldsymbol{E}_{01}$ are measured with relations of $I_5 = \frac{1}{2}(I + S_3)$ and $I_6 = \frac{1}{2}(I - S_3)$. $S_3 = I_5 - I_6$ is obtained.

In practice, the loss of polarizer and wave plate should be taken into account. The measurement of Stokes vector is the basis of polarimeter.

APPENDIX 4: SPECIFICATIONS OF RELATED MATERIALS AND DEVICES

Table A4.1 ITU-T Communication Fibers

	Single mode fiber ITU-T G.652	Nonzero dispersion shifted ITU-T G.655	Bending loss insensitive ITU-T G.657	Multimode fiber (Corning)
Mode field diameter	8.6–9.5 μm ± 0.6 μm @1310 nm	8.0–11.0 μm ± 0.7 μm @1550 nm	8.6–9.5 μm ± 0.4 μm @1310 nm	Core diameter 62.5 ± 2.5 μm
Cladding diameter	125.0 ± 1.0 μm	125.0 ± 1.0 μm	125.0 ± 0.7 μm	125 ± 2 μm
Coating diameter				245 ± 5 μm
Cut-off wavelength	≤1260.0 nm	≤1450.0 nm	≤1260.0 nm	–
Macrobend loss	≤0.1 dB @1550.0 nm For r = 30 mm	≤0.5 dB @1265.0 nm For r = 30 mm	≤0.25 dB @1550.0 nm For r = 15 mm	–
Proof stress	≥0.69 GPa	≥0.69 GPa	≥0.69 GPa	≥0.7 GPa
Zero dispersion wavelength	1312.0 ± 12.0 nm	1547.0 ± 17.0 nm	1312.0 ± 12.0 nm	1332–1354 nm
Attenuation coefficient	≤0.5 dB/km @1310.0 nm ≤0.4 dB/km @1550.0 nm	≤0.35 dB/km @1550.0 nm ≤0.4 dB/km @1625.0 nm	≤0.4 dB/km @1310–1625 nm ≤0.3 dB/km @1550.0 nm	≤2.9 dB/km @850 nm ≤0.6 dB/km @1300 nm
Dispersion coefficient	17.0 ps/[nm.km] @1550.0 nm	1.0≤D≤10.0 ps/[nm.km] @1530.0–1565.0 nm	/	/

Quoted from references [23, 25, 26], and [27] of Chapter 2.

Table A4.2 Erbium-Doped Fibers[a]

	Er 1500C	Er 1500C3 LC	Er 1600L3
Peak absorption range @1530 nm	4.0–8.0 dB/m	5.0–10.0 dB/m	18.0–29.0 dB/m
Peak absorption range @980 nm	≥2.5 dB/m	≥3.0 dB/m	–
Cutoff wavelength	≤1300 nm	≤980 nm	≤1400 nm
Mode-field diameter @1550 nm	6.25 ± 0.75 μm	5.4 ± 0.4 μm	5.5 ± 0.3 μm
Cladding outside diameter	125 ± 1 μm	125 ± 1 μm	125 ± 1 μm
Coating outside diameter	245 ± 10 μm	245 ± 10 μm	245 ± 10 μm
Operating temperature	–40 to 85°C	–40 to 85°C	–40 to 85°C
Numerical aperture	0.21	0.22	0.23
Polarization mode dispersion	≤4 fs/m	≤4 fs/m	≤5 fs/m
Splicing loss to SMF	0.10 dB	0.10 dB	0.10 dB
Splicing loss to 980 Fiber	0.10 dB	0.10 dB	0.10 dB

[a] Provided by Corning.

Table A4.3 PANDA Polarization Maintaining Fibers[a]

	PM1550	PM1300	PM980
Wavelength	1500 nm	1300 nm	980 nm
Mode-field diameter	10.5 ± 1 μm	9.5 ± 1 μm	6.6 ± 1 μm
Beat length range	3.0–5.0 mm	2.5–4.0 mm	1.5–2.7 mm
Maximum cross-talk	≤–30 dB/100 m	≤–30 dB/100 m	≤–30 dB/100 m
Cutoff wavelength	1290–1450 nm	1100–1290 nm	800–950 nm
Maximum attenuation	0.5 dB/km	1 dB/km	2.5 dB/km
Coating outer diameter	245 ± 15 μm	245 ± 15 μm	245 ± 15 μm
Cladding outer diameter	125 ± 1 μm	125 ± 1 μm	125 ± 1 μm
Coating type	UV/UV acrylate	UV/UV acrylate	UV/UV acrylate
Core-to-cladding offset	≤0.7 μm	≤0.7 μm	≤0.7 μm

[a] Provided by Corning.

Table A4.4 Typical Parameters and Specifications of LD[*]

	LD(Tc=25°C)			LED	SLD
Wavelength	980 nm	1310 nm	1550 nm	1550 nm	1310 nm
Optical Power	10 mw	10 mw	6 mw	2 mW	1–15 mW
Spectral Width	/	/	1.5 nm	100 nm	60 nm
Threshold Current	20 mA	6 mA	10 mA	/	/
Operating Current/Voltage	50 mA/2V	16 mA/1.1V	30 mA/2V	20 mA	100 mA
Slope Efficiency	0.5 mW/mA	0.5 mW/mA	0.5 mW/mA	/	/
Beam divergence	10°(‖) 30°(⊥)	25°(‖) 30°(⊥)	25°(‖) 30°(⊥)	15°	/
Package Style	TO-18 (5.6 mm Ø)	TO-CAN (5.6 mm Ø)	TO-CAN (5.6 mm Ø)	/	Butterfly or DIL

[*] Provided by THORLABS.

Table A4.5 Typical Parameters and Specifications of PD[a]

	PD	PIN-FET	APD
Wavelength	1310 nm, 1550 nm	1110–1650 nm	1260–1575 nm
Responsivity	0.75 A/W@1310 nm	0.85 A/W@1310 nm	0.85 A/W@1550 nm
	0.70 A/W@1550 nm		
Dark current	0.3 nA	<0.5 nA	1 nA
Capacitance	1.1 pF	–	<0.6 pF
Bandwidth*	1.5 GHz	1.3 GHz	3 GHz
Rise time	250 ps	–	–
Active diameter	100 μm	–	153 μm
Forward current	<10 mA	–	<10 mA
Breakdown voltage	–	–	33–53 V@10 μA
Gain	–	–	10@2 V and 1 MHz

[a]Provided by JDSU.

Table A4.6 Main Specifications of LN-MZM[a]

Specifications	Condition	Specification
Wavelength range		1528–1564 nm
Insertion loss		2.5–5 dB
Return loss	Input an output ports	35 dB
DC V_π	Differential drive	< 2 V
MZ extinction ratio		25 dB
RF bandwidth		12 GHz
RF voltage	At 2 GHz	< 2.6 V

[a] JDSU 10 Gb/s double drive MZM.

A4.1 Fiber Connectors

Fiber connectors are used widely in optical communication and fiber sensor systems. The fiber connector fixes the fiber at the middle of a ceramic ferrule, and can be inserted coaxially and precisely into an adapter with a ceramic tube. Most widely used connectors are divided into two types: PC (physical contact) and APC (angled physical contact). PC makes the gap between two end facets of the fibers to be connected as thin as possible to eliminate almost all Fresnel reflection, which would occur at the interface between silica and air. The end surface is polished into a spherical form with its top coincided with the fiber axis, as illustrated in Figure A4.1a. To further reduce the reflection R, the facet is lapped and polished in an angle of ∼8° off the fiber axis, called APC, as shown in Figure A4.1b. In APC connectors, the residual reflected beam will be deflected off the fiber axis to be lost in a short propagation. Two specification items are used to

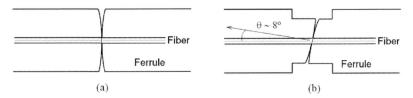

Figure A4.1   Configurations of (a) PC and (b) APC fiber connectors.

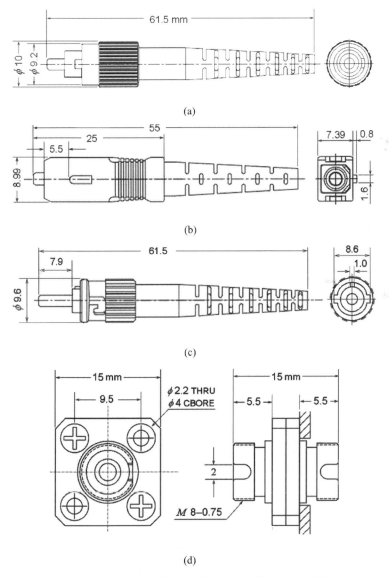

Figure A4.2   Connectors: (a) FC/PC; (b) SC; (c) ST; and (d) adapter.

characterize the performance: the insertion loss and the return loss; the latter is defined as

$$L_R = -10 \log(1 - R). \tag{A4.1}$$

Figure A4.2 shows the structures of commercial connectors and a typical adaptor used widely today.

The typical insertion loss of FC/PC connection is 0.15 dB for single mode fibers, and 0.3 dB for multimode fibers; and the typical return loss is 50 dB for PC connection, and 65 dB for APC connection.

INDEX

Fundamentals of Optical Fiber Sensors, First Edition.
Zujie Fang, Ken K. Chin, Ronghui Qu, and Haiwen Cai.
© 2012 John Wiley & Sons, Inc. Published 2012 by John Wiley & Sons, Inc.

WILEY SERIES IN MICROWAVE AND OPTICAL ENGINEERING

KAI CHANG, Editor
Texas A&M University

FUNDAMENTALS OF MICROWAVE TRANSMISSION LINES • *Jon C. Freeman*

OPTICAL SEMICONDUCTOR DEVICES • *Mitsuo Fukuda*

MICROSTRIP CIRCUITS • *Fred Gardiol*

HIGH-SPEED VLSI INTERCONNECTIONS, Second Edition • *Ashok K. Goel*

FUNDAMENTALS OF WAVELETS: THEORY, ALGORITHMS, AND APPLICATIONS, Second Edition • *Jaideva C. Goswami and Andrew K. Chan*

HIGH-FREQUENCY ANALOG INTEGRATED CIRCUIT DESIGN • *Ravender Goyal (ed.)*

RF AND MICROWAVE TRANSMITTER DESIGN • *Andrei Grebennikov*

ANALYSIS AND DESIGN OF INTEGRATED CIRCUIT ANTENNA MODULES • *K. C. Gupta and Peter S. Hall*

PHASED ARRAY ANTENNAS, Second Edition • *R. C. Hansen*

STRIPLINE CIRCULATORS • *Joseph Helszajn*

THE STRIPLINE CIRCULATOR: THEORY AND PRACTICE • *Joseph Helszajn*

LOCALIZED WAVES • *Hugo E. Hernández-Figueroa, Michel Zamboni-Rached, and Erasmo Recami (eds.)*

MICROSTRIP FILTERS FOR RF/MICROWAVE APPLICATIONS, Second Edition • *Jia-Sheng Hong*

MICROWAVE APPROACH TO HIGHLY IRREGULAR FIBER OPTICS • *Huang Hung-Chia*

NONLINEAR OPTICAL COMMUNICATION NETWORKS • *Eugenio Iannone, Francesco Matera, Antonio Mecozzi, and Marina Settembre*

FINITE ELEMENT SOFTWARE FOR MICROWAVE ENGINEERING • *Tatsuo Itoh, Giuseppe Pelosi, and Peter P. Silvester (eds.)*

INFRARED TECHNOLOGY: APPLICATIONS TO ELECTROOPTICS, PHOTONIC DEVICES, AND SENSORS • *A. R. Jha*

SUPERCONDUCTOR TECHNOLOGY: APPLICATIONS TO MICROWAVE, ELECTRO-OPTICS, ELECTRICAL MACHINES, AND PROPULSION SYSTEMS • *A. R. Jha*

TIME AND FREQUENCY DOMAIN SOLUTIONS OF EM PROBLEMS USING INTEGTRAL EQUATIONS AND A HYBRID METHODOLOGY • *B. H. Jung, T. K. Sarkar, S. W. Ting, Y. Zhang, Z. Mei, Z. Ji, M. Yuan, A. De, M. Salazar-Palma, and S. M. Rao*

OPTICAL COMPUTING: AN INTRODUCTION • *M. A. Karim and A. S. S. Awwal*

INTRODUCTION TO ELECTROMAGNETIC AND MICROWAVE ENGINEERING • *Paul R. Karmel, Gabriel D. Colef, and Raymond L. Camisa*

MILLIMETER WAVE OPTICAL DIELECTRIC INTEGRATED GUIDES AND CIRCUITS • *Shiban K. Koul*

ADVANCED INTEGRATED COMMUNICATION MICROSYSTEMS • *Joy Laskar, Sudipto Chakraborty, Manos Tentzeris, Franklin Bien, and Anh-Vu Pham*

MICROWAVE DEVICES, CIRCUITS AND THEIR INTERACTION • *Charles A. Lee and G. Conrad Dalman*

ADVANCES IN MICROSTRIP AND PRINTED ANTENNAS • *Kai-Fong Lee and Wei Chen (eds.)*

SPHEROIDAL WAVE FUNCTIONS IN ELECTROMAGNETIC THEORY • *Le-Wei Li, Xiao-Kang Kang, and Mook-Seng Leong*

COMPACT MULTIFUNCTIONAL ANTENNAS FOR WIRELESS SYSTEMS • *Eng Hock Lim and Kwok Wa Leung*

ARITHMETIC AND LOGIC IN COMPUTER SYSTEMS • *Mi Lu*

OPTICAL FILTER DESIGN AND ANALYSIS: A SIGNAL PROCESSING APPROACH • *Christi K. Madsen and Jian H. Zhao*

Printed and bound by CPI Group (UK) Ltd, Croydon, CR0 4YY

17/04/2025

14658882-0001